The Literature of Agricultural Engineering

Literature of the Agricultural Sciences
WALLACE C. OLSEN, Series Editor

Agricultural Economics and Rural Sociology: The Contemporary Core Literature edited by Wallace C. Olsen

The Literature of Agricultural Engineering edited by Carl W. Hall and Wallace C. Olsen

THE LITERATURE OF AGRICULTURAL ENGINEERING

EDITED BY

Carl W. Hall and
Wallace C. Olsen

Cornell University Press

ITHACA AND LONDON

This book was typeset from disks supplied by the staff of the Core Agricultural Literature Project, Albert R. Mann Library, Cornell University. Sharon Van De Mark prepared the machine-readable text. The research was financially supported by Cornell Agricultural Experiment Station; National Agricultural Library, United States Department of Agriculture; and the Rockefeller Foundation.

Copyright © 1992 by Cornell University

All rights reserved. Except for brief quotations in a review, this book, or parts thereof, must not be reproduced in any form without permission in writing from the publisher. For information, address Cornell University Press, 124 Roberts Place, Ithaca, New York 14850.

First published 1992 by Cornell University Press.

International Standard Book Number 0-8014-2812-2
Library of Congress Catalog Card Number 92-24516
Printed in the United States of America
Librarians: Library of Congress cataloging information appears on the last page of the book.

⊗ The paper in this book meets the minimum requirements of the American National Standard for Information Sciences—Permanence of Paper for Printed Library Materials, ANSI Z39.48-1984.

Contents

Preface   vii

SECTION I. *From Beginnings to Worldwide Influence*

1. Introduction   3
   IR. F. COOLMAN
2. The Influence and Scope of the Agricultural Engineering
   Profession   11
   CARL W. HALL
3. Agricultural Engineering Education Programs   25
   DONALD M. EDWARDS
4. Agricultural Engineering Literature in Developing Countries   43
   MAKOTO HOKI, HISASHI HORIO, AND GAJENDRA SINGH

SECTION II. *Reviews of Emerging Areas in Agricultural Engineering*

5. Energy in Agriculture   59
   BILL A. STOUT AND TERRI G. HUFF
6. Literature in Instrumentation, Control, Electronics and
   Automation   96
   G. C. ZOERB
7. The Current Core Literature of Food Engineering   117
   R. PAUL SINGH
8. Development Forest Engineering and Its Literature   126
   J. L. FRIDLEY, J. A. MILES, AND F. E. GREULICH
9. Aquacultural Engineering   144
   JAW-KAI WANG

SECTION III. *Publishing and the Classical Literature*

10. Textbook Publishing in Agricultural Engineering   163
    GEORGE E. MERVA

11.	Publishing of the American Society of Agricultural Engineers JAMES A. BASSELMAN AND WILLIAM CHANCELLOR	175
12.	The Place of Standards in Agricultural Engineering JIMMY L. BUTT	197
13.	Characteristics of Agricultural Engineering Literature WALLACE C. OLSEN	213
14.	Citation Analysis and the Core Monographs WALLACE C. OLSEN	226
15.	Primary Journals and Reports WALLACE C. OLSEN	318
16.	Reference Collection Update CYNTHIA S. KAAG	332
17.	Primary Historical Literature, 1850–1950 SUSAN J. THOMPSON AND CARL W. HALL	339
	Appendix	406
	Index	413

Preface

The first book in English which had the words "agricultural engineering" in the title was published in England in 1852. The profession in the United States will soon be a century old. A growing profession depends for the dissemination of its principles and standards on scholarly activities, the most important of which is the writing of documents that come to form the literature. The development of agricultural engineering in many countries of the world is thus chronicled, often initially in reports and articles, then later in books. After surveying nearly one hundred authors and users of the literature, the editors made judgments and compiled listings that identify the dominant works in the field. *The Literature of Agricultural Engineering* therefore lists, analyzes, and summarizes the books and journals that have made important contributions to the evolution of the subject.

A thorough assessment of this literature takes on added significance with the increasing globalization of the economy. Books and journals originating in the developed countries are valuable sources of information for Third World countries, and those sources are identified in this book. But the Third World countries also have much to offer the rest of the world. Although their literature may not be voluminous and some of it is in languages not widely known, many of their books should be in the agricultural engineering libraries of the developed countries. Scholars and literature collectors throughout the world will find assistance here in evaluating their collection strengths, in measuring their journal holdings, and in making decisions about the preservation of pertinent historical literature.

The Literature of Agricultural Engineering builds on the traditional subject areas of power and machinery, soil and water, structures and environment, and electric power and processing. Sections I and III cover these traditional subject areas in detail. Although the names of these areas have endured over the years, the subject matter has changed tremendously, as new topics and techniques have been incorporated into their activities and publications. Thus, subjects such as erosion, soil transport, chemical drift, precision application, filtering pollutants, conservation, materials handling, waste management, and safety, to name a few, are now important aspects of the traditional areas.

At the same time some new fields of major significance have developed which are not easily absorbed into the traditional areas of agricultural engineering. A Steering Committee that counseled the Core Agricultural Literature Project on this work recommended that the emerging or growth areas of the discipline would be well served by individual literature reviews. Section II therefore includes review chapters that highlight the developments in food engineering, forest engineering, and aquacultural engineering, as well as cross-cutting subjects such as energy, instrumentation, electronics, and automation.

Athough agricultural engineering predated the founding of the American Society of Agricultural Engineers in 1907, that professional society has had a tremendous impact on the literature of the field. Similar organizations in other countries have likewise contributed to the development of an important and expanding knowledge base. As we move toward the twenty-first century, the pace of scholarship and the need for a record of scholarly activities will only increase, and the media will be expanded. Comprehensive databases, coupled with distribution of information through electronic networks, high-speed telecommunications and satellite transmission not restricted by national borders, will draw agricultural engineers around the globe together as better-informed people and thus will improve their service to society.

The Literature of Agricultural Engineering is the second book in the Literature of the Agricultural Sciences series. The first is *Agricultural Economics and Rural Sociology: The Contemporary Core Literature*. A total of seven books is anticipated. Volumes to follow are Animal Science and Health; Soil Science; Crop Protection and Improvement; Food Science and Human Nutrition; and Forestry and Agroforestry.

Members of the Steering Committee advising the Core Agricultural Literature Project on this volume, in addition to myself, were James Basselman, former manager of publications for the American Society of Agricultural Engineers; Joseph Campbell, Cornell University; William Chancellor, University of California, Davis; and Elizabeth Roberts, Washington State University.

CARL W. HALL
Senior Editor

Arlington, Virginia

SECTION I

From Beginnings to Worldwide Influence

1. Introduction

Ir. F. Coolman
Director Emeritus, I.M.A.G.
Dutch Research Institute of Agricultural Engineering
Wageningen, Netherlands

Each development in agricultural engineering depends on a flow of knowledge from an original source to a field of application. The sources can be scientific workers and institutions, commercial enterprises, fieldworkers or inventors. As well as a demand for an item of knowledge in a field of application, there should be a flow back from the users to the source. Good feedback sparks new ideas and contributes to further development. However, rapid developments may expose shortcomings and even gaps in the communication links. This introduction analyzes the main causes of these gaps, which are particularly likely to be present when the source is in a developed country and the field of application is in the Third World.

Agricultural engineering, in the most general terms, is the use of engineering science in agriculture. Both engineering and agriculture today cover very broad areas. Engineering offers a rapidly increasing number of possibilities for solving problems. Agriculture identifies a rapidly increasing number of production and processing problems to be solved. Engineering extends from simple tools to complex computer programs and robotic machines; agriculture extends from simply rotating crops and grazing sheep on poor soils, to intensive production with many technical aids and scientific support. Agricultural engineering contributes to progress in varying ways across the producer-to-consumer system. It contributes not only in increasing production, but in improving a number of aspects of rural life such as public facilities, roads, environment, landscape, and rural health. The quality of human life has to be kept in mind by those making plans with regard to the environment, energy production, health, food and feed, and landscape. This is true not only for industrial countries, but also for developing Third World countries, where agricultural engineering is still in its early stages.

A. Stages of Development and the Role of Agricultural Engineering

In agricultural and rural development nearly every step is based on a foregoing one. The present stage is the starting level for the next stage of progress; development is step-by-step. Natural factors and often human factors prevent too big a jump in development and restrict agricultural or engineering possibilities. The human factors include individual capacities, rural customs, social and religious rules, infrastructure, economics, and market situations. Developmental planning and projects have to be balanced on the basis of feasibility, taking into account all these aspects of a situation. A few examples with an increasing level of development show this:

(1) Blades or knives of rotary hoes need regular sharpening or welding. As a first step, local infrastructure has to be brought into being; local repair shops need know-how, equipment, and training to do the repair or to replace the parts. At a later stage of development, on larger farms, farmers can perform these tasks, having gained that ability through special courses and broad experience. The use of better materials leads to higher levels of plant production.

(2) Introducing engines into an area's farms requires extensive training of drivers and other users, as well as strong guidance to the distributor and the local repairshop. The first tractor introduced in a neighborhood requires considerable negotiation and much thought—even more than the first truck service to the nearest town. Tractors must match field sizes and roads. The total agricultural community becomes involved, e.g., in the repair operation. Strong sentiments and social circumstances often play a major role in whether a new way of carrying out jobs is accepted. Literature on experiences in similar projects, especially on why expected goals were not reached, is a valuable aid in planning, instruction and guidance.

(3) Typical of far-reaching technological decisions is the introduction of electricity in a village and on farms. Electricity changes life and work completely, e.g., in the way water is provided for humans, livestock, and irrigation, and in the living facilities of the home. Moreover, public life changes and offers more opportunities for progress, especially in the field of processing and storing of agricultural goods and products. Most of the first networks built in Third World countries have already been renewed, giving more power than was foreseen initially. Reports are available about increasing demands in several areas. A thorough study of these beforehand can prevent or reduce disappointments.

(4) Raised agricultural output requires a purchaser for the increased production. Especially in developing countries one cannot always find a dependable, trustworthy, or stable marketing system. Technological progress is often hampered in such circumstances. Market information must be taken into account at an early stage; market research studies need to be published and made available. Case studies of past projects can be used in working out the masterplan of a project directed toward more output through improved production technologies.

(5) Enterprise management of today's farms with a high output per worker (high wage level) means confronting complex problems of decision-making. Proper computer programming and adequate information are needed. Engineering is now delivering sensors and programs, suitable for following in detail, what a plant, an animal and the soil need. Information is also available for retrieval; so that programming for individual situations can be carried out with greater detail and ease.

Many more examples could be given to illustrate the relation between the simple and the complex in agricultural engineering knowledge. The steps in development require intensive planning, each step in connection with the foregoing one.

B. Hampering Influences

The adoption of new methods in agriculture has always met with a certain resistance. This resistance often has a direct relationship to the level of complexity or the degree of unknown consequences. Farmers all over the world usually have very strong feelings about what can go wrong, what can eventually lead to a lower crop yield or income. Some experiences in the past support those anxious feelings. A well-known saying is: "They may say (or write) so, but. . . ." Thus, information has to be properly directed and well presented.

New technologies need a certain period for gaining acceptance. In the mind of the farmer, using a new technology could mean taking a financial risk for a period during which the level of yield is uncertain. Information should be clear as to the risks involved, including possible profit risks. Convincing a farmer or a group of farmers is often a difficult task. Extension officers have to talk the language in which the farmers are accustomed to think. Written material should explain the risks and benefits in an understandable way. In many cases library or research information has to be translated into understandable language. Extension officers, instructors, trainers, teachers, and lecturers should have experience in doing this. If they do it well, a connection will be made in the technology transfer process. Even well-educated farmers often do not quite understand the literature on advanced developments.

In addition to the personal factors hampering adoption of new technologies, the community may resist for different reasons. Most common is the general opposition to changing a custom that appears to be working well. This type of opposition is usually found in the first stages of development. When a new technology requires a different type of community cooperation, resistance often develops. Giving information about how the new technology was worked out in comparable communities is often the best way to educate the participants before a decision is made. After having experience with a technology, people tend to become more technically oriented in their thinking and open to advancements. A higher investment level often leads to more community interest.

Religious beliefs may also be involved in resistance to a modern technology. Such resistance is more difficult to solve. Either the technology must be adapted to such resistance or the beliefs must be modified. There are no general rules on how to handle such situations. The experiences of those who have faced similar problems can be helpful, although accounts of such experiences are rarely published.

Technologies may meet with resistance because they disturb existing social relationships. A technology may require modifying the way tasks are divided between husband, wife, sons and daughters on a family farm, or it may mean a change in the way the whole community cooperates. Examples:

— A simple thresher replaces grain threshing by hand.
— Man-operated milking machines are introduced in a community where women have been doing the milking by hand.
— A small dairy plant opens and takes over butter-making, formerly done on individual farms.
— A bulk delivery system and shop sale of products replace traditional farm-based sales.

As soon as the community reaches a higher technical level, new techniques are more readily accepted. But at a higher technical level with more community involvement, more rules and regulations have to be made and accepted. Development plans often do not incorporate these aspects although evaluation reports sometimes give attention to them.

Economic forecasts of development and plans for Third World projects particularly are often too optimistic. Plans are set up for the mindset found in a developed country on a higher technical level. In many cases the plans lead to disappointments. FAO and the World Bank have quantified such experiences and pitfalls. Many national foreign aid organizations in developed countries are familiar with them, and many private enterprises have also experienced them. They all result from plans made with care, but with less than complete knowledge of Third World reality. The real limitations were underestimated. If exchanges of knowledge and experience can help to overcome the shortcomings and to fill a part of the gap, a very valuable contribution to the effectiveness of aid programs is realistic.

C. *Advanced Technologies, High Technologies*

Developments in microelectronics have opened a wide range of possibilities in agricultural engineering. Automation, robotization, computer-directed planning and management are some of the key areas of high-level technology. Knowledge transfer of mainly technical and physical data has a new biological aspect at this advanced level. Sensors make it possible to get information about the production aspects of living material, animals, plants, soils, and the microclimate. Biological sciences are contributing knowledge of relationships about growth, health, and production shortcomings. Not all of them are yet quantifiable but estimations, if necessary, make them useful in computer programs or in microchips for controlling, e.g., feed, fertilizer, moisture, and air. These have led to a number of advanced technologies in agriculture as well as in some aspects of rural life.

At present one can view these technologies as the highest stage of agricultural engineering, applied in high level production situations. The next phase seems to be a much more detailed approach to and control of quality in its different aspects: health of animals and plants, environmental relationships, ease and exactness of carrying out jobs, management information on the spot and at the computer disk, harvest expectations and market forecasts, and adequate reactions to climate influences. Agricultural engineering will be expected to make major contributions to an improved quality control of reproduction in the whole system of plant and animal breeding. A wide spectrum of sciences is involved, and a well-functioning system of knowledge transfer is essential. Such transfer has a multidisciplinary character. High-level technology offers the possibility of considering fine details along with more general aspects. These can be of a broadly occurring nature but can also be tied to local production circumstances. If either direct or indirect measurement is possible, the application of technologies can be partly automated, but a part of the application will always be left to the persons who control the work. This means that they must understand and be able to apply and guide the characteristics of the high-level technology. It means that complete information is required on the various levels of technology and development involved. One of the aims of *The Literature of Agricultural Engineering* is to contribute to an understanding of the multidisciplinary aspects of the consequences and possibilities of modern agricultural engineering.

D. *Technical Levels and Feasibilities*

As discussed previously, there exists a strong link between technical progress and development in general. This is true not only for the lower or beginning stages of technology, but also at the higher levels. The different limitations to technical progress that have been mentioned are mainly based on human individual and social characteristics. General economical and political circumstances should be added when feasibility is considered.

Governmental promotion of modernization can be quite effective in motivating progress.

Once a development is started, the next steps will have to be established, and if the first experiences went down well, continued progress will occur. Whereas nearly all local progress is expressed in the local money unit, the parameter for measuring or expressing large-scale development is overall economical and social value. It should not be forgotten that the main goal of development is the satisfaction or the happiness that the people concerned derive from their activity. Each human being, each community, each population determines its own values. Each also chooses the literature that is read or received on development possibilities.

The higher the level of technological development the less subject it is to human limitations. Scientific approaches should be objective, and usually they are. But the more closely that scientific approaches can be linked to practical application, the more likely they are to be used. This means that someone must be able to interpret the information received and often adapt the technology described in the literature. The higher the scientific level or the more detailed the subject treated, the greater the chance of a gap developing between the author and the reader. Also, local circumstances in agriculture have a great influence. The bigger the difference in circumstance from what the technology assumes as a norm, the more the chance that the user will not understand the literature.

Technical aids used worldwide can also be misunderstood. Some examples at different levels illustrate the point:

(1) Excellent plowing can be done only with a plow bottom adapted to local soil conditions. Plow factories, research workers in this field, and clever farmers know this. The distance, however, between the plow bottom designer and those with the knowhow about the soils is often too great.
(2) Milking machines, especially the size and shape of the teatcup, have to be adapted to cow breed and behavior. General identifications of udder type and milk yield are not sufficient to design the right teatcup. The dairy cow of the person doing the milking is seldom the same as the dairy cow the technical designer has in mind. Robot-milking increases this gap.
(3) Management programs must include a lot of local input. The influence of each local factor is significant. A standard program may miss the goal. Even non-linear relationships have to be considered, and the way a project is carried out can be the most important factor in its success.

These three examples show that knowledge transfer should be as complete as possible so that the receiver of the information can make full use of it. Many publications do not take this into account. Authors state their results or opinions on the basis of local or regional or national applications, and often do not foresee an international or wider use. On the one hand such a use stimulates progress; on the other hand, international service can be made easier if an author correctly chooses and comments on points that have to be considered in broadening the application. Extension workers are accustomed to doing this, whereas those who are not working in the field often forget it. Scientists or engineers may believe in a much wider potential use of their results than is justified. Feasibility discussions may help assure that appropriate consideration is given to the range of application of results. Such discussions often take place when a team of researchers is tackling complex problems in a multidisciplinary way. They have available literature as an important base to assess the feasibility of applying new results in their area and often "redress" published results, making them suitable for their own circumstances and goals.

Feasibility studies are required for plans and projects in developing countries. If advanced technologies are incorporated, training and education must be part of the project. Special courses provide a way to become acquainted with a new technology. The real learning, however, takes place as the new technology is put into practice. The more advanced the technology, the more time required to get it understood and applied.

There exists a distinction between anticipated results and benefits actually achieved. Scientific information contributes mainly to the planning phase of a project and sometimes to the final reporting on it, especially if new facts or findings come to light. Also, hitherto unknown circumstances can influence the value of scientific information. Perhaps an old technology is completely worn out, but maybe a local factor makes its renewal possible. If any information provides a real input that makes it possible to advance to a higher technology, a positive contribution is made. At the lower development levels a clear view of feasibility is essential; on the higher levels such a view is more difficult to achieve because of many uncertainties.

E. Methods of Knowledge Transfer

How do those who want to know or learn and those who can deliver the knowledge or practical know-how find each other and communicate? The most direct way is the personal contact, but it is also the most expensive. During scientific meetings useful contacts can often be made. The Commission Internationale du Génie Rural (CIGR), the worldwide association of agricultural engineering organizations, adopted a subject taxonomy which includes advanced techniques and high technology in 1989/1990 in order to be of more service to its members. The growth of multidisciplinary approaches was one reason for changing the structure of the organization. In the mid-1980s *Who Is Who: A Directory of Agricultural Engineers Available for Work in Developing Countries* was published.[1] Negotiations are underway to computerize this information for the entire field of agricultural engineering. Such attempts are being made because personal knowledge is often regarded as the most important source of scientific and engineering information. But the main flow of such information comes from the literature and marketed technical aids.

Written information is available worldwide in well-equipped libraries. Keywords make it easier to find facts about subjects and to find the names of those who have published on various subjects. But information is not easily available to all, and that gap needs to be filled. In addition to microfilm, microfiche, indexes, and abstracts, library compact disc services will be an important aid in disseminating a broad view of what is known and done. An internationally accepted system of keywords and documentation codes would be a big step forward. Constant renewing and adaption will be necessary to keep information current.

F. Looking Backward and Forward

I have described a number of shortcomings in knowledge transfer that may lead to misunderstandings of literature and documentation. Misunderstandings are caused both by those who are looking for more knowledge and also by those who write and publish information. The former often do not know the circumstances influencing the published results; the latter often generalize too easily. Both participants in a communications link should be aware of these characteristics. In many cases a personal contact by phone or letter can clear up or prevent misunderstandings and thus avoid disappointments.

One of the tendencies of human beings is to believe that the other person is the one who is often not correct. In business the seller and the buyer have to agree in order for progress to occur, so that both can benefit. Scientific information is not clearly sold or bought, but for both parties the exchange has a material value. The exchange of scientific knowledge has another sort of value as well. The owner of such knowledge should also possess a willingness to make it known to those who can use it. The benefits are manifold when the exchange of scientific information works well and the receiver and the deliverer advance their mutual understanding.

1. F. Coolman, compiler, *Who Is Who: A Directory of Agricultural Engineers Available for Work in Developing Countries* (Paris: Commission Internationale du Génie Rural, 1985).

2. The Influence and Scope of the Agricultural Engineering Profession

CARL W. HALL

Engineering Information Services, Arlington, Virginia

A. Early Years

The industrial revolution set the stage for a dramatic change in farming and agriculture. Developments during the industrial revolution enabled business and industry to increase productivity, and were a precursor to the mechanization of farming. In open societies, one segment of society does not change without affecting other segments. Thus, the development of iron and steel, materials, fuels and lubricants, power units and communications were integral to changes in the agricultural sector.

Before 1900, the principal sources of power on the farm were oxen, horses, and mules, and, of course, human beings. Horses were used for transportation in the military, in large cities, and on canals, as well as by individuals. In 1900 there were about 20 million horses and mules in the United States; their number reached a peak in 1917 with about 27 million, although they were being replaced by steam engines as early as the 1890s.[1]

The inventions of those early years—the plow, the reaper, the steam engine, the cotton gin, the electric power systems, the trucks and automobiles—along with the discovery of oil set changes in motion that affected all segments of the economy. That motion continues today.

The most significant event of the nineteenth century for United States agriculture, beginning the revolution still in progress, was the Morrill Act, or the Land-Grant Act, that was passed in 1862 in the midst of the Civil War. Both agriculture and the mechanic arts (engineering and technology) were identified as targets of opportunity for the United States. The Morrill

---

1. United States Department of Agriculture, *Agricultural Statistics*, published annually, various issues.

Act firmly set the stage for development of the profession of agricultural engineering. Although the Morrill Act focussed on higher education, additional legislation, the Hatch Act of 1887, followed to strengthen agricultural research. The Smith-Lever Act in 1914 established the extension function. This new set of land-grant institutions struggled for fifty years or more to meet its charge and to be accepted by other education institutions. In the process other institutions were impacted and changed to a more open, relevant educational system at a lower cost. Sons and daughters of the industrial class and farm population became students of this new approach to advanced education.

The mechanization of industry and agriculture was not without its shortcomings. Often the health and safety of workers were compromised to reach new levels of productivity. Thus, legislation was passed to eliminate child labor, to restrict hours of work per day and per week, and to provide an improved working environment. These or similar laws usually did not apply to farm workers until many years later.

The driving force for mechanization was primarily the desire to reduce the drudgery of clearing the land, draining the swamps, preparing the land for production, and harvesting the crops. Although labor productivity played an important role, production of the land was more important in those days.

The driving force for the development of agricultural engineering was provided by devices and processes, either available or perceived, rather than from a scientific theory. Practitioners of the day—John Deere, Cyrus Hall McCormick, and Henry Ford—in the United States, could be considered the early agricultural engineers based on the devices they developed and the impact they made. Even Thomas Jefferson carried out many activities that today could be considered in the domain of agricultural engineers. Biographies of these people are replete with examples of significant contributions. Some fields of study start with a base in theory, and include not only undergraduate study, but also work at graduate levels such as in physics, astronomy, mathematics; basic instruction then moved to the lower levels of education. Agricultural engineering started in an opposite manner with movement from the practitioner, the blacksmith who made tools and equipment, the craftsman, and the manufacturer. To improve upon these, the knowledge base of science was added later.

Man has always been seeking methods and developing devices to reduce drudgery. It wasn't until the 1830s with the plow, seeder and reaper, and the early 1900s with the steam and gasoline engines, that major advances with considerable visibility gave exposure to the emerging profession of agricultural engineering. Books in the field document these developments.[2]

2. Robert E. Stewart, *Seven Decades That Changed America* (St. Joseph, Mich.: American Society of Agricultural Engineers, 1979).

People such as Thomas Edison, Eli Whitney, Guglielmo Marconi, Charles Kettering and Werner von Siemans produced technologies that later had a tremendous impact on agricultural engineering. These technologies were adopted and later studied and improved to provide major advancements. It was not until post-World War II that the basic theories gained importance in improving tools, devices, and processes in agricultural engineering. New technologies served an important role in driving the theoretical developments. It is often said that "the steam engine did more for thermodynamics than thermodynamics did for the steam engine," which is true with many agricultural engineering applications.

B. 1900–1920

From 1900 to 1920, the horsepower available to each farm worker increased from 2.0 to 3.3, not a rapid but a significant change, primarily as a result of tractor and steam power. By 1960, each farm worker had thirty-five horsepower available in mechanical power, twice that available to each industrial worker.[3]

Gasoline and kerosene tractors began to replace the steam tractors with 10,000 power units produced per year in 1915, increasing to 200,000 per year in 1920. Stationary operations such as threshing, grinding, and sawmills, continued to use steam power. Steam power units were available for using byproducts such as straw and fodder for fuel.

The design, building and testing of these power units was a very active field for agricultural engineers. In 1919, the Nebraska tractor test was established by law and became the model for testing and evaluation, and for providing information on tractors on a world-wide basis, a role continuing today.

The application of the new technologies in industry for use in agriculture and manufacturing serving agriculture was not as direct as one might expect. Agriculture was usually seasonal; weather played an important role, and education of farmers was not extensive, although they greatly respected education. To assist the farmer by providing information on the latest practices for the farmstead and in the home, most states established extension services to meet those needs. The 4-H movement for youth grew out of that program in which there were machinery, tractor, tool, carpentry and electricity projects. Youths were made aware of the opportunities provided by additional education and were stimulated to pursue a college education. Rural schools and rural free delivery mail routes expanded, improving

3. M. L. Esmay and C. W. Hall, *Agricultural Mechanization in Developing Countries* (Tokyo: Shin-Norinsha Co., 1973).

information flow and strengthening education and adding grades nine through twelve to the previous basic eight-year school. The rural population had access to and utilized education and training to make use of technological developments.

The demise of horses and mules for mobile power decreased the need for land to provide feed for those animals, changed the design of structures for housing and feeding, and increased the demand for fuel. On some university campuses, the vacated horse barn even became the facility for developing agricultural engineering.

Agricultural engineering built on existing engineering fields. In the late 1800s and early 1900s, civil and mechanical engineers were the primary participants who developed agricultural engineering by forming the American Society of Agricultural Engineers (ASAE) in 1907 in the United States. At that time, to be a member of ASAE required an engineering degree in a related field.

Most institutions initiated studies in agricultural engineering in sections or departments of agricultural mechanization, agricultural mechanics, and agricultural machinery. These were usually established as a part of a department of agronomy, in a college of agriculture. Later, when staffed with people who had engineering degrees, these departments were established in colleges of engineering, or jointly with colleges of agriculture. The leading departments of agricultural engineering, consisting primarily of engineers, that become best known for providing the foundation for the profession were at Kansas State University, University of Nebraska, Iowa State University, University of Missouri, and Utah State University in 1919, although many other departments were in existence and made significant impacts.

It is believed that the first Ph.D. in agricultural engineering in the United States was granted in 1917, at Cornell University, awarded to E. A. White.[4] His major professor and chairman of the doctoral committee was a member of the Department of Rural Engineering. The dissertation topic was tillage in which he worked on a mathematical equation to relate the surface of a plow and the nature of soil movement over the plow.

Rural engineering was established in the U.S. Bureau of Public Roads (BPR) in 1915; the BPR had been established in the Department of Agriculture in 1893.

Early agricultural engineering was devoted to meeting immediate needs with little attention given to long-range problems. Thus, horseshoeing, har-

---

4. E. A. White, "A Study of the Plow Bottom and Its Action upon the Furrow Slice," *Transactions of the American Society of Agricultural Engineers* 12 (1918): 42–50; and in the *Journal of Agricultural Research* 12(4) (1918): 149–182. These publications were accepted in lieu of a separately written dissertation for the Ph.D., a common practice at the time.

ness making, hitch design, drainage, building construction, rural roads, sanitation and waste, and steam power were early topics, some aspects of which continue. Much of the effort in teaching and research was devoted to the practice rather than the theory of agricultural engineering. Industry was not close to and was not dependent upon university and government research. Inventors and entrepreneurs provided the driving force for the new tools, equipment and structures. Experiment stations, some in engineering, most in agriculture, began to participate in projects related to the field. The industries were dependent on the universities and colleges for personnel, usually limited to the bachelor's level, and not much for research.

J. B. Davidson of Iowa State University authored one of the first books in the United States which included all of the subjects important to agricultural engineering. His *Agricultural Engineering* (St. Paul: Webb Publishing Co.) was issued in 1913 and intended for the general reader and students of secondary schools, not for the professional engineer.

C. 1920–1945

Immediately following World War I, the trend to increase the production of crops accelerated. Windmills driving electric generators or pumps were widely used, particularly in areas where electricity was not available. The focus on gasoline and kerosene farm tractors as stationary power units (belt-driven) and for breaking and preparing the soil continued. The first popular row crop, all-purpose tractor was the Farmall, in 1924, which was followed by many other makes. Agricultural engineers were involved in many of the developments such as providing power to the belt, power takeoff, and wheels. The need for these units is demonstrated by the fact that some people put a pulley on the rear wheel of their automobile which, when jacked up, would provide power through a belt to a feed grinder, sausage grinder, or water pump. Also people would cut down a heavy truck to make a tractor. These "doodlebugs" were quite common devices that were used not only in the field, but for farmstead operations and for travel to nearby locations. These ingenious strategies attest to the creativity and inventiveness of the farmer.

Early tractors had steel wheels with steel lugs which severely damaged macadam rural roads. Riding a tractor on steel lugs gave considerable discomfort to the driver operating on these roads. Steel rims, placed over the lugs, were used as a less than satisfactory solution, often requiring changing rims as tractors were moved from field to field. "Tractors with lugs prohibited," signs were posted and enforced to protect the roads. Hard rubber tires with zero pressure appeared in 1931, but by 1935 about 15% of the tractors had pneumatic or inflated rubber tires. The claim that rubber "poisoned the soil" was quickly abandoned as the comfort and wide applicability overrode criticisms. The need for weight to provide traction was met by wheel weights or the addition of water that had to be drained in winter or protected with antifreeze to prevent freezing. Many agricultural engineers were involved in the development of appropriate tires and treads through this and later periods.

The tractor became a movable power plant to which pneumatic and hydraulic controls were added, and later electrical generators for driving and controlling attached units. Units for seeding, cultivating and harvesting became integral with the tractor rather than pulled by a team or tractor.

Farm implements, originally horse-drawn, were modified to be used with the tractor. Then implements were designed specifically for the tractor; these implements were not usually interchangeable between power units. Lifting, adjustment, and use of implements could usually be manipulated from the driver's seat.

The next phase was that power units were designed to accommodate the attachments, a reverse of previous approaches. Agricultural engineers were particularly influential in merging the needs of farmers and the needs of the manufacturers. This approach was particularly true for specialized units used for production, harvesting, and processing of fruits and vegetables and for work around the farmstead.

The rapid increase in the use of electric power and its distribution throughout the United States largely missed the rural areas until the 1930s.[5] The low density of farms as compared to industries, the impression that farmers would use electricity only for light rather than power applications, did not make extension of electricity to rural areas an attractive investment. The Rural Electrification Act (REA) of 1935 stimulated the development of rural electric power distribution lines so that electricity became readily available to many rural areas. This new source of energy began a dramatic change not only on the farmstead, but in the farmhouse which now had the potential for the same working conditions for rural and city folks. The telephone and running water were not far behind. And the radio brought news, commodity price information, and entertainment into the farm home. Household appliances were included as part of the field of agricultural engineers working with home economists. Except for the depression years of the 1930s, the trends were for increasing the use of fossil-related fuels and electricity. With the coming of World War II, these changes accelerated to the extent that materials and people were available to do the job.

5. Marquis Childs, *The Farmer Takes a Hand—The Electric Power Revolution in Rural America* (New York: Doubleday, 1952).

The United States dust bowl of the 1930s raised a popular concern for the natural resources of the future. The need for conservation measures was demonstrated by the Soil Conservation Service, established to promote the protection of the soil and water. Engineering was involved in many of the soil and water conservation efforts.

During times of prosperity people in rural areas moved to industrial jobs as a result of easier work and usually a cleaner environment, leaving a void for labor on the farm. Also the high demand for military personnel put a tremendous drain on labor available for working on the farm. The door for increased mechanization and electrification was opened and helped replace the working hands who moved elsewhere. In these depression years, jobs were scarce and people yearned for those jobs they had left. But most would not return to the labor intensive jobs on the farm.

Sentiments developed against mechanization. These views were reflected in numerous discussions pertaining to the movement of people off the farm. The pull-or-push of rural migration to the urban areas was discussed widely. Did the mechanical cotton picker push people off the farm? Or were people pulled from the farm by the lure of an eight-to-five, well-paying industrial job while living in an urban area or in a rural area close to a city? This controversy arose again after World War II particularly in connection with providing jobs for migrants in sugar beet, cotton, fruit and vegetable fields.

In the 1920s, the technical sections of ASAE were

> Farm Power and Equipment
> Farm Buildings
> Reclamation
> Rural Electric

An E & R Section was formed to represent the education and research interests of ASAE, which persisted as a section and later as a division until 1954. The most recent organization (1989) has an education department and a research department, both under a major Professional Department group of the Society. During this period, graduate degrees were granted in cooperation with other academic departments, either in engineering or agriculture or a combination of these, depending upon the specialty area. The emphasis was primarily on production and on-farm processing.

D. After 1945

The critical need for production of food was recognized during World War II, and draft deferments were generally available for people in food-related occupations, but most people served in the military anyway, making farm labor scarce. Those agricultural engineers who completed their education helped to meet a variety of United States needs in war production plants and in the military service. Few returned to the farm, although many continued in other agricultural pursuits including equipment manufacture, processing, distribution, and marketing of farm products. Although the number of people on farms decreased, the total number of people in agriculturally related enterprises has remained fairly constant from 1920 to the present. As the civilian industrial production got back in gear, new larger pieces of equipment became available, thus increasing the productivity of labor.

Whereas previously the emphasis of mechanization was to increase production, land and resource productivity now became paramount. Irrigation was a way of increasing land productivity. To conserve water as compared to channel or ditch irrigation, overhead irrigation was used. Later, the center pivot overhead system was introduced as a means of increasing labor productivity as well as increasing efficiency in water use on some soils.

Whereas mechanization could improve precision and reduce drudgery, labor productivity increased rapidly also with larger power units and implements operated at high speeds. In fact, labor productivity improved more rapidly than land productivity. Larger farm units, with nearly ten times the area per worker evolved after 1945. Likewise farm equipment and tractors greatly increased in size—from 15–30 hp to 100 hp, from 2–4 row to 12 row equipment, and hay and storage bales from 40–100 lb. to 500, and 5 ft. combines to 20 ft.—all of which generally increased worker productivity.

The question as to whether labor was pulled or pushed from the farm is still debated, although both phenomena are involved. Without the ability of one person to produce enough food for fifty to seventy-five others, the industrial capacity of the United States could not have been reached. The issue of replacing workers with mechanization, particularly in heavily populated regions of the world, continues.

After World War II, considerable emphasis was placed on increasing the productivity of livestock. Materials handling, structures, environmental controls, and storages were designed and built, often making previous facilities obsolete. Devices for lifting and handling materials in or on pallets were widely used after 1945. Labor-serving (as contrasted to labor-saving) systems were sought to handle many operations for seed and fertilizer, grain and fodder, manure, forage, and produce. Agricultural engineers had a major role in conceiving, designing, manufacturing, and using these devices.

The use of chemicals to increase land productivity for weed and insect control, soil treatment, and product treatment put stringent specifications on chemical distribution and use to assure safety to the environment—a new challenge for the engineer. The point of diminishing return appears to have been reached for chemical fertilizer application on several crops, particularly cereals. Concerns for the chemical pollution of the environment caused new consideration on the amount, type and precision of application of chemicals. Agricultural engineers have designed new equipment to improve accuracy of distribution, uniformity of coverage, and low-volume distribution devices.

Vegetable, fruit, and horticultural crops, including those in greenhouses, also are the province of the agricultural engineers from the standpoint of the greenhouse structure and its environmental control. Handling the products has been greatly improved as a result of research and development. Quality control of products developed through electronic sorting and handling devices, expert systems, light transmission and other radiation measurements for quality, and CAD/CAM systems. Food engineering has become an integral part of the profession. Much of the on-farm food processing moved to industry sites and is now classified as industrial food processing. And new fields are emerging in agricultural engineering such as catfish farming, aquaculture, hydroponics, sod and turf management, and energy production.

Biotechnology in the 1970s with the new biology based on DNA and genetic engineering has become an important field for agricultural engineers. Molecular biologists collaborated with agricultural engineers in some cases as members of agricultural engineering departments.

With the rapid increase in the price of oil, the use of agricultural and forestry biomass as a fuel, opened up new areas of interest and opportunities. The production of biomass provides a means of capturing and concentrating solar energy. Considerable research and development are being done on various biomass treatments to produce fuel and feedstocks.

Computers are now widely used by teacher, researcher, producer and processor. These tools are as important in agriculture as in other industries.

Responding to these trends, several University departments are currently identified by different names: agricultural and biological engineering; agricultural and chemical engineering; agricultural and irrigation engineering; biological and agricultural engineering; bio-resource engineering; civil and agricultural engineering; forest engineering; natural resource engineering; and food engineering. These project stronger biological commitments for the future and the need to more closely identify with biologists is clear.

During the 1950s, agricultural engineering became truly a multi-disciplinary field. Dozens of disciplines have become partners for research and problem-solving including economics, other fields of engineering, and the sciences (agricultural, biological, chemical, electrical, physics). Problems are studied on the basis of the system involved and an attempt is made to consider smaller problems or subsystems on the basis of the larger system. Agricultural engineering represents an amalgamation of parts of several fields of engineering plus an understanding of the plant, animal, product and food systems, often with the human factors involved. And all must be done with respect for the natural resources—soil, plants, water, and air.

Following World War II, there was considerable activity throughout the world to develop undergraduate and graduate programs. Every continent had at least one major project to develop programs in agricultural engineering with people from Europe and North America contributing money and personnel. Curricula of different styles and contents were fashioned by the needs of the country. Many faculty members from developing countries studied and received advanced degrees at North American and European universities, probably in greater proportions than other fields of engineering and agriculture.

Educational institutions greatly increased agricultural engineering Ph.D. programs beginning around 1950 (see Chapter 3). These programs consisted primarily of courses in mathematics, physics, and other engineering fields with the dissertation directed towards an agricultural engineering subject. In this mode of operation, many people with Ph.D.s in agricultural engineering have the equivalent of a degree in another field where most of the course work was taken. Accreditation of engineering degrees became commonplace and agricultural engineering was recognized on the same basis as other engineering degrees with a appropriate undergraduate curriculum. And, professional registration in the states provided a separate category of agricultural engineers, although the registration was usually "professional engineer" without specialty designation. The same fundamentals of engineering examination was required for all engineers. Another measure of being accepted as an engineer by the United States government is a listing in the *Dictionary of Occupational Titles*. In the 1950s, agricultural engineers received a standard occupational classification, SOC 1632. This was significant in that agricultural engineers were recognized as engineers, and it was not necessary to give them a title outside their field. There was high interest by students, particularly as the teaching and research emphasis shifted to principles using examples for practice. Research projects usually were done cooperatively with people in the related areas such as plant and animal scientists, machine designers, materials specialists, mathematicians, and physicists. The *Journal of Agricultural Engineering Research* in England became recognized as one of the leading journals to publish fundamental findings and a place to locate the latest literature, in English, on the subject. In selected areas, German (machine design and structures), French (mechanization and soil and water) and Russian (tractors, tillage and grain handling) literature were widely cited.

In the midfifties, the following technical divisions made up the ASAE indicative of the major professional interests:

> Power and Machinery
> Soil and Water
> Farm Structures
> Electric Power and Processing

Through the efforts of many professionals, the need was envisioned for a series of books based on science and engineering laws and principles, with greater emphasis on principles than previous books. Thus, during the fifties, the *Ferguson Foundation Agricultural Engineering Series*, published by John Wiley, provided the standard textbooks designed for those studying in the accredited curriculum in agricultural engineering.

Barre, Henry J., and L. L. Sammet. *Farm Structures*. New York: Wiley, 1950. 650p.
Barger, Edgar L., et al. *Tractors and Their Power Units*. New York: Wiley, 1952. 496p.
Bainer, Roy, R. A. Kepner, and E. L. Barger. *Principles of Farm Machinery*. New York: Wiley, 1955. 571p.
Frevert, Richard K. et al. *Soil and Water Conservation Engineering*. New York: Wiley, 1955. 479p.
Henderson, S. M. and R. L. Perry. *Agricultural Process Engineering*. New York: Wiley, 1955. 402p.
Schwab, Glenn O. et al. *Elementary Soil and Water Engineering*. New York: Wiley, 1957. 296p.
Heinton, Truman E., Dennis E. Wiant, and Oral A. Brown. *Electricity in Agricultural Engineering*. New York: Wiley, 1958. 393p.

These books, some now in their third and fourth editions, continue as major undergraduate references. Although designed primarily as undergraduate texts, these books have provided tremendous improvement in instruction, and helped increase the interest in the field, stimulated and improved research, and helped attract people with a greater inclination for academic and research interests. McGraw-Hill published a parallel series in agricultural engineering, generally serving a broader audience, with Quincy C. Ayres as consulting editor. There are twelve books in the McGraw-Hill agricultural engineering series.

The 1960s, and beyond, witnessed a maturing of the profession with the publication of the *Agricultural Engineers' Handbook* (1961), by C. B. Richey, P. Jacobson, and C. W. Hall (McGraw-Hill). This paralleled other engineering handbooks. The *Agricultural Engineering Index*[6] and its successors and the *Dictionary of Agricultural Engineering* provided those basic references available to most fields. Agricultural engineering joined other professions in the academic structure from the standpoint of a complete range of programs from undergraduate to graduate and research. Initiative for research came primarily from industry and experiment stations and later from a wide range of government/industry foundations, associations and commodity groups. Ph.D. degrees began to be offered as an identified A.E. degree by the department as a part of the graduate school programs of major universities. Supporters of research in agricultural engineering became more tolerant of theoretical considerations, putting less demand on researchers to work on problems of immediate concern.

In the last half of the sixties, the AVI Publishing Co., now a part of Van Nostrand Reinhold, published several engineering books devoted to the food processing industry, along with books specifically for agricultural engineers emphasizing the environmental aspects of plants, animals, and products. An *Encyclopedia of Food Engineering*, the first in the field, was published by AVI in 1970 and later revised (1984).

Although relevancy continued to be an important driving force for agricultural engineering research, the emphasis on the application subsided, leaving that aspect or approach to the industries involved. And along with society in general, departments became more interested in research and teaching and turned to working with energy and the environment.

The developments in agricultural engineering have paralleled developments in technology, particularly including such areas as robotics, materials, properties, energy, manufacturing, and expert systems, with books written to incorporate the research on the subjects.

Many of these developments were chronicled by Robert E. Stewart in a 1979 book, *Seven Decades That Changed America*, published by ASAE. Bryan A. Morgan, head of the Library at Silsoe College in Cranfield, wrote *Keyguide to Information Sources in Agricultural Engineering*, published by Mansell Publishing, Ltd., London. FAO has published a comprehensive listing worldwide of instructional research, and related laboratories in agricultural engineering. The era closed with a three-volume *Handbook of Engineering in Agriculture*, edited by R. H. Brown, published by CRC Press.

The 1980s witnessed a change to the following ASAE divisions to represent the major fields in agricultural engineering:

---

6. Carl W. Hall, *Agricultural Engineering Index, 1907–1970* (Reynoldsburg, Ohio: Agricultural Consulting Associates, 1961–72).

Electrical and Electronic Systems
Food and Process Engineering
Power and Machinery
Soil and Water
Structures and Environment
Emerging Technologies (to identify new technologies; mistakenly considered by some as a home for new divisions, although that is not the objective)

The change in name of academic units parallels economic changes and those taking place in the profession. Thus, at Cornell University, for example, the department names in succession are:

>Division of Rural Engineering and Architecture, 1907
>Department of Farm Mechanics, 1910
>Department of Rural Engineering, 1913
>Department of Agricultural Engineering, 1930
>Department of Agricultural and Biological Engineering, 1988

E. Summary of Major Trends Influenced by Agricultural Engineering[7]

(1) Labor productivity (production per worker-hour) increased much more rapidly (four times) than land productivity (production per hectare; 1.5 times) particularly from 1935 to the present.
(2) Tractors rapidly replaced horses and then with the larger implements and greater speed increased labor productivity.
(3) Electrification helped improve both animal productivity and labor productivity on the farmstead and in the home. The appliances of the farm home including telephone, radio, television, and bathroom and laundry facilities, are equivalent to those of an urban home.
(4) Development of steel, fuel and lubricants, and rubber has greatly influenced the design and use of power units and their attachments.
(5) Control of soil erosion and efficient use of water were important developments that depended on agricultural engineers for leadership. These areas were first important as a part of the conservation efforts and later were elements in minimizing tillage of soil to decrease energy, and to decrease soil and water loss while increasing production.
(6) Need for labor for military service and for operating industrial services during World War II created a demand for agricultural systems that needed to be more labor efficient.
(7) Use of biomass as a means of capturing solar energy and as a fuel to replace fossil fuels for agriculture and industry remains a potential alternative to those who would unreasonably increase the price of fossil fuels. Use of ethanol produced from biomass for reducing pollution of combustion systems became practical.

---

7. For more background, see: United States Department of Agriculture, *The Yearbook of Agriculture* (Washington, D.C.: USDA, 1960).

(8) Recent units, devices and systems incorporate electronics, sensors, controls, expert systems, instrumentation and computers to provide a near parallel to what is available in other industries.
(9) Reduction and elimination of pollution of water bodies, streams, and ground water from chemicals used for fertilizer, pest control and plant regulators, and from wastes became an important part of agricultural engineering.
(10) Biological systems are becoming more important in improving productivity, so engineers are recognizing the need to strengthen the biological component of their education, research and practice.

3. Agricultural Engineering Education Programs

DONALD M. EDWARDS
College of Agriculture and Natural Resources
University of Nebraska

Employment in the United States food and agricultural system has changed significantly over the past 200 years. In 1800, over 95% of the five million people in the United States were involved in agriculture (Figure 3.1). The production unit, the farm, also served as the place for processing and storing the raw materials. By 1870, the population had grown to about 40 million, of which nearly 90% were engaged in agriculture. For approximately the past 125 years, employment in the American agricultural and food system has remained nearly constant at about 35 million people.

Since the mid-1900s the composition of the employment has changed significantly. In the 1990s about 15% of the American population will be employed in the total agricultural and food systems with about 2–3% only in production agriculture. While the 2–3% represents about 5 million people, the needs of these people and the others employed in the total food and agriculture system have changed significantly. The need for specially educated people has increased.

In the early 1800s, higher education was restricted primarily to theological and classical studies. During the mid-1800s, there was a major reform movement toward a more democratic quality of education focusing on the practical application of physical and natural sciences for provision of skilled farmers and factory workers. Schools were created to advance agriculture, industrial and mechanical arts, and the trades. The underlying theme was that agriculture and science should be essential elements in the curriculum complemented by mechanical arts. Prior to the 1860s, higher education provided little access for members of working-class families. Because most of the universities were private, the legislators moved to establish universities for the masses. This led to the passage of the First Morrill Land-Grant Act

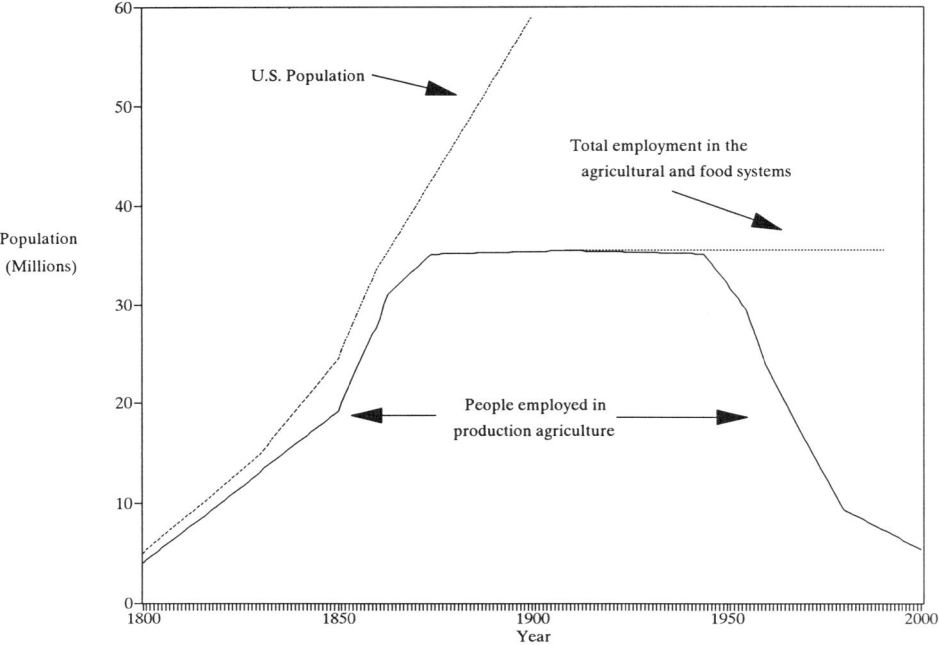

Figure 3.1. People employed in the United States agricultural and food system. (Sources: D. M. Edwards et al., *The Future Role of Engineering in the United States Department of Agriculture*, Washington, D.C.: USDA Cooperative State Research Service, 1988 [Special Report for a USDA S & E Opportunity Evaluation Study], pp. 1–30; also in D. M. Edwards, *Agricultural Engineering in North America in the 21st Century*, Proceedings of the Eleventh International Congress on Agricultural Engineering, Dublin, Ireland: Rotterdam, Netherlands: A. A. Balkeman, 1989.)

of 1862,[1] which provided that all Americans should have equal access to higher education and occupations in agriculture and the industrial and mechanical arts. With the emancipation of slaves in the late 1860s, Congress moved to enact the Second Morrill Act in 1890 to insure that black Americans be included in the mainstream of the United States educational system.[2]

During the last half of the 1800s, colleges of agriculture and engineering were being developed. The Land-Grant Act provided the rapid development of these colleges at public institutions. Typical agriculture and engineering curricula during this time had science occupy one-half of the student's time. English, philosophy, history, and geography comprised one-quarter of the

---

1. J. B. Kendrick, *The Past Is but Prologue for the Future: Or Is It?* (Washington, D.C.: USDA Cooperative State Research Service, 1986), *Agriculture Leadership in Science.*
2. W. P. Hytche, *A National Resource—A National Challenge: The 1890 Land-Grant Colleges and Universities* (Washington, D.C.: USDA Cooperative State Research Service, 1989), *Justin Smith Morrill Memorial Lecture.*

student's time with the remainder devoted to practical courses that were augmented by performing daily tasks on a farm or in a factory.

During the late 1800s, departments of civil and rural engineering became commonplace at land-grant colleges. A typical description in a college catalog would read:

> This study embraces the application of the engineering art to agriculture as seen in improved machinery, well-constructed buildings, conveniently arranged fields, etc. The special subjects will be construction, use and repair of farm machinery. The principles of building include farming, bridge work, limes, mortars, and cement. Drainage will consist of surveying and construction, road building, and repair. The student shall be required to submit for criticism neatly drawn plans of buildings, maps of farms, and drawings or models of useful mechanical contrivances. An engineering lathe or some other machinery tools will be at the service of the class.[3]

As the quiet revolution from human to animal to mechanical power was occurring between the late 1800s and early 1900s, the need for engineers was recognized. Engineers, by virtue of their ability to solve problems, were called upon to respond to increased demands placed upon the American food and agricultural system. Due to the rapid development of the country, few engineers of the established engineering professions of civil, mechanical, mining, and electrical engineering were interested in serving agriculture. In fact, many people were more interested in getting away from agriculture as represented by farming.

In 1906 J. Brownlee Davidson went from the University of Nebraska to Iowa State University to head the first agricultural engineering department in the United States.[4] Shortly after he arrived at Iowa State, the first undergraduate degree program in agricultural engineering was started. The thrust of the program was to prepare graduate engineers who could direct their talents to meet the needs and solve the problems of agriculture. The initial graduates were similar in educational pursuits to the civil and rural engineers of the late 1800s, except they focused on the needs of agriculture. Courses in agricultural engineering were practical and skills oriented. These courses did not change significantly throughout the first half of the 1900s. Representative course offerings in agricultural engineering departments are shown in Table 3.1. Although the courses appeared to provide practical "hands-on" experiences, there was a concern that the college graduates were conceited and had a know-it-all attitude as they began jobs in a practi-

---

3. Accreditation Board for Engineering and Technology, *Annual Reports* (New York, 1932–89).

4. R. E. Stewart, *Seven Decades That Changed America: A History of the American Society of Agricultural Engineers, 1907–1977* (St. Joseph, Mich.: American Society of Agricultural Engineers, 1979).

Table 3.1. Representative courses taught by an Agricultural Engineering department, 1906–40

Year	Course Title
1906–1919	Wood Shop
1906–1919	Forge Shop
1906	Power Machinery
1906	Farm Machinery
1910–1938	Farm Structures
1914–1938	Farm Conveniences (1931 called Farm Home Utilities)
1916–1937	Farm Drainage and Land Clearing
1920–1929	Cement, Concrete, and Materials
1922	Farm Structures (Graduate Course)
1922–1929	Farm Conveniences (Graduate Course)
1922	Power and Machinery (Graduate Course)
1925–1932	Poultry House Construction
1925	Farm Equipment
1926–1940	Explosives
1928	Rural Electrification
1928	Technical Problems
1928–1930	Testing Farm Equipment
1929	Farm Mechanics
1929	Explosives (2nd course of two)
1929	Rural Electrification (Graduate Course)
1930	Land Development (Graduate Course)
1932–1938	Farm Structures (1st course of two)
1938	Power Machinery (2nd course of two)
1941	Farm Shop Practices

(Source: A. W. Farrall et al., *From the Ground Up—75 Years of Progress in the Agricultural Engineering Department, 1906–1981*, Michigan State University, E. Lansing, Mich., 1986.)

cal world among people who had grown into special jobs by the route of the "school of hard knocks."

In the very early 1900s, concern was addressed to the definition of agricultural engineering and what constituted an agricultural engineering curriculum. Agricultural engineering was often referred to as rural or farm engineering and instruction was considered to have the two objectives, first, of providing a practical working knowledge of the ordinary mechanical operations of the farm, and second, of preparing the student for a professional career in some field of engineering applied to agriculture. This was the first expression of the distinction between service and professional teaching in agricultural engineering. This distinction was not resolved until after agricultural engineering programs became accredited as engineering programs beginning in the mid-1930s. Even in the 1990s, agricultural engineering departments provide both service and professional teaching usually with two sets of different courses.

An early recognition for the professional development of a student beyond the confines of the classroom was the establishment in 1919 at Iowa State University of a campus-based agricultural engineering student branch of the recently formed American Society of Agricultural Engineers. The purpose of the student branch was to develop professionalism, provide an early entry into the national organization, and to tie academic instruction with professional practice.

Throughout the 1910s and 1920s, debate continued between deans of agriculture and deans of engineering on the role of instruction in agricultural engineering. Agricultural deans felt more at home in discussing the contributions of agricultural engineers to the field of farm mechanics. By contrast, engineering deans, especially at Iowa State University, University of Nebraska, and Kansas State University, generally saw agricultural engineers as professional engineers who brought engineering expertise directly to solving agricultural problems. The departments continued to debate the role of service versus professional instruction. Most agricultural engineering instruction remained practical and applications oriented.

The American Society of Agricultural Engineers (ASAE) established a College Division in 1920.[5] The purpose of the Division was to further the interests of agricultural engineering in land-grant colleges, to promote research and better teaching methods, and to better correlate these activities with those of the U.S. Department of Agriculture. One outcome was to develop a set of standard methods in education. The College Division continued until it evolved into the Education and Research Division in 1968. Agricultural engineering developed strong technical directions during the 1920s that began to impact the curricula. Also, during the 1920s, agricultural engineers began to participate actively in engineering educational activities of the Society for Promotion of Engineering Education. (The name was later changed to the American Society for Engineering Education.)

The first attempt by the College Division of ASAE at standardizing courses and curricula occurred in 1922. Debates on course content for agricultural engineers centered on animal versus mechanical power, farm motors and farm machinery to be taught as separate or integrated courses, or quality of engineering courses as compared to other college courses, and the role of agricultural engineering in teaching farm mechanics in rural high schools.

5. *Agricultural Engineering, Journal of the American Society of Agricultural Engineering* (St. Joseph, Mich.: American Society of Agricultural Engineers, 1919–20).

As of 1921, the professional agricultural engineering courses had produced 157 B.S. degrees, and 14 post-graduate degrees, with 173 students enrolled (Table 3.2). The students were instructed by 140 teachers. Discussion focuses on the need for textbooks.

By 1928 fourteen colleges in the United States and one in Canada had established degree programs in agricultural engineering. C. W. Smith, University of Nebraska, suggested that ASAE appoint an inspection committee to visit each college trying for an accredited standing by following the plan worked out by a standardization committee. The first agricultural engineering program to be accredited did not occur until 1936.

In the late 1920s, agricultural engineers began to participate in international consultation and visitations to other countries. This lead to major international activities in the 1930s resulting in more internationalism of agricultural engineering curricula. A major involvement in international professional development occurred in 1930 with the holding of the first International Congress of Agricultural Engineering in Belgium. The congress was organized by the Commission Internationale du Génie Rural (CIGR). Until recently these congresses have been devoted to research reports, with little emphasis on curricula.

Table 3.2. Student enrollments in majors offered by United States Agricultural Engineering departments

Years	Agricultural engineering enrollment	Years	Mechanized agriculture enrollment
1910	0	1979	1960
1921	173	1980	2169
1931	210[a]	1983	2270
1941	200[a]	1985	1839
1959	2002	1986	1214
1962	2265	1987	907
1975	1945	1988	896
1980	3364		
1982	3328		
1984	2674		
1986	1822		

(Sources: *Agricultural Engineering, Journal of the American Society of Agricultural Engineering*, St. Joseph, Mich.: American Society of Agricultural Engineers, 1919–90; W. A. Maley, *ASAE Data*, St. Joseph, Mich.: American Society of Agricultural Engineers, 1990; *Engineering Education, 1983–1990*, publication of the American Society for Engineering Education, Washington, D.C.: ASEE.)

[a]Estimated

Twenty-one departments graduated forty B.S. degree majors in agricultural engineering in 1930. A major concern for the very low enrollments was a lack of aggressive recruitment of "white farm boys." For the first time, recruitment efforts were initiated nationally. Throughout the 1930s, more departments were changing from "farm mechanics" titles to "agricultural engineering" titles, but the curricula was little changed.

Ohio State University implemented the first five-year professional agricultural engineering curriculum in 1935 with joint administration between agriculture and engineering. Students were becoming actively involved in ASAE professional activities and were recognized in 1933 with the establishment of the National Council for Student Branches.

The agricultural engineering curricula experienced an increased emphasis on soil and water conservation during the dustbowl years, and an increased emphasis on power and machinery during the 1940–50 decades. The dustbowl problems of the 1930s created much of the momentum on soil and water engineering. The high demand for farm machinery following World War II increased emphasis on power and machinery engineering. With growing concern of the agricultural engineering curriculum being recognized as a professional engineering program that included standardization of programs, agricultural engineering requested accreditation from the Engineering Council for Professional Development (currently Accreditation Board for Engineering and Technology). In 1936 agricultural engineering programs at Iowa State University, University of Nebraska, and Kansas State University were accredited (Tables 3.3, 3.4, 3.5).

Agricultural engineering continued to seek its uniqueness among the engineering professions. C. O. Reed from Ohio State University introduced and promoted the concept of agricultural engineering as the "engineering of agricultural biology" in 1937. To him it seemed that agricultural engineering was simply the service of mechanical, civil, electrical, architectural and industrial engineering applied to the industry of agriculture. His efforts received some support but nothing really happened. Stewart writes in his *Seven Decades That Changed America*:

> Had the concept been adequately matured and widely accepted, certain troubles lying in wait for agricultural engineering could possibly have been avoided or minimized. Accreditation of professional curricula and recruitment of students into them might have been made easier by a philosophy which clearly differentiated from those of other branches of engineering. The role and structure of ASAE itself might also have been clarified in relation to other technical societies.[6]

6. Stewart, *Seven Decades That Changed America*.

Table 3.3. ABET (Accreditation Board for Engineering and Technology)—accredited engineering programs in United States Agricultural Engineering departments (or similarly named departments)

Year Accredited	Institution	Program
1932	ECPD (Currently ABET)	Agricultural Engineering
1936	Iowa State University	Agricultural Engineering
	Kansas State University	Agricultural Engineering
	University of Nebraska	Agricultural Engineering
1949	Oregon State University	Agricultural Engineering
1950	University of Idaho	Agricultural Engineering
	University of Illinois	Agricultural Engineering
	Louisiana State University	Agricultural Engineering
	Michigan State University	Agricultural Engineering
	University of Minnesota	Agricultural Engineering
	Oklahoma State University	Agricultural Engineering
	Purdue University	Agricultural Engineering
	Texas A & M University	Agricultural Engineering
	Washington State University	Agricultural Engineering
1951	Virginia Polytechnic Inst.	Agricultural Engineering
1952	University of Missouri	Agricultural Engineering
	Rutgers University	Agricultural Engineering
1953	Auburn University	Agricultural Engineering
	Clemson University	Agricultural Engineering
	North Dakota State University	Agricultural Engineering
1954	Mississippi State University	Agricultural Engineering
	Ohio State University	Agricultural Engineering
1955	University of Arkansas	Agricultural Engineering
1956	Pennsylvania State University	Agricultural Engineering
1958	Colorado State University	Agricultural Engineering
	Cornell University	Agricultural Engineering
	University of Maine	Agricultural Engineering
	North Carolina State University	Biological & Agricultural Engineering
1959	Louisiana Tech University	Agricultural Engineering
1960	New Mexico State University	Agricultural Engineering
1961	University of Georgia	Agricultural Engineering
	South Dakota State University	Agricultural Engineering
1962	University of Florida	Agricultural Engineering
1964	University of Tennessee	Agricultural Engineering
1965	University of California, Davis	Agricultural Engineering
1967	University of Kentucky	Agricultural Engineering
	Texas Tech University	Agricultural Engineering
	Utah State University	Agricultural Engineering
1968	University of Wyoming	Agricultural Engineering
1969	University of Arizona	Agricultural Engineering
1970	University of Maryland	Agricultural Engineering
1971	University of Wisconsin	Agricultural Engineering
1972	Mississippi State University	Biological Engineering

1973	California Polytechnic State University, San Luis Obsipo	Agricultural Engineering
1974	Montana State University	Agricultural Engineering
1975	Utah State University	Agricultural Engineering & Irrigation Engineering
1977	University of Maine	Forest Engineering
1979	Arkansas State University	Agricultural Engineering
1983	State University of New York, Syracuse	Forest Engineering
1985	California State Polytechnic University, Pomona	Agricultural Engineering Agricultural Engineering
1986	Purdue University	Food Processing Engineering

(Sources: Accreditation Board for Engineering and Technology (ABET), *Annual Reports*, New York: 1932–90; W. A. Maley, *History Chronology 1906–1986*, St. Joseph, Mich.: American Society of Agricultural Engineers, 1986.)

Similar thrusts to gain a uniqueness for agricultural engineering by emphasizing biological engineering did not occur until the 1960s, and again in the late 1980s. Each time some changes occurred, especially in shaping the curricula.

World War II brought the agricultural engineering enrollments almost to zero. Students and many faculty were called to war duty. Following the war, agricultural engineering programs placed greater emphasis upon power and machine applications and the immediate "training" of students to enter the industries and farms serving agriculture. A concern was that students were being trained and not provided an education. This concern lead to an

Table 3.4. CEAB—accredited engineering programs in Canadian Agricultural Engineering departments (or similarly named departments)

Year Accredited	Institution	Program
1965	University of Saskatchewan	Agricultural Engineering
	University of British Columbia	Agricultural Engineering
1971	University of Manitoba	Agricultural Engineering
1972	University of McGill (Macdonald College)	Agricultural Engineering
	University of New Brunswick	Forest Engineering
1973	University of Guelph	Agricultural Engineering Biological Engineering Water Resources Engineering
	University of Laval	Rural Engineering
1974	Technical University of Nova Scotia	Agricultural Engineering
1979	University of British Columbia	Bio-Resources Engineering
1983	University of Alberta	Agricultural Engineering

(Source: ABET, *Annual Reports*, New York.)

Table 3.5. Non-ABET—accredited engineering programs in United States Agricultural Engineering departments (or similarly named departments)

Institution	Program
Auburn University	Forest Engineering
University of Hawaii	Agricultural Engineering
University of Southwestern Louisiana	Agricultural Engineering
University of Massachusetts	Food Engineering
North Carolina A & T University	Agricultural Engineering
Prairie View A & M University	Agricultural Engineering
University of Washington	Forest Engineering

(Sources: *Agricultural Engineering, Journal of the American Society of Agricultural Engineering*, St. Joseph, Mich.: American Society of Agricultural Engineers, 1919–90; W. A. Maley, *History Chronology 1906–1986*, St. Joseph, Mich.: American Society of Agricultural Engineers, 1986.)

aggressive promotion of professionalism among agricultural engineers and the encouragement of professional engineering accreditation. Prior to 1949, only three agricultural engineering programs were accredited, but by 1960, twenty-five additional agricultural engineering programs became accredited. The 1950s became the decade that established agricultural engineering as a recognized branch of engineering.

Doctoral programs in agricultural engineering were actively established during the last half of the 1940s. Michigan State University and Iowa State University were the major producers of Ph.D. degree graduates. Only eight doctor's degrees were granted prior to 1950. Between 1950 and 1956, forty-three Ph.D. degrees were granted, twenty-six by Michigan State University.[7]

The first agricultural engineering department outside North America was developed in 1921 at Allahabad Agricultural Institute in India.[8] Under the leadership of Mason Vaugh, a graduate in the United States, and the first department head, an undergraduate agricultural engineering curriculum was established in 1942. Fourteen institutions in India now offer undergraduate agricultural engineering degree programs; fourteen offer M.S. agricultural engineering degree programs (first established in 1956), and five offer Ph.D. agricultural engineering degree programs, first established in 1962.

7. A. W. Farrall et al., *From the Ground Up—75 Years of Progress in the Agricultural Engineering Department* (E. Lansing, Mich.: Michigan State University, 1986).

8. A. Kumar and S. Tiwary, "Agricultural Engineering Teaching, Research, and Extension in India: Present State and Future Need" and A. Kumar, S. Tiwary, and R. K. Rai, "Response of Undergraduate Agricultural Engineering Students to Systems of Education and Evaluation," in *Proceedings of the Eleventh International Congress on Agricultural Engineering* (Dublin, Ireland; Rotterdam, Netherlands: A. A. Balkeman, 1989).

Unlike other branches of engineering in India, the agricultural engineering profession aims at helping the agricultural industry, directly and indirectly.

In many countries of South America, Africa, and Asia where educational assistance for higher education was provided, curricula in agricultural engineering are often similar to programs in the United States and Canada. These are usually five years in length for bachelor's degrees with an extra 1–2 years for master's. Usually the recipient country chooses a university to provide the curriculum and support that they believe would be most appropriate.

There is little similarity of agricultural engineering curricula in Europe. To date there has been no accreditation organization. Few institutions have the combined engineering and agriculture expertise at one location. However, regional meetings have been held in England, Ireland, and Western Europe as a result of EC92 (European Community) increased coordination and collaboration can be expected.

To gain employment opportunities for agricultural engineering graduates within the U.S. government, considerable effort was put forth in the 1950s to have the U.S. Department of Labor include agricultural engineering on the Critical Occupations List. In 1956 the title was listed.

ASAE had been active in the American Society of Engineering Education (ASEE), and was elected to membership in 1954. This is significant in that it recognized agricultural engineering as having a major interest in engineering education. This membership led agricultural engineering to have its own division within ASEE. The name of the division was changed to Biological and Agricultural Engineering Division in 1988.[9]

While registration of agricultural engineers as professional engineers was available as early as 1929, ASAE became more closely associated with the National Council of State Boards of Engineering Examiners in 1952 in promoting greater uniformity among the states in registration of engineers. This action led to registration of engineers as Professional Engineers, the last step in recognizing agricultural engineers as professionals.[10] With accreditation of agricultural engineering curricula and the opportunity to become registered professional engineers, agricultural engineering students were provided all professional recognition opportunities of other engineering students.

The debate continued throughout 1940s and 1950s on the role of agricultural engineering departments offerings of academic curricula. In 1948 at

9. *Engineering Education, 1983–1990*, publication of the American Society of Engineering Education (Washington, D.C.: ASEE).
10. *Professional Engineers 1934–1990*, publications of the National Society of Professional Engineers (Alexandria, Va.: NSPE).

a meeting of the Association of Land-Grant Colleges and Universities, a resolution was passed that set the future pattern for curricula offerings and departmental administration of departments of agricultural engineering. Under the leadership of A. A. Potter, Dean of Engineering at Purdue University, the deans of agriculture and engineering adopted this resolution:

> Resolved that the principles approved by the Resident Instruction Sections of the Divisions of Agriculture and Engineering at their joint conference on Tuesday, November 7, 1948, be used as a guide in formulating curricula leading to degrees of Bachelor of Science in Agricultural Engineering and to Bachelor of Science in Agriculture with a major in Mechanized Agriculture, and that land-grant institutions which have sound programs of study leading to the B.S. degree in Agricultural Engineering apply promptly to the Engineers' Council for Professional Development (ECPD) for inspection and accreditation.[11]

After recognition of the two distinct undergraduate curricula, the need for graduate programs was laid out. Finally, it was recommended that mechanized agriculture (or agricultural mechanization) programs be administered by the agricultural dean, and that the agricultural engineering programs be jointly administered by the engineering and agricultural deans, presently a common practice.

The decade after the Soviet Sputnik went into orbit in 1957 saw a growing emphasis on electronics, environmental issues, energy, human productivity and efficiency, safety, and natural resources utilization. A renewed interest in bio-engineering surfaced. Several agricultural engineering departments considered broadening the agricultural engineering base to include all aspects of biological engineering. ASAE was encourage to change its name to the American Society of Biological Engineers. These issues had significant impacts upon the agricultural engineering curricula. Greater emphasis was placed on the incorporation of the basic biological sciences in the curricula.

A. W. Farrall at Michigan State University stated that the 1960s were the decade that initiated the biological age. He expressed his views in: "The question might be asked, what is the difference between agricultural engineering and the other principal disciplines of engineering. The answer might be: 'Agricultural engineering is unique in that it involves specifically biological and environmental factors, since it deals with engineering applied to biological matter—food, feed, natural fiber, animals and humans.' Note that the unique feature, the one we have which is different from any other is the emphasis on the biological factor."[12] Many agricultural engineering cur-

---

11. *Engineering Education*, op. cit.
12. Stewart, *Seven Decades That Changed America*.

ricula made significant changes in course content and names. Only a few curricula and department names were changed to reflect this redirection. North Carolina State University, Mississippi State University, and Rutgers University were the first to change department and curricula names to Agricultural and Biological Engineering, or to Biological Engineering.

The ASAE membership was not ready to make a name change. The industrial members of ASAE were reluctant to encourage the idea that all agricultural engineers were some kind of bioengineers, or that ASAE was composed only of bioengineers. These decisions did not signify rejection of biology as a major component of the agricultural engineering practice, but the decisions postponed the needed definition of the uniqueness of the agricultural engineering profession.

The agricultural mechanization programs began to increase during the 1960s (Table 3.6) as agricultural efficiency and profitability claimed greater attention. These programs increased emphasis on management and on working with a complete production "system." Equipment became more available and, at times, appeared to be the answer to increased agricultural productivity.

One program, Agricultural Engineering Technology at the University of Delaware, became accredited as an ABET Technology Program in 1984.[13] Maximizing yields was a major consideration in curricula content. This changed during the 1970–80 decades with greater emphasis on optimizing yields and profit efficiency through systems analysis and engineering. Additionally, greater consideration of environmental factors was given in agricultural engineering studies.

In addition to changes in university curricula content and a few name changes, the only visible change within ASAE was the establishment of the food engineering division in 1967, and the bioengineering committee in the newly formed Education and Research Division in 1968.[14]

With greater emphasis on the quality of education in the 1960s, engineering sciences were expanded. Many professional engineers were concerned that the focus on the engineering sciences was detracting from design. In recognition of quality professional engineering education, ASAE became a full member of ECPD in 1966.[15] The ASAE Curriculum and Course Content Committee established in the 1940s became active in the 1960s in assisting departments to develop quality agricultural engineering curricula, to encourage accreditation, and to delineate between accredited and nonaccredited programs in publications. Certification of non-engineering programs was initiated by ASAE for those programs in agricultural engineering

13. ABET, *Annual Reports*.
14. *Agricultural Engineering*.
15. ABET, *Annual Reports*.

Table 3.6. Non-engineering programs in United States and Canada Agricultural Engineering departments (or similarly named departments), 1989

Institution	Program
UNITED STATES	
University of Arkansas	Agricultural Mechanization
University of Arizona	Irrigation
California Polytechnic State University, San Luis Obispo	Agricultural Mechanization and Agricultural Engineering Technology
California State University, Chico	Agricultural Engineering Technology
California State University, Fresno	Agricultural Mechanization Agricultural Engineering Technology
University of Connecticut	Resource Engineering Technology
Cornell University	Biosystems Technology
University of Delaware	Agricultural Engineering Technology
University of Florida	Agricultural Operations Management
University of Georgia	Mechanized Technology
University of Hawaii	Mechanized Agriculture
University of Idaho	Agricultural Mechanization
Southern Illinois University	Agricultural Mechanization
University of Illinois	Agricultural Mechanization
Purdue University	Agricultural Mechanization
Iowa State University	Agricultural Mechanization
Kansas State University	Agricultural Mechanization
Louisiana State University	Agricultural Mechanization
Michigan State University	Agricultural Technology and Systems Management Building Construction Management Electrical Technology
Mississippi State University	Agricultural Engineering Technology and Business
University of Missouri	Agricultural Mechanization
Montana State University	Mechanized Agriculture
University of Nebraska	Mechanized Agriculture
Rutgers University	Biological/Agricultural/Environmental Sciences
North Carolina State University	Agricultural Systems Technology
North Dakota State University	Agricultural Mechanization
Ohio State University	Agricultural Mechanization and Systems
Oklahoma State University	Mechanized Agriculture
Oregon State University	Agricultural Engineering Technology
Pennsylvania State University	Agricultural Mechanization
University of Puerto Rico	Mechanized Technology in Agriculture
Clemson University	Agricultural Mechanization and Business
South Dakota State University	Mechanized Agriculture
Texas A & M University	Agricultural Systems Management
University of Vermont	Agricultural and Energy Technology
Washington State University	Agricultural Mechanization
University of Wisconsin, Madison	Agricultural Mechanization and Management

University of Wisconsin, Platteville	Agricultural Engineering Technology
University of Wisconsin, River Falls	Agricultural Mechanization
CANADA	
University of Alberta	Agricultural Mechanization
University of Saskatchewan	Agricultural Mechanization

(Sources: *Agricultural Engineering, Journal of the American Society of Agricultural Engineering*, St. Joseph, Mich.: American Society of Agricultural Engineers, 1919–90; W. A. Maley, *History Chronology 1906–1986*, St. Joseph, Mich.: American Society of Agricultural Engineers, 1986.)

departments not accredited by ECPD. During this period, ASAE took an aggressive role along with the universities and industries in career guidance. Following major publicity and evaluation of results, it was concluded that the most effective recruitment and career guidance was performed at the local and state levels.

The early 1970s was a period of student protests. The outcome resulted in increased emphasis upon non-military use of engineers, social effects of technology, ethical implications of new technologies and biotechnologies, environmental issues, greater use of renewal energy resources, and less dependency upon replacing people with machines. While the youthful idealism asked for too much too quickly, notable changes were incorporated into both agricultural engineering and agricultural mechanization programs. To accommodate many of the social, ethical, environmental and energy topics, biological sciences were reduced in the course content.

Whereas the 1960s appeared to place emphasis on the beginning of a biological era, the next decade threatened to change this focus to the beginning of a social and ethical era. Curricula began to emphasize accountability of technology, the humanities and social sciences, ethics, and a pastoral approach to agricultural production. Pastoral was defined as production in keeping with constraints on available natural resources. More emphasis was placed on the utilization and conservation of natural resources, especially soil, land, forests, and water. Food process engineering gained attention during the 1970s, and agriculture was beginning to be viewed as a total food system.[16]

During this period, several departments of agricultural engineering began reconsidering the search for their unique identity among the engineering professions. As noted in Tables 3.3 and 3.4, several engineering programs changed names and became accredited. Forest Engineering, Bio-resources

---

16. Ibid., *Engineering Education*.

Engineering, Biological Engineering, and Food Engineering became accredited programs during the 1970–80s.

The events of the past twenty years have begun to force agricultural engineering departments to identify clearly their uniqueness among academic programs. Budgets became short and programs had to be reduced, merged, or dropped. Due to the lack of a clear and unique identity, several departments were reduced in size and scope, merged with other engineering departments, or eliminated. Other agricultural engineering departments will face these same decisions during the 1990s, especially if they do not identify their uniqueness and justify their existence.[17]

Beginning in 1960, expanding in the 1970s, and becoming commonplace in the 1980s, electronic computers made changes in curricula. Computers were used to reorder the solutions to complex problems. The inexpensive microcomputer of the last decade brought computing power capabilities to all individuals, classrooms, and curricula. From word processing to computer-generated graphics to complex design, computers made an impact that continues at a rapid pace.

Agricultural engineering heads from throughout North America came together in 1987 to discuss the future roles of the agricultural engineering profession. The result was an accelerated effort to reevaluate their academic programs. A consensus was formed on the purpose of agricultural engineering, or a similarly named department. The purpose states: "Departments of Agricultural Engineering (name may change depending on local preferences) are engineering and agriculture—based academic units of professionals. These professionals develop, adopt, and disseminate knowledge and technologies that focus on the engineering and management of biological and agricultural systems and natural resources. Their goals are to formulate efficient and effective production, processing, storage, distribution and utilization of agricultural and other biological products for food and non-food needs of society."[18]

Engineering curricula have been completely redesigned (Table 3.7), and programs established in Food Processing Engineering, Biological Systems Engineering, Agricultural and Biological Engineering, and Biochemical Engineering. Several departments in North American have changed or will be changing names of departments and curriculum to address the purpose stated above. Non-engineering curricula have been changed in content (Ta-

---

17. (a) D. M. Edwards, ed., *Project 2001: Engineering for the 21st Century* (Columbus, Ohio, 1987), Proceedings of the International Conference on Strategic Planning for North American Agricultural Engineering Departments. (b) A. Johnson et al., *A Model Biological Engineering Curriculum* (Washington, D.C.: National Science Foundation, 1989).

18. Edwards, *Project 2001*.

Table 3.7. Typical courses taught in Agricultural Engineering departments, 1990

Undergraduate	Graduate
Introduction to Agricultural Engineering	Land Locomotion
Physical Principles of Biological Processes	Finite Element Method
Physical Principles of Plant Environment	Bio-Processing Engineering
Environment of Biological Systems	Physical Properties of Agricultural Products
Electric Power and Controls	Instrumentation for Agricultural Engineering Research
Food Process Engineering	
Systems of Agricultural Machines	Research Methods in Agricultural Engineering Seminar
Meteorology	
Water Movement in Biological Systems	Advanced Power and Machinery Design
Microclimatology	Biological Waste Management
Communication Techniques for Agricultural Engineers	Environmental Engineering
	Bioengineering
Principles of Structures and Machines	Non-Newtonian Flow
Agricultural Structural Design	Special Problems
Processing Biological Products	Master's Thesis Research
Operations Research	Doctoral Dissertation Research
Special Problems Management	
Soil and Water Engineering	
Power Systems	
Fundamentals of Design and Design Project	

(Source: University Academic Catalogs, 1990.)

ble 3.8) and in name (Table 3.6) to Agricultural Engineering Technology, Agricultural Technology and Systems Management, Biosystems Technology, and Agricultural Systems Management. During this period of time, ASAE continued to make insignificant changes in organization to reflect the changing trends.

The changing trends in America's agricultural and food system, Figure 3.1, mandate that changes occur in agricultural engineering departments' curricula, both in content and name. Anything less than a serious and thorough review and implementation of new thrusts will result in numerous current programs to cease to exist. As can be seen in Tables 3.1, 3.7 and 3.8, considerable changes have occurred since 1900. Significant changes equal to the accumulative changes since 1900 must be made in the next few years if the agricultural and food systems are to be served. Departments of agricultural engineering must change curricula to address the changes resulting from biotechnology, high technology, renewable biological systems, renewable energy sources, bioprocessing, natural resources utilization and management, environmental management, food and personal safety, human ergonomics, information and management systems, ethics, social issues, internationalism and computing.

Table 3.8. Typical courses taught in Agricultural Mechanization or Agricultural Engineering Technology, 1990

Undergraduate	Graduate
Introduction to Agricultural Engineering Technology	Agricultural Mechanization in Developing Countries
Computer and Information Processing in Agriculture and Natural Resources	Environmental Measurements
	Analysis of Agricultural Systems
Technical Agricultural Mechanics Skills	Man-Machine Relationships
Materials and Processes	Research Methods in Agricultural Engineering Technology Seminar
Commercial Food Processing Systems	
Housing Conservation	Special Problems
Automotive and Recreational Engines	Master's Thesis Research
Microclimate: The Biological Environment	Doctoral Dissertation Research
Technical Skills	
Systems Analysis in Agricultural Production	
Mechanical Systems in Agriculture and Natural Resources	
Processing Systems for Biological Products	
Energy in the Food System	
Teaching Agricultural Mechanics	
Agricultural and Natural Resources Safety	
Light Structural Systems	
Electrical Energy Utilization	
Irrigation, Drainage, and Erosion Control Systems	
Machinery and Tractor Systems	
Special Problems	

(Source: D. M. Edwards, ed., *Project 2001: Engineering for the 21st Century*, Proceedings of the International Conference on Strategic Planning for North American Agricultural Engineering Departments, Columbus, Ohio, 1987.)

4. Agricultural Engineering Literature in Developing Countries

MAKOTO HOKI
Department of Bioprodution Machinery, Mie University

HISASHI HORIO
Department of Agricultural Engineering, Kobe University

GAJENDRA SINGH
Division of Agricultural and Food Engineering, Asian Institute of Technology

A. Introduction

Most of what are developing countries today made a transition after World War II from colonial status to independence. Agricultural engineering research, development and education started after independence and grew with the nation.

Agricultural engineering in developing countries must deal with wide-ranging circumstances of agricultural development depending on regional, environmental and historical conditions. The agricultural and socio-economic systems in most developing countries before World War II were formed under the economic framework of the mother countries. Therefore, socio-economic conditions were at a stage of primitive capital accumulation. In these times, technology transfer from the developed countries was seldom done since there was no incentive for the transfer as viewed from the mother countries. Only after independence did the new nations start their own agricultural development, and agricultural engineering considerations.

Most agricultural engineering innovations have been based on labor saving tools and mechanization. These machines have served to promote intensive farming. This is only true, however, where an economic structure is well developed and sufficient social capital are available. The limited success experienced by many developing countries in establishing agricultural engineering technologies has been partly due to insufficient investment on research and development.

Agricultural innovation or mechanization cannot by achieved simply by the introduction of advanced technologies. Mechanization has often been misunderstood as being synonymous with labor saving technology for even small-scale farming in rice production in many developing countries. Labor saving concepts should be viewed from two sides: enlargement of cultivated land, and capital-saving technology.

An influential fact is that mechanization often has been synonymous with labor saving technology thereby misleading many agricultural scientists and their decision-makers who are concerned about labor displacement and unemployment. This situation has resulted in less attention to or neglect of mechanization and agricultural engineering.

B. Reference Books Relevant to Developing Countries

Agricultural engineering in developing countries requires the engineering discipline as a base which is discussed in another section. Books and reference materials pertaining to agricultural technologies and systems useful for developing countries are reviewed here.

The literature of agricultural engineering for developing countries includes diverse subjects. Due to various reasons, the literature dealing with agricultural engineering or mechanization in the developing countries is short of the demand, especially to meet the growing needs of research and education programs for universities. Books dealing with agricultural engineering's role in the total biological production systems are of urgent need in order to cope with rapidly increasing population and environmental concerns.

Specialized agricultural engineering books and journals for the developing countries have appeared considerably later when compared to other agricultural fields.

Esmay and Hall[1] are credited with one of the earliest books in the field, *Agricultural Mechanization in Developing Countries* (1973), which was reviewed by Marilyn Car in *Appropriate Technology* (3 (1): 18); the main part of the review is quoted here.

> Over the last decade, economists and technologists have concentrated to such an extent on the role of tractors in agricultural production, that the terms

1. M. L. Esmay and C. W. Hall, eds., *Agricultural Mechanization in Developing Countries* (Tokyo: Shin-Norinsha Co., 1973).

"tractorization" and "agricultural mechanization" have become almost synonymous. It is refreshing, therefore to find a book which covers agricultural mechanization in its fullest sense, and gives extensive coverage to such neglected technologies as improved hand tools, animal-drawn equipment, and storage facilities. Mechanization is defined, in fact as "any devices, tools and machines which extend the hand of man, and reduce drudgery." The first four chapters cover the principles of agricultural mechanization, and the current patterns of mechanization in Equatorial Africa, Asia, and Latin America. The information included in these chapters provides a very useful introduction for anyone interested in both the theoretical and practical aspects of the subject. These are followed by chapters relating specifically to irrigation, drying, sorting and handling of food grains, and the extremely important topics of ownership patterns of machinery, and education and training for agricultural mechanization. Thus, the book indicates not only the wide range of technologies available for increasing levels of food production in developing countries, but also the range of alternative ways of introducing such technologies to smaller farms, and the ensuring efficient utilization of equipment.

For the economic stability of developing countries, more attention must be paid to small-holdings farmers who are a majority. They should be able to receive more benefits and better income by virtue of technological improvement or mechanization of any segment of the agricultural production and processing system. Mechanization aimed at large farms in the major world granary areas entails labor saving technologies as a major element. The technologies adopted for large-scale farming are not effective for the farms having small land holdings and limited investment capabilities. Small farm mechanization is a more difficult and important subject from the Third World point of view.

A challenge was made by Crossley and Kilgour in their *Small Farm Mechanization for Developing Countries*.[2] The book focuses attention to small farm requirements, their environment, and the suitability of alternative technical solutions. It provides information on the design, development, testing, selection, operation, and maintenance of small scale machinery and systems, in the economic, social, and physical environments of developing countries. This is a unique textbook for those students, teachers and practicing engineers who apply engineering principles and tackle practical problems of agricultural mechanization on small farms in developing countries. Crossley and Kilgour placed an emphasis on small-scale upland farming which makes this book suitable to deal with specific prevailing conditions in some areas of Africa, the Middle East, and India.

2. P. Crossley and J. Kilgour, *Small Farm Mechanization for Developing Countries* (New York: John Wiley, 1983).

As previously mentioned, agricultural engineering is an integrated discipline based on the engineering and agricultural sciences. With the advancement of modern technologies an understanding of their impact on the complex socio-economic, cultural and environmental factors involved in agricultural development becomes increasingly important.

Arnon wrote *Modernization of Agriculture in Developing Countries* based upon his substantial experience in agricultural research, teaching and program implementation in many developing countries in Asia, Africa and Latin America.[3] The book contains detailed descriptions and a review of present knowledge on constraints encountered in modernizing agriculture in developing countries and the problems that arise after new technologies have been adopted. The information provided is useful to agricultural engineers who need to deal with diversified other fields related to their own activities.

Agricultural development accompanied by mechanization has been understood by many people as a Western method of large scale farming. Yet the majority of farmers of the world practice small-scale farming where mechanization has been unsuccessful in many cases in the past. Although not numerous, examples of successful small-scale mechanization exist in some Asian, rice-producing countries including Japan, Korea, and Taiwan. Although their ways of mechanization cannot be applied directly to other countries having different backgrounds, a thorough review will be useful in understanding the processes and significant factors leading to successful small-scale rice mechanization.

Ohkawa et al. wrote *Agriculture and Economic Growth: Japan's Experience*.[4] The book provides relevant information for a comprehensive appraisal of Japan's agricultural development based upon small-scale and intensive rice mechanization over the last 100 years. There is discussion of economic and historical aspects of mechanization and their relevance to developing countries. The book can be an excellent text or reference book for a graduate level course dealing with small-scale mechanization processes and their economics.

Thailand and Malaysia started mechanization in the 1960s and are now making a rapid transformation from the traditional into modern farming. Chancellor provided an excellent interim report entitled *Survey of Tractor Contractor Operation in Thailand and Malaysia*.[5] Although country-spe-

---

3. I. Arnon, *Modernization of Agriculture in Developing Countries*, 2d ed. (New York: John Wiley, 1987).

4. K. Ohkawa, B. F. Johnston, and H. Kaneda, eds., *Agriculture and Economic Growth: Japan's Experience* (Tokyo: University of Tokyo Press, 1969; and Princeton, N.J.: Princeton University Press, 1970).

5. W. J. Chancellor, *Survey of Tractor Contractor Operations in Thailand and Malaysia* (Davis, Calif.: University of California, Department of Agricultural Engineering, 1970).

cific, the report contains both quantitative and qualitative engineering and economic material relevant to the mechanization process for other developing countries. The contrast of Thai and Malaysian mechanization is also presented. The study results which contain a wide range of field data will be of great help in comprehending an intermediate phase of rice mechanization applicable to other developing countries.

Farm Mechanization in Japan by Kaburaki deals with small-farm mechanization characteristics and the historical process of development and technical advancement which follow.[6] This book is a useful information source for practicing engineers in developing countries because of its descriptive technical information and engineering data relevant to small-scale rice production and upland crop production machinery including forage, vegetable, fruits, livestock, pulse and industrial crops.

After extensive work on small-scale mechanization, the International Rice Research Institute (IRRI) compiled *Consequences of Small-Farm Mechanization*.[7] The book contains perspectives and issues of the subject in Asian countries including Pakistan, Philippines, Thailand, Bangladesh, Indonesia and Nepal based on a research project: The Consequences of Small Farm Mechanization on Production, Income and Rural Employment in Selected Countries of Asia undertaken jointly by IRRI and the Agricultural Development Council Inc.

IRRI and International Crops Research Institute for the Semi-Arid Tropics (ICRISAT) have agricultural engineering research sections which are active in mechanization research and extension programs with particular emphasis for developing countries. They also provide technical information and publications of practical use which are available upon request.

The Agricultural Engineering Service of the Food and Agricultural Organization of the United Nations (FAO) is engaged in the development work of agricultural mechanization, rural storage, handling and drying, and rural buildings. Technical information pertaining to these subjects is published as *FAO Agricultural Service Bulletins* and in other formats. The United Nations Industrial Development Organization (UNIDO, P.O. Box 300, A-1400 Vienna, Austria) is also a source of publications for industrial development and rural technology transfer. The *UNIDO Newsletter* is sent free of charge to approved readers; inquiries should be made to the Vienna headquarters.

Agricultural Mechanization in Development by Gifford, published by FAO, defines and puts in perspective the relationships between agricultural

---

6. H. Kaburaki, A. Hosokawa, and K. Maeda, eds., *Farm Mechanization in Japan* (Tokyo: Association of Agricultural Relations in Asia, 1982).

7. International Rice Research Institute, and Agricultural Development Council, *Consequences of Small-Farm Mechanization* (Manila: IRRI, 1983).

mechanization and national development objectives in developing countries.[8] This brief FAO bulletin provides guidelines for appropriate mechanization strategy formulation. Another *FAO Agricultural Service Bulletin* is *Elements of Agricultural Machinery* in two volumes by Wilkinson and Braunbeck.[9] The book provides guidelines for teachers and technicians in the developing countries about the role of agricultural mechanization and management. The content deals with technical issues and economics of relatively large-scale upland crop production machinery.

Equipment for Rice Production by Stout is a concise introductory book for small-scale rice machinery published by FAO in 1966.[10]

Considering current diversified conditions and a wide range of technical levels of mechanization, the most needed books today should deal with small rice machinery, and management applicable to specific levels and conditions.

Engineering technology advancements provide the tools for agricultural engineers to be involved in all aspects of agricultural production and environmental protection. These tools are often needed to grasp the implications of the agricultural or bioproduction system as well as to monitor details of individual operations and processes. A systems approach to agriculture is becoming a powerful tool to cope with problems of developing countries because of the many different and complex agricultural configurations.

Spedding wrote an excellent textbook, *An Introduction to Agricultural Systems*, which defines and describes a multidisciplinary systems approach, the need for it, and the ways it can be used.[11] Chapters include those on subsistence farming and shifting cultivation which still prevail in some developing countries, as well as industrial food production systems. This is a good introductory undergraduate textbook for agricultural engineering, but can also serve as an introduction to the systems approach for practicing engineers who need to cope with complex agricultural situations.

Some practical reference materials dealing with appropriate agricultural technology include *Regional Catalogue of Agricultural Implements* prepared by the Regional Network for Agricultural Machinery;[12] *Tools for Agriculture: A Buyer's Guide to Appropriate Equipment* by Volunteers in

---

8. R. C. Gifford, *Agricultural Mechanization in Development: Guidelines for Strategy Formulation*. (Rome: FAO, 1981). (*FAO Agricultural Service Bulletin* no. 45)

9. R. H. Wilkinson and O. A. Braunbeck, *Elements of Agricultural Machinery*, Vol. 1 and Vol. 2. (Rome: FAO, 1977). (*FAO Agricultural Service Bulletin* no. 12)

10. B. A. Stout, *Equipment for Rice Production* (Rome: FAO, 1966).

11. C. R. W. Spedding, *An Introduction to Agricultural Systems* (London: Elsevier Applied Science Publishers Ltd., 1988).

12. Regional Network for Agricultural Machinery (RNAM), *Regional Catalogue of Agricultural Implements* (Los Baños, Philippines: RNAM, 1983).

Technical Assistance (VITA);[13] *Harnessing and Implements for Animal Traction: An Animal Traction Resource Book for Africa* by Starkey;[14] *Directory of Agricultural Machinery and Manufacturers* by Singh and Bhardwaj;[15] and *Agricultural Engineering in Development: Guidelines for Establishment of Village Workshops* by Pothecary et al.[16]

C. Journals and Proceedings Relevant to Developing Countries

Technical articles and papers dealing with agricultural engineering related specifically to developing countries have appeared seldom in primary professional journals such as *Transactions of the ASAE* and *Journal of Agricultural Engineering Research*. These prominent journals have been more concerned with academic research for industrialized countries, although this pattern is changing. Recent articles in these journals dealing with the subjects pertaining to developing countries are reviewed by Hoki et al.[17]

There are two major journals that specifically deal with agricultural engineering in developing countries: *Agricultural Mechanization in Asia, Africa and Latin America (AMA);* and *Appropriate Technology* (International Technology Development Group, Parnell House, 25 Wilton Road, London SW1V 1JS, U.K.).

The AMA journal covers all aspects of agricultural mechanization and has opened a valuable channel of information exchange in developing countries. Information in this quarterly journal for the past 18 years is summarized in Table 4.1. Recent years of the AMA are also indexed by the ASAE.

Appropriate Technology is a unique quarterly that specifically addresses appropriate rural technologies with emphasis on the poorer people. It provides practical and effective self-help techniques using various indigenous materials. There is an emphasis on technologies which are labor-intensive, capital saving, and simple to use and maintain by people without mechani-

13. Volunteers in Technical Assistance (VITA), *Tools for Agriculture: A Buyer's Guide to Appropriate Equipment* (Mt. Rainier, Md.: VITA, 1985). 4th ed., 1992, issued by Intermediate Technology Publ., London.
14. P. Starkey, *Harnessing and Implements for Animal Traction: An Animal Traction Resource Book for Africa* (Braunschweig/Wiesbaden, Germany: Friedr. Vieweg, 1989).
15. G. Singh and K. C. Bhardwaj, *Directory of Agricultural Machinery and Manufacturers* (Bhopal, India: Central Institute of Agricultural Engineering).
16. B. Pothecary, S. U. Skarp, and S. A. Dembner, *Agricultural Engineering in Development: Guidelines for Establishment of Village Workshops* (Rome: FAO, 1988). (*FAO Agricultural Service Bulletin* no. 71)
17. M. Hoki, H. Horio, and G. Singh, *Agricultural Engineering in Developing Countries;* Annual Report of Bioproduction Systems Research Institute, no. 3. (Tsu, Japan: Mie University Department of Bioproduction Machinery, 1990). p. 143–152.

Table 4.1. Articles published in *Agricultural Mechanization in Asia, Africa, and America*

Engineering areas	No. of articles	Primary countries of authors	Primary subjects
Human power	10	Bangladesh, India, Philippines	Ergonomics of field operations; Manually operated machines; Anthropometry.
Animal power	16	Bangladesh, India, Mexico, Nigeria, Thailand, UK	Potential in developing countries; Harness designs; Performance under varying load and climatic conditions.
Handling & processing	200	Bangladesh, Bolivia, Canada, Chile, China, Denmark, Ghana, India, Italy, Japan, Kenya, Korea, Malaysia, Nepal, Netherlands, New Zealand, Nigeria, Pakistan, Papua New Guinea, Philippines, Saudi Arabia, Sri Lanka, Sudan, Taiwan, Thailand, Turkey, UK, USA, West Germany, West Indies	Drying of rice, peanut, rapeseed; Rice parboiling techniques; Post-harvest losses; Grain storage; Mechanized fruit harvesting; Reapers for different crops; Multi-crop threshers.
Mechanization & management	450	Almost all countries of Asia; Australia, Brazil, Canada, Chile, Denmark, Egypt, Ghana, Italy, Kenya, Lebanon, Libya, Mexico, New Zealand, Netherlands, Nigeria, Papua New Guinea, Sierra Leone, Spain, Sudan, Switzerland, Turkey, West Germany	Historical development, status, and prospects of mechanization; Socio-economic, institutional and energy problems; Mechanization strategies; Economics of machinery operations; Optimum machinery selection; Farm machinery industry.
Land & environment	60	Bangladesh, China, Colombia, Ethiopia, Ghana, Guyana, India, Iran, Iraq, Japan, Lebanon, Malaysia, Netherlands, New Zealand, Nigeria, Pakistan, Philippines, Saudi Arabia, Sri Lanka, Sudan, Syria, Taiwan, Thailand, USA	Evaluation of soils for tillage characteristics; Canal irrigation systems; Socio-economic aspects of irrigation; Drainage; Zero tillage; Runoff & soil erosion.
Field implements	200	Bangladesh, Brazil, Cameroon, Canada, Chile, Denmark, Hungary, India, Iran, Iraq, Ireland, Japan, Netherlands, New Zealand, Nigeria, Pakistan, Philippines, Sudan, Tanzania, Thailand, Turkey, UK, USA, West Germany, Zambia	Traditional and modern hand, animal, and power tillage; Plant protection and seeding implements and equipment; Tillage practices in developing countries; Comparative performance of traditional and modern implements; Soil bins; Tractive performance of

> cage wheels and pneumatic tires; Manual and power operated sprayers; Direct sowing methods; Seed-cum-fertilizer drills; Oscillatory tools; Transplanters; Mechanics of puddling operation with different implements.

cal training. The journal covers all aspects of living necessaries including food, clothing, housing and small village industry, and is useful for small subsistence farms in some African countries, semiarid tropics and other areas with similar conditions.

An important information source for agricultural engineering technology relevant to developing countries is the proceedings of international conferences and workshop. These contain the most current information and in many ways supplement the present insufficiency of proper books and journal articles. Such proceedings include substantial information on experimental, theoretical, and field research results from around the world. Conferences and workshops pertinent to developing countries, held in the past fifteen years, are presented below.

One of the proceedings compiled by Moens and Siepman entitled *Development of the Agricultural Machinery Industry in Developing Countries* is based on The Second International Conference on Agricultural Mechanization in Developing Countries held in Amsterdam, January 1984.[18] This volume includes a large amount of information from countries worldwide, about actual status, recommendations and relevant issues in dealing with the development of the agricultural machinery industry as well as mechanization.

The *Journal of the Japanese Society of Agricultural Machinery* contains a wide variety of highly technical information about crop production, management, handling and processing machinery for small-scale farming. The information is very helpful in solving specific problems in developing countries. The articles are mostly in Japanese, but the English summaries of recent article are available as *JSAM Abstract 1989* from the Japanese Society of Agricultural Machinery (c/o Seiken Kikoh, Nisshin-cho 1-40-2, Omiya, Japan 331) at 500 Yen per copy including postage.

---

18. A. Moens and A. H. J. Siepman, eds., *Development of the Agricultural Machinery Industry in Developing Countries* (Wageningen, Netherlands: Pudoc Publishing, 1984).

List of Valuable International Conferences, Workshops, Seminars, Meetings and Symposia

1974. Expert Consultation Meeting on the Mechanization of Rice Production, June 10–14, International Institute of Tropical Agriculture, Ibadan, Nigeria. *Report* of 280 pages with papers on numerous aspects of rice production mechanization. Published in Ibadan by IITA, 1975.

1977. International Conference on Rural Development Technology: An Integrated Approach, June 21–24, Asian Institute of Technology, Bangkok. *Proceedings* (847 pages) edited by G. Singh and J. H. De Goede entitled *An Integrated Approach*. Papers on rural development strategies; technological changes in agriculture; agricultural mechanization; solar energy applications; bio-gas technology; rural industrialization; transport and marketing; educational and social development; infrastructure and housing; environmental aspects; and case studies on integrated rural development were included in the conference. Published by A.I.T., 1977.

1978. Special International Conference on Agricultural Technology for Developing Nations: Farm Mechanization Alternatives for 1–10 Hectare Farms, May 23–24, University of Illinois at Urbana-Champaign. The *Proceedings* (105 pages) cover papers on socio-economic aspects of small farm mechanization and development, manufacturing and marketing of agricultural equipment.

1978. International Agricultural Machinery Workshop, International Rice Research Institute, Manila. The *Proceedings* (203 pages) cover papers on the status of agricultural mechanization in various Asian countries. Published by IRRI at Los Banos, 1978.

1979. International Conference on Agricultural Engineering in National Development, September 10–16, Universiti Pertanian Malaysia Serdang, Selangor, Malaysia. *Proceedings* edited by Choa S. Lin, M. Z. Bardaie, N. C. Saxena and V. V. Tran cover papers on various aspects of agricultural engineering in development. Published by the University at Selangor, 1981.

1979. UNIDO First Consultation Meeting on the Agricultural Machinery Industry, 15–19 October, Stresa, Italy. New York; United Nations Industrial Development Organization, 1979. 38p. [Report ID/239 (ID/WG.307/9/Rev. 1)]

1981. Regional Seminar on the Appropriate Mechanization of Rural Development with Special Reference to Small Farming in the ASEAN Countries, January 26–31, Jakarta, Indonesia. *Proceedings* (298 pages) include papers on land development; prospectus and manufacturing of agricultural equipment; harvest and post-harvest technology; management aspects of agricultural mechanization; and mechanization policy.

1981. International Conference on Agricultural Engineering and Agro-Industries in Asia, November 10–13, Asian Institute of Technology, Bangkok. *Proceedings* (489 pages) edited by V. K. Jindal, G. Singh, D. Gee-Clough and J. R. Jensen cover soil and water; farm machinery; processing and storage; energy and education.

1982. Regional Seminar on Farm Power and Employment in Asia: Performance and Prospects, October 25–29, Agrarian Research and Training Institute, Colombo, Sri Lanka. *Proceedings* (437 pages) edited by John Farrington, Fredrick Abeyratne and Gerard J. Gill on issues in agricultural mechanization; farm power in Asia; colonization, irrigation and mechanization in Sri Lanka; and future prospects of mechanization.

1983. Second Consultation on the Agricultural Machinery Industry, October 17–21, Vienna. Published in New York; United Nations Industrial Development Organization. 45p. [Report ID/307 (Id/WG.400/8/Rev. 1)]

1984. The Second International Conference on Agricultural Mechanization in Developing Countries, January 23–26, Amsterdam. *Proceedings* (400 pages) compiled by A. Moens and A. H. J. Siepman, entitled *Development of the Agricultural Machinery Industry in Developing Countries* include the information on the actual status, recommendations and relevant issues dealing with the development of agricultural machinery industry as well as mechanization, from countries worldwide. Published in Wageningen, Netherland by Pudoc Publishing, 1984.

1984. Expert Consultation on Irrigation Water Management, July 16–22, Yogyakarta and Bali, Indonesia. *Proceedings* (200 pages) entitled *Participatory Experiences in Irrigation Water Management*, published by Food and Agriculture Organization in Rome (1985) include case studies from Indonesia, Philippines, China, India and Sri Lanka.

1985. International Conference on Small Farm Equipment for Developing Countries: Past Experiences and Future Priorities, September 2–6, International Rice Research Institute, Manila. *Proceedings* (629 pages) cover general mechanization issues in developing countries, agricultural equipment developed and manufactured in various countries, and future R & D needs. Published in Manila by IRRI, 1986.

1986. Third Consultation on the Agricultural Machinery Industry, September 29–October 3, Belgrade. Published in New York by United Nations Industrial Development Organization, 1986. 39p. [Report ID/346 (ID/WG.462/11)]

1987. International Symposium on Agricultural Mechanization and International Cooperation in High Technology Era, April 3, University of Tokyo, Tokyo. *Proceedings* (574 pages) cover papers on energy and automation, biomass and post-harvest technology, and international cooperation and technology transfer.

1987. International Conference on Agricultural Systems Engineering, August 11–14, Changchun, China. *Proceedings* (624 pages) cover review of agricultural systems engineering, theory of agricultural systems, applications of systems engineering for agricultural production techniques, and applications of microcomputer for agricultural production and management. Published by China Machine Press, 1, Nanli, Baiwanzhuang, Beijing.

1988. International Commission of Agricultural Engineering Inter-Sections Symposium, September 5–10, Ilorin, Nigeria. *Proceedings* (463 pages) edited by K. C. Oni contain papers on production machinery, processing machinery and storage systems for small farmers; technology for small-scale irrigation systems; and livestock and domestic water supply for rural communities. Published in Ilorin by NSAE, Department of Agricultural Engineering, Faculty of Engineering & Engineering Technology, University of Ilorin.

1988. The Second Asia-Pacific Conference of the International Society for Terrain Vehicle Systems, December 6–10, Asian Institute of Technology, Bangkok. *Proceedings* (635 pages) edited by David Gee-Clough and Vilas Salokhe cover papers on measurement of soil properties, soil-tool interaction, soil-wheel interaction, tires and tracks, tractive performance and steering, and equipment and machine performance.

1989. International Research Symposium on Draught Animals in Rural Development, July 3–7, Cipanas, Indonesia. *Proceedings* (347 pages) edited by D. Hoffman, J. Nari and R. J. Petheram cover papers on regional emphasis and progress; biological aspects in draught animal power; farming systems research; and engineering aspects and economics of draft animal power. Published and distributed by Inkata Press Pty. Ltd., 13/170 Forster Road, Mount Waverley, Victoria 3149, Australia.

1989. Regional Expert Consultation on Agricultural Implements and Machinery, September 5–8, Regional Office for Asia and the Pacific (FAO), Bangkok. *Report* (51 pages) reviews policy developments relating to agricultural mechanization in the region; progress in the use of improved agricultural implements and machinery; existing arrangements for supply of agricultural implements and machinery to farmers; and the role of government agencies, cooperative/private sector in the promotion, utilization and development of agricultural implants and machinery in the region.

1989. International Symposium on Agricultural Engineering, September 12–15, Beijing. *Proceedings* (1246 pages) entitled *Potentiality of Agricultural Engineering in Rural Development*, edited by Wang Mao-hau contain papers on agricultural mechanization; farm power and machinery; post-harvest engineering and processing engineering; soil and water engineering; farm structure and environment; farm electronic, electrification, automation and remote sensing; rural energy engineering; and agricultural systems and management engineering. Published in 1989 by International Academic Publishers, Xizhimenwai Dajie, Beijing Exhibition Center, Beijing 100044.

1990. Asia-Pacific Regional Conference on Engineering for the Development of Agriculture, June 5–7, Universiti Pertanian Malaysia, Serdang, Selangor, Malaysia. *Proceedings* (555 pages) edited by M. N. Ibrahim, K. C. Yan, A. Yahya and N. Endut cover agricultural machinery development, mechanization, irrigation and drainage, land and water management, waste management, agricultural processing, food engineering, information technology, education, and energy in agriculture.

1990. International Agricultural Engineering Conference and Exhibition, Dec. 3–6, Asian Institute of Technology, Bangkok, Thailand. *Proceedings* (4 vols.) edited by V. M. Salokhe and S. G. Ilangantileke. Volume subjects are: V.1: Farm Power and Machinery; V.2: Post Harvest Technology and Bio-Technology; V.3: Soil and Water Engineering; V.4: Agricultural Systems, Agricultural Waste Management, Energy in Agriculture, Ergonomics, Extension and Training; Structures and Environment. Published in 1991 by Asian Institute of Technology (at U.S. $110., available from V. M. Salokhe)

D. *Concluding Remarks and Recommendations*

The literature reviewed includes a wide variety of the books, journals, reports and proceedings pertaining to the promotion of agricultural engineering technology for the developing countries.

Many developed countries throughout the world have initiated research, education, extension or development programs in agricultural engineering. These programs have often operated over a relatively short period of time or intermittently so there is often a weak trail of work done. Literature resulting from these activities usually does not get distributed, catalogued or indexed in an adequate manner so that others learn of it. Such publications are difficult to retrieve even if they are known. The literature identified in this chapter consists of internationally recognized publications and is not intended to include the numerous important publications with limited circulation and site-specific contents.

The literature is insufficient in a sense to cope with diverse and complex problems; there is also a lack of a proper referencing system suitable for the subjects of developing countries. This literature review at this time forms the foundation for a more thorough review and data consolidation from which textbooks can be published in the future.

A literature review shows that future efforts need to be exerted by agricultural engineers in the following areas.

(1) A network should be implemented to identify the agricultural engineering literature of developing countries through appropriate international organizations and member governments. FAO for example might be one of the possible organizations appropriate for implementing this type of network.
(2) Field survey projects should be initiated to collect current data pertaining to agricultural technology in select developing countries. Emphasis should be placed upon the small-scale production systems of rice and major crops.

The engineering technology which has been regarded as economically valuable should be now considered in combination with socio-economical conditions. This is to create a new technological perspective concerning not only civil and mechanical engineering but also appropriate mechanization and processing technology for developing countries.

SECTION II

Reviews of Emerging Areas in Agricultural Engineering

5. Energy in Agriculture

BILL A. STOUT AND TERRI G. HUFF
Department of Agricultural Engineering
Texas A & M University

A. Introduction

Agricultural engineering began to emerge as a branch of engineering in the early years of the twentieth century. From the beginning, agricultural engineers were concerned with efficiency in the application of energy in agriculture. While energy efficiency was not always an explicit goal, it was often a major driving force as improved machinery, power units, water systems, and other technologies evolved. For example, the steel plow reduced draft, and rubber tires on tractors improved traction. Both developments saved fuel and reduced operating costs. More recently, agricultural engineers have worked to improve the efficiency of electrical energy use in grain dryers, food processing systems, and a host of other applications. Thus, agricultural engineers have always been energy engineers in agriculture.[1]

Fred Buelow began teaching a course on energy concepts in agriculture at Michigan State University in 1957. Topics included world energy, atomic and nuclear energy, solar energy, human energy, and control of energy. Dr. Buelow's energy course was discontinued after a few years because it was ahead of its time. However, his students included Jacob Pos, who years later conducted research in Ontario on anaerobic digestion of livestock wastes; Andy Deshmukh, who helped modernize the milk processing industry in India to make it more energy-efficient; and James Butler, who directed the USDA Southern Biomass Energy Research Center at Tifton, Georgia.

---

1. Robert E. Stewart, *Seven Decades That Changed America* (St. Joseph, Mich.: American Society of Agricultural Engineers, 1979).

United States agriculture has evolved throughout the past half-century in an era of inexpensive and plentiful energy. Consequently, production, processing, marketing, and utilization of food has become heavily reliant on fossil fuels and electricity. Dramatic gains in agricultural productivity have resulted from technological advances vitally dependent on energy. Energy has been utilized primarily to (1) reduce labor requirements; (2) increase the quantity and variety of food produced, while using less land; (3) lower the risk of crop failure or food spoilage; (4) manufacture fertilizers for increased crop production; and (5) reduce waste, while maintaining food quality preferred by consumers as indicated by their willingness to pay.[2]

However, a series of embargoes, strikes, and wars have occurred. These, coupled with shortages resulting from depletion of finite energy resources, have alerted us to the realization that the era of cheap energy is gone and energy supplies for the future are uncertain. Recall, for example, the following events: Oil embargo of 1973, natural gas crisis of 1975, coal strike of 1976, Iranian oil cutoff in 1978–79, gasoline shortages of 1979, and uncertain diesel fuel supplies in 1979. Similar mini-crises are likely to become a part of our future as the demand on finite world oil, gas, and other resources increases.

On April 20, 1977, U.S. President Jimmy Carter gave his famous energy policy speech which outlined a series of energy conservation measures and declared the nation's energy crisis the "moral equivalent of war." So what is the United States energy situation a decade or two later? Have we "won the war"? Hardly. The price of oil dropped from a peak of $35 per barrel in 1981 to around $15 in 1990. The United States oil and gas industry has collapsed. The number of domestic drilling rigs dropped from about 4,000 in 1981[3] to less than 1,000 in 1990. In addition, the United States is importing about 50% of its oil at a cost of more than $50 billion per year,[4] a major factor in our imbalance of trade.

While agricultural energy use is less than 3% of the nation's overall energy consumption, it is a vital 3%. The high yields of modern agriculture have occurred primarily by substituting energy from oil and natural gas for human and animal energy expenditures. Even a brief shortfall or interruption of energy supply can have a severe impact on agricultural production and farm income because the timeliness of farm operations is crucial.

For example, planting and harvesting operations cannot be delayed with-

---

2. R. T. Van Arsdall and P. J. Devlin, *Energy Policies: Price Impacts on the U.S. Food System* (Washington, D.C.: Economic Research Service, USDA, 1978 [*USDA Agricultural Economics Report* no. 407]).
3. *The Houston Post*, Dec. 28, 1989.
4. *The Houston Chronicle*, Sept. 3, 1989.

out adversely affecting crop yields. Many perishable commodities require cooling or refrigeration immediately following harvest to maintain quality and prevent losses. Wet grain must be dried promptly to prevent mold growth and spoilage. Greenhouse operations cannot tolerate even one night of near-freezing temperatures. All of these activities require a reliable supply of energy in the proper form. A factory can be shut down for a few days if an energy shortfall occurs, and production will be lost for the down time. However, an energy shortfall in farming operations has much more serious implications and can dramatically reduce crop yield or lead to a total crop loss.[5]

The purpose of this chapter is to provide a guide to significant sources of information relating to energy in agriculture. Only a small fraction of the vast array of references have been cited and fewer still are discussed in the text because of length constraints. Three publications in particular contain more comprehensive reference lists related to agricultural energy. *Handbook of Energy for World Agriculture*[6] contains more than 600 references, *Energy for World Agriculture*[7] lists 2,613 different titles, and *Bibliography of Biomass Energy*[8] contains 1,550 titles of books, bulletins, and dissertations.

While this review of literature is limited primarily to books published between 1973 and 1989, it would be incomplete without some important conference and workshop proceedings and energy plans. For example, the Council for Agricultural Science and Technology[9] produced two energy reports which called public and legislative attention to the energy needs of agriculture and highlighted energy conservation opportunities in agriculture.[10]

In addition, the American Society of Agricultural Engineers (ASAE) has sponsored many important energy conferences. In 1980, ASAE sponsored the National Energy Symposium at which dozens of papers were presented on energy in crop and livestock production and food processing. Renewable alternative forms of energy, such as solar heating and biomass fuels, were

---

    5. B. A. Stout, "Energy and Agricultural Productivity," Presented at U.S. Department of Energy Regional Hearing, Omaha, Nebr., Dec. 8, 1989.
    6. B. A. Stout, *Handbook of Energy for World Agriculture* (London: Elsevier, 1989).
    7. B. A. Stout et al., *Energy for World Agriculture* (Rome: Food and Agriculture Organization, 1979 [*FAO Agriculture Series* no. 7]).
    8. C. W. Hall, *Bibliography of Biomass Energy* (St. Joseph, Mich.: American Society of Agricultural Engineers, 1986 [Publication 6-86]).
    9. Council for Agricultural Science and Technology, *Potential for Energy Conservation in Agricultural Production* (Ames, Iowa: Council for Agricultural Science and Technology, 1975 [*CAST Report* no. 40]).
    10. (a) B. A. Stout et al., *Energy Use in Agriculture: Now and for the Future* (Ames: Iowa State University, Council for Agricultural Science and Technology, 1977 [*CAST Report* no. 68]).
(b) B. A. Stout et al., *Energy Use and Management in Agriculture* (Boston: Breton, 1984).

also discussed.[11] ASAE also published two public policy issue reports, one addressing the cost and policy impacts on agriculture and the consumer,[12] and another focusing on biological liquid fuels alternatives.[13]

This chapter is intended to serve as an introduction to energy in agriculture for educators, scientists, engineers, and administrators who are seeking a brief overview of the subject. It is not intended to be a substitute for examination of the literature itself. The chapter is organized into six main sections: (1) Energy Flow in the Food System, (2) Energy Management in Agriculture, (3) Renewable Energy, (4) International Dimension, (5) Environmental Implications, and (6) Energy Alternatives and Policies.

B. Energy Flow in the Food System

Total Energy Flow

The total United States food system consumes energy in the industrial (5.5%), commercial (3.2%), residential (3.4%), and transportation (4.4%) segments of the economy. According to a study prepared for the Federal Energy Administration, the total food system uses 16.5% of the nation's energy.[14] Other studies report values from 12–20%, depending on the boundaries given the food system and the extent to which indirect energy usage, such as machinery and buildings, is charged to the food system.[15]

Energy Required for 1 Kilocalorie (Kcal) of Food

Of every 16 kcal of potential plant energy in food, feed, and fiber, only eleven actually enter the United States food chain, with 1.2 kcal being used directly for human food and 9.8 for animal feed. After processing, transport, marketing, and final food preparation, 62% is consumed as vegetable products and 38% as animal products. Of the original 16 kcal, only 1/16 is actually eaten by the United States population.[16]

---

11. B. F. Clary et al., *Selected Papers and Abstracts from the 1980 ASAE National Energy Symposium*, Kansas City, Mo., Sept.–Oct. 1980 (St. Joseph, Mich.: American Society of Agricultural Engineers, 1980).
12. B. A. Stout et al., *Energy—A Vital Resource for the U.S. Food System: Cost and Policy Impacts on Agriculture and the Consumer* (St. Joseph, Mich.: American Society of Agricultural Engineers, 1978).
13. M. L. Miller et al., *The Biological Liquid Fuels Alternative: Technology Status and Engineering Considerations* (St. Joseph, Mich.: American Society of Agricultural Engineers, 1981).
14. Federal Energy Administration, *Energy Use in the Food System* (Washington, D.C.: Office of Industrial Programs, U.S. Government Printing Office, 1976).
15. Stout, *Energy Use and Management in Agriculture*.
16. Stout, *Energy for World Agriculture*.

Figure 5.1 shows the total energy flow needed to feed the entire United States population for one year. For a population of 215 million people with a daily caloric intake of 3,000 kcal per caput, 236 x 10^{12} kcal per year are required.[17]

In 1968, agricultural production in the United Kingdom required 378.4 x 10^{15} joules (90 x 10^{12} kcal) of energy, or 4.6% of that country's primary energy consumption.[18] For this investment, 130 x 10^{15} joules (31 x 10^{12} kcal) of human food (¼ cereals, ¼ roots and vegetables, ¼ meat, and ¼ milk) were delivered, or enough to feed half the population of over 5 billion people.

Table 5.1 provides a detailed breakdown of the energy inputs to agriculture for primary food production. These inputs are only a minor fraction of the total energy needed to produce, process, transport, package, and sell food in the United Kingdom and to import all the animal feed and human food that cannot be or are not grown domestically.[19]

A total of nearly 1,300 x 10^{15} joules of energy for food production to the point of retail sale took 15.7% of the primary energy utilized by the United Kingdom. Agriculture plus domestic fisheries accounted for 31.6%; food processing, 36.6; food and fish imports, about 21%; and food sales (including certain transport inputs), 10.7%.[20]

To illustrate the allocation of the energy flow in agriculture in the United Kingdom, Leach presented a detailed breakdown of the energy inputs for a 1 kilogram (kg) loaf of white bread, sliced and wrapped (Figure 5.2). Just under 20% of the total energy is consumed in growing the wheat, and all but 3% of the remainder is used in processing, packaging, and transport.

There cannot be more energy output from any process than input. Energy is neither created or destroyed. Thus, if all input energy is considered, the output energy is exactly the same, but with reduced availability (increased entropy). The energy ratio measures the edible energy output divided by the energy input.[21] An energy ratio of 1 indicates that the energy content of the fuel produced is exactly equal to the cultural energy input to grow the crop and convert the biomass to a liquid fuel. Many other studies of energy output have been conducted with similar results.[22]

Nearly all pre-industrial farmers and food gatherers achieve very large energy returns on the work they put in.[23] A 1:1 energy ratio is not neces-

17. Ibid.
18. G. Leach, *Energy and Food Production* (Guildford, Surrey, U.K.: IPC Science and Technology Press, 1976).
19. Ibid.
20. Ibid.
21. Ibid.
22. Stout, *Handbook*.
23. Leach, *Energy and Food Production*.

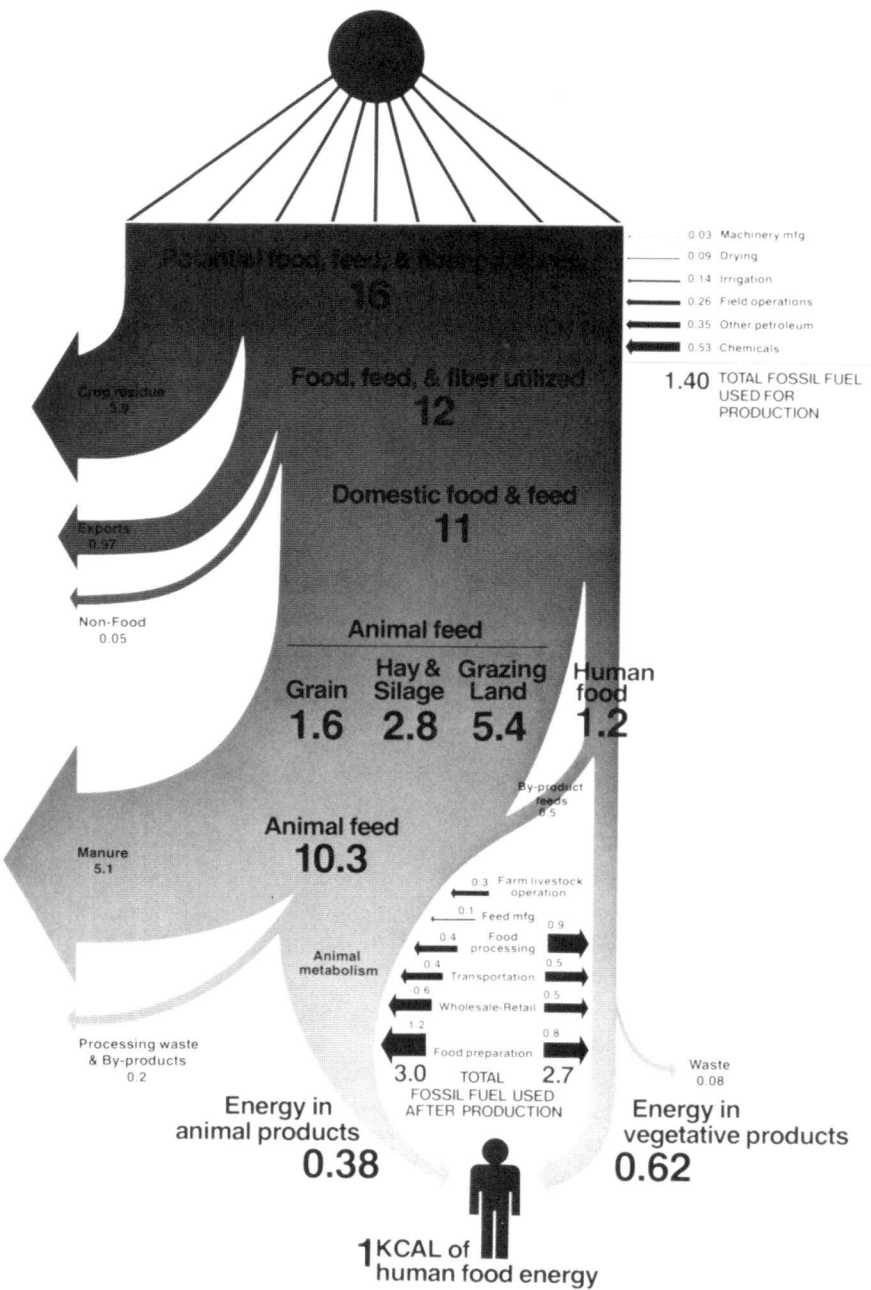

Figure 5.1. Energy flow in the United States food chain to produce a year's supply of food for 215 million people at 3,000 kilocalories per day. (Source: F. C. Stickler, W. C. Burrows, and L. F. Nelson, *Energy from Sun, to Plant, to Man*, Deere and Company, 1975, p. 4.)

Table 5.1. Energy inputs to agriculture in the United Kingdom, 1968

		Quantity	Energy (MGJ)
Coal		0.172 Mt	5.62
Coke		0.10 Mt	3.31
Electricity		(3.444 TWh)	(49.59)
	60% for non-domestic power units	2.066 TWh	29.75
Petroleum:	diesel or gas-oil	0.645 Mt	33.79
	fuel oil	0.034 Mt	1.70
	vaporizing oil	0.060 Mt	3.18
	lubrication	0.011 Mt	0.91
	50% motor spirit	0.111 Mt	6.01
Petroleum:	heating, drying, etc.	0.484 Mt	24.15
(Total petroleum)		(1.345 Mt)	(69.74)
Total direct energy			*108.42*
Fertilizers:	N	0.783 Mt	62.64
	P	0.482 Mt	6.75
	K	0.459 Mt	4.13
	lime	4.20 Mt	8.4
Total fertilizers		*5.924 Mt*	*81.92*
Machinery:	capital	111.0 M£	21.29
	non-capital	46.8 M£	10.48
Total machinery		*157.8 M£*	*31.77*
Chemicals		26.0 M£	8.48
Buildings		87.6 M£	22.77
Miscellaneous		26.1 M£	4.28
Transport, services, etc.		194.1 M£	16.28
Total other inputs		*333.8 M£*	*51.81*
Feedstuffs:	food industries	15.13 Mt	51.3
	grow imports	(7.01 Mt)	35.2
	ship imports	(7.01 Mt)	18.0
Total feedstuffs		*15.13 Mt*	*104.5*
OVERALL TOTAL (rounded)			378
Total output: 2380.2 £M.			
Input/Output: 159 MJ/£			

(Source: G. Leach, *Energy and Food Production*, Guildford, Surrey, U.K.: IPC Science and Technology Press, 1976, p. 48.)

sarily adequate; instead, it appears that a ratio of 15:1 to 20:1 must be achieved to ensure adequate food for these societies.[24] Figure 5.3 shows energy ratios for both subsistence primitive agricultural and industrialized agricultural societies. *Agricultural Energetics* provides a detailed discussion of the uses and limitations of energy ratios.[25]

24. R. C. Fluck and C. D. Baird, *Agricultural Energetics* (Westport, Conn.: AVI, 1980).
25. Ibid.

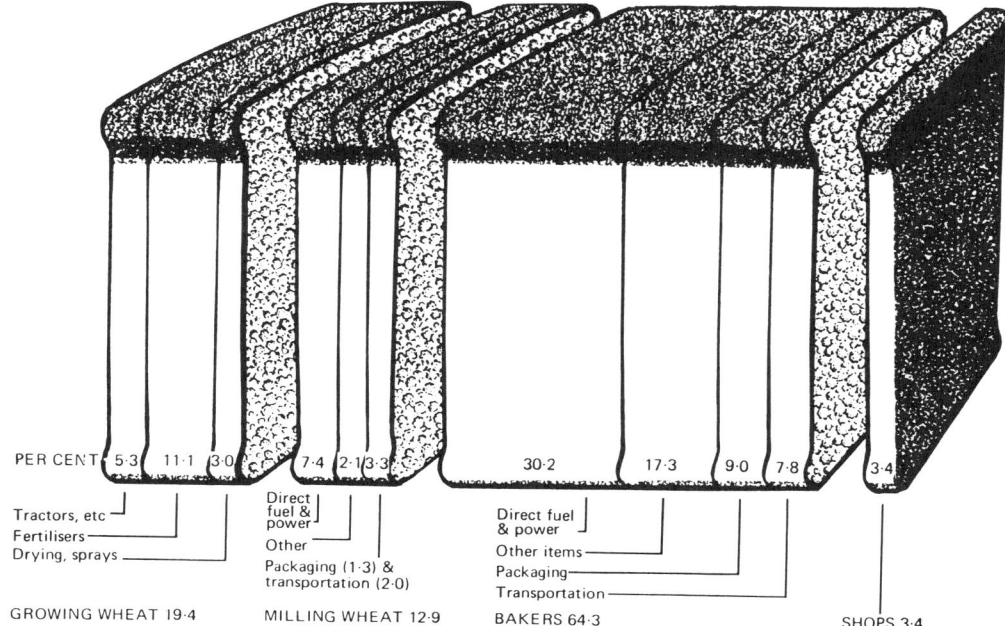

Figure 5.2. Percentage breakdown of the energy required to produce a 1-kilogram loaf of white bread and deliver it to a retail store in the United Kingdom. The total amount of energy required is 20.7×10^6 Joules. (Source: G. Leach, *Energy and Food Production*, Guildford, Surrey, U.K.: IPC Science and Technology Press, 1976, p. 29.)

Both the United States type of agriculture and the "green revolution" type of agriculture have been eminently successful in increasing crop yields through improved technology. In the *Science* article, "Food Production and the Energy Crisis," Pimentel investigated the relationship of energy inputs to crop production by using data on corn. While corn yields increased about 240% from 1945 to 1970, the labor input per acre decreased more than 60%. Intense mechanization reduced the labor input and, in part, made possible the increased corn yield. From 1945 to 1970, mean corn yields increased from about thirty-four bushels per acre to eighty-one bushels per acre (2.4–fold); however, mean energy inputs increased from 0.9 million kcal to 2.9 million kcal (3.1–fold) (Table 5.2). Hence, the yield in corn calories decreased from 3.7 kcal per one fuel kcal input in 1945 to a yield of about 2.8 kcal from 1954 to 1970, a 24% decrease.[26]

26. D. Pimentel et al., "Food Production and the Energy Crisis," *Science* 182 (1973): 443–49.

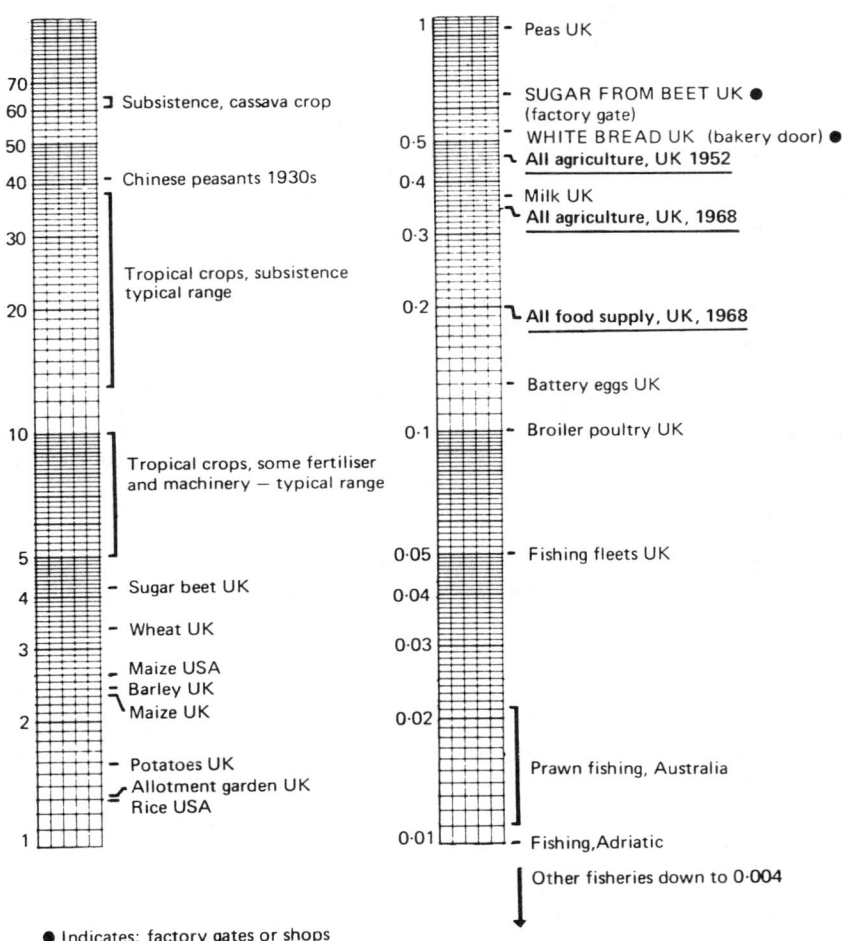

Figure 5.3. Energy ratios for food production. (Source: G. Leach, *Energy and Food Production*, Guildford, Surrey, U.K.: IPC Science and Technology Press, 1976, p. 8.)

These trends in energy inputs and corn yields confirm several agricultural evaluations which conclude that the impressive agricultural production in the United States has been gained through large inputs of fossil energy. This means that the principal raw material of modern United States agriculture is fossil fuel, whereas the labor input is relatively small (about nine hours per crop acre). As agriculture becomes more dependent upon fossil energy, crop production costs will also soar when fuel costs increase two- to five-fold. Then a return of 2.8 kcal of corn per 1 kcal of fuel input may

Table 5.2. Energy inputs (kilocalories) in corn production

INPUT	1945	1950	1954	1959	1964	1970
Labor	12,500	9,800	9,300	7,600	6,000	4,900
Machinery	180,000	250,000	300,000	350,000	420,000	420,000
Gasoline	543,400	615,800	688,300	724,500	760,700	797,000
Nitrogen	58,800	126,000	226,800	344,400	487,200	940,800
Phosphorus	10,600	15,200	18,200	24,300	27,400	47,100
Potassium	5,200	10,500	18,900	36,500	30,400	63,000
Seeds for Planting	34,000	40,400	50,400	60,400	68,000	68,000
Irrigation	19,000	23,000	27,000	31,000	34,000	34,000
Insecticides	0	1,100	3,300	7,700	11,000	11,000
Herbicides	0	600	1,100	2,800	4,200	11,000
Drying	10,000	30,000	60,000	100,000	120,000	120,000
Electricity	32,000	54,000	100,000	140,000	203,000	310,000
Transportation	20,000	30,000	45,000	60,000	70,000	70,000
Total inputs	925,500	1,206,400	1,548,300	1,889,200	2,241,900	2,896,800
Corn yield (output)	3,427,200	3,830,400	4,132,800	5,443,200	6,854,400	8,164,800
Kcal return/input kcal	3.70	3.18	2.67	2.88	3.06	2.82

(Source: D. Pimentel et al., "Food Production and the Energy Crisis," *Science* 182 (1973): 443–49.)

be uneconomical.[27] However, Heichel shows that primitive agriculture produces a higher energy ratio than high-input, modern agriculture, but yields are low. Clearly, high yields are necessary to feed a growing world population, and reduced energy ratios may be necessary to maximize production.[28]

In *Handbook of Energy Utilization in Agriculture*,[29] energy ratios for a variety of crops are discussed. Fossil energy input and food energy output analyses have been calculated and the amount of protein produced is also listed. In assessing fossil energy inputs and crop yields, it becomes evident that the marginal return in increased production per input kcal decreases as a particular input continues to increase. The yield of corn, for example, continues to increase, but at a decreasing rate, with each additional increment of nitrogen fertilizer applied until an input of about 200 kg per hectare (ha) is reached. If the quantity of nitrogen applied increases further, corn yields often decline, particularly without additional water. As a result, the

27. Ibid.
28. G. H. Heichel, *Comparative Efficiency of Energy Use in Crop Production* (New Haven, Conn.: The Connecticut Agricultural Experiment Station, 1973 [*CAES Bulletin* no. 739]).
29. D. Pimentel, *Handbook of Energy Utilization in Agriculture* (Boca Raton: CRC Press, 1980).

energy output/input also declines with the further increment of nitrogen becoming less favorable.

USDA 1974 Agricultural Energy Data Base

With legislative, conservation, production, and research needs in mind, a comprehensive investigation of energy use in agricultural production was undertaken by the Economic Research Service (ERS) of the U.S. Department of Agriculture in cooperation with the U.S. Federal Energy Administration (FEA).[30] The results were published in 1976.[31] An agricultural energy accounting model was designed with five dimensions or categories: energy form, state within the United States, commodity, month, and operation. Figure 5.4 provides a summary of energy use in United States production agriculture in 1974 by operation.

The USDA Farm Enterprise Data System (FEDS) contains detailed cost budgets for production inputs and operations over a wide range of commodities and regions and is continuously updated. The FEDS energy cost data were converted to physical energy units. Validation procedures included comparing total energy use in agriculture with estimates from other sources and energy use in fertilizers with known fertilizer tonnages. Based on all available controls and checks, the aggregate estimates appear to closely represent actual energy consumption.[32]

Between 1974 and 1978, most major crops experienced a substantial increase in production. The increasingly intensive use of liquid fuels was accompanied by an increase in the usage of embodied energy in pesticides and fertilizer. The total energy in United States agriculture in 1978 was 2.2 EJ—up from 2.0 EJ in 1974 (revised data)—an increase of about 7%. During the same period, farm outputs rose about 14%.

Other Energy Audits

Numerous state and commodity farm energy analyses have been conducted.[33] An energy analysis is the study of energy flows, energy input/output, energy accounting or other energy systems investigation. In 1974,

---

30. United States Department of Agriculture and Federal Energy Administration, *Energy and U.S. Agriculture*, 1974 database, Vol. I, FEA/D-76/459 (Washington, D.C.: U.S. Government Printing Office, 1976).
31. Stout, *Energy Use and Management in Agriculture*.
32. Ibid.
33. C. A. Myers, "A Review of Literature on Agricultural Energy Consumption," unpublished paper for AE840, Agricultural Engineering Department, Michigan State University, East Lansing, 1980.

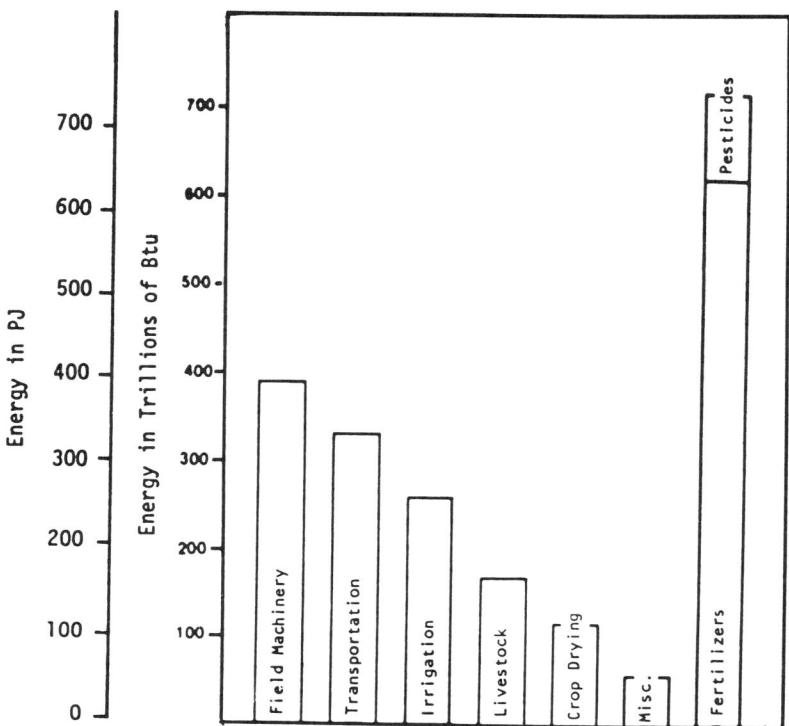

Figure 5.4. Energy use in United States production agriculture, 1974, by operation. (Source: United States Department of Agriculture and Federal Energy Administration, *Energy and U.S. Agriculture*, 1974 data base, Vol. I, FEA/D-76/459, Washington, D.C.: U.S. Government Printing Office, 1976.)

two similar studies were reported: California[34] and New York.[35] Each study presented agricultural energy use in their respective states and derived their numbers by using three basic methods: (1) engineering analysis, (2) ques-

---

34. (a) V. Cervinka et al., *Energy Requirements for Agriculture in California*, Joint study by California Dept. of Food and Agriculture and Agricultural Engineering Dept. Davis, Calif.: University of California-Davis, 1974). (b) "Methods Used in Determining Energy Flows in California Agriculture," *Transactions of the ASAE* 18 (2) (1975): 246–51. (c) W. J. Chancellor et al., *Energy Requirements for Agriculture in California (1978 data base)*, Vols. 1 and 2, Joint study by California Dept. of Food and Agriculture, and Agricultural Engineering Dept. (Davis, Calif.: University of California-Davis, 1981).

35. (a) W. W. Gunkel et al., *Energy Requirements for New York State Agriculture: Food Production* (Ithaca, N.Y.: Cornell University, 1974 [*Agricultural Engineering Extension Bulletin* no. 405]). (b) W. W. Gunkel et al., *Energy Requirements for New York State Agriculture: Indirect Energy Inputs* (Ithaca, N.Y.: Cornell University, 1976 [*Agricultural Engineering Extension Bulletin* no. 406]).

tionnaires or surveys, and (3) conversion of enterprise cost budgets to energy terms.

The California Department of Food and Agriculture and the Agricultural Engineering Department at the University of California-Davis determined the fuel and electricity requirements of California agriculture on a commodity basis. Later, the study was extended in more depth for one commodity—wheat. In Michigan, data were analyzed and energy usage standards were determined for the following areas: field operations, crop enterprises, and livestock operations. Also, a set of energy management tools was developed to enable farmers to analyze problem areas and implement conservation programs. The objectives of a New York study were to determine the fuel and electricity needs of agricultural production in New York State and to establish an index or model for use in projecting agricultural energy requirements. The Kansas/Nebraska fuel use survey was part of a broader program that also included a component on energy management and conservation, while the Ontario study included an analysis of the ratio of primary energy to delivered energy.

These energy audits are examined in detail in *Energy Use and Management in Agriculture*.[36] This book analyzes the energy needs of United States agriculture, documents energy use on farms, provides guidelines for conservation and more efficient energy management, and considers alternatives to petroleum and natural gas. Individuals concerned about energy use and management in agriculture will find this publication to be a helpful resource.

Roles of Various Energy Forms

A 1984 reexamination of the data on machinery stocks and sales indicated that by 1990, farmers would have 88% of their tractors diesel powered. Motorized combines were apparently also shifting to diesel. It was then estimated that by 1990 half of the farm trucks would be diesel powered.[37]

Reduced tillage practices were expected to be adopted in some form on 50% of the cropland area to be harvested in 1990. These practices range from eliminating one of the conventional tillage practices to no-till, in which there is no land preparation before planting and no cultivation of the crop. Changes in fuel use for alternate tillage practices were assumed to be directly related to the extent of reduction in tillage. A likely farm demand

36. Stout, *Energy Use and Management in Agriculture*.
37. Ibid.

for fuel in 1990 was estimated at 17.8 billion liters of diesel fuel and 4.5 billion liters of gasoline.[38]

The primary use of LP gas is in crop drying, although some tractors, combines and irrigation pumps are powered by this fuel. With the expanded production of grain crops and with no trend toward the older practice of harvesting ear corn, the demand for LP gas was projected to rise 13% from 1974 to 1990, when 6.4 billion liters of this fuel would be needed.[39]

Whatever the future, the high priority for agriculture dictated that farmers continue to use natural gas in 1990 much the same as they did in 1974. The increased use of agricultural chemicals, principally fertilizers and pesticides, will cause the indirect demand for natural gas in farm production to increase substantially.[40]

National energy policy looks to our vast domestic resources of coal for greatly expanding the use of electricity in the future. This energy source was expected to play an increasing role in agricultural production by 1990. Projections for 1990 were for electricity to provide 33.6 billion kwh of power, or 5% above the 1974 level—an estimate that may have been on the low side. As energy costs from fossil fuels increase, the trend toward converting irrigation pumping to electric power will continue. However, the expected shift to diesel-powered irrigation systems will not occur; the shift instead will be to electric-powered units. Further, the trend toward use of center-pivot and other pressurized water systems also increases energy needs.[41]

Embodied energy is the indirect energy used to manufacture farm equipment, fertilizers, and other production inputs. Transportation of farm products, processing, trade, and finally household energy used in food storage and preparation are all sometimes included in the energy budget charged against agriculture.[42]

The efficient use of electric energy on American farms during the past half-century has been a primary factor affecting productivity and the reduction of human drudgery. This is especially true with on-farm processes and operation related to livestock and poultry enterprises in which housing, feeding, environmental conditioning, waste control, product refrigeration, and similar requirements must be met on a regular basis. However, electric energy is of equal importance to farms that grow and store only crops. It is used as a primary power source to drive crop irrigation machines; for con-

38. Ibid.
39. Ibid.
40. Ibid.
41. Ibid.
42. Fluck, *Agricultural Energetics*.

veying, drying, and removing from storage; and/or for processing these crops on the farmstead.[43]

C. Energy Management in Agriculture

Fertilizer and Chemicals

Fertilizer is the greatest single energy input in United States agriculture, as shown in Figure 5.5.[44] Fertilizers represent the greatest amount of energy involved in world crop production as well, accounting for 40–50% of the total input. While fertilizers are used to a lesser extent in developing countries, their use is increasing each year. Pesticides use less than 5% of the energy invested in world agriculture, but they require the greatest amount of energy for manufacture on a per unit basis of any agricultural input.[45]

Energy in Plant Nutrition and Pest Control discusses the significant role plant nutrition and fertilizers, and pest control and pesticides play in energy in world agriculture. This volume details the use of fertilizers and pesticides, examines the energy required for their manufacture and distribution, and explores the potential to conserve these inputs.[46] The authors evaluate the energy balance from the use of plant nutrients and pest control and seek to determine their monetary, social, and economic implications. *Energy from Sun, to Plant, to Man*[47] discusses the role of fossil fuels in American agriculture. In *Energy Use and Management in Agriculture*,[48] Stout devotes a section to fertilizer management, which discusses the energy invested in producing and transporting fertilizer, as well as fertilizer placement and efficient use.

Irrigation

The energy used for pumping and distributing irrigation water has an impact on a significant portion of the irrigated land in the United States. In those areas where irrigation is required for continuous agricultural production, the energy use for irrigation may approach 75% of the total energy demands of crop production. The relatively large energy demands of irriga-

---

43. K. L. McFate, ed., *Electrical Energy in Agriculture* (Amsterdam: Elsevier Science Publishers, 1989).
44. Stout, *Energy Use in Agriculture*.
45. Z. R. Helsel, ed., *Energy in Plant Nutrition and Pest Control* (Amsterdam: Elsevier Science Publishers, 1987). (Energy in World Agriculture Series)
46. Ibid.
47. F. C. Stickler, W. C. Burrows, and L. F. Nelson, *Energy from Sun, to Plant, to Man* (Deere and Company, 1975).
48. Stout, *Energy Use and Management*.

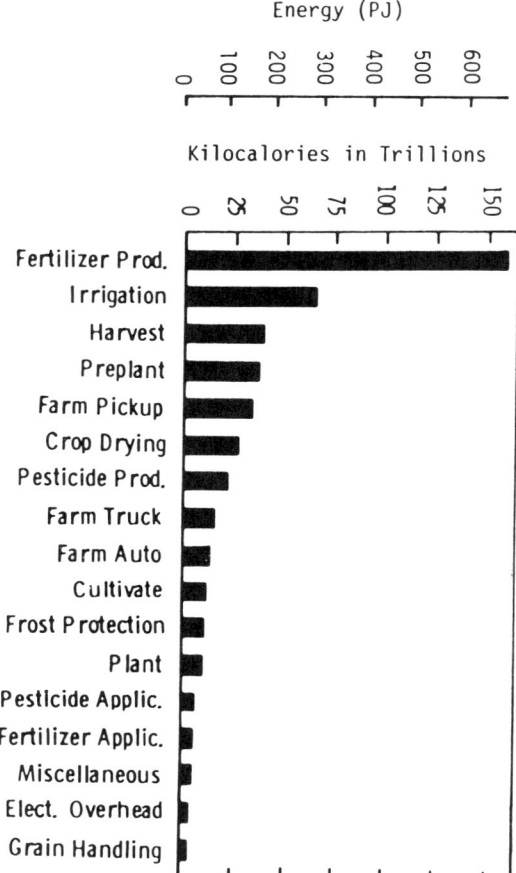

Figure 5.5. Consumption of energy in different aspects of crop production in the United States in 1974. (Source: B. A. Stout et al., *Energy Use in Agriculture: Now and for the Future*, Ames: Iowa State University, Council for Agricultural Science and Technology, 1977 (CAST Rpt. no. 68); B. A. Stout et al., *Energy Use and Management in Agriculture*, Boston: Breton, 1984.)

tion systems supplied with pumped groundwater and the rapid expansion of pressurized irrigation systems have provided an impetus to the development of improved energy considerations in the design of new systems.[49]

Potential energy-conserving practices include reductions in the volume of water pumped, reductions in total dynamic head, and improvements in the

49. J. R. Gilley and R. J. Supalla, "Energy Requirements and Management for Pumping Irrigation Water," in *Proceedings of the International Conference on Water and Water Policy in World Food Supplies*, Texas A. & M. University, May 26–30, 1985.

efficiencies of the pump, drive, and power units. Many potential areas of improvement are not realistically available to existing systems, and not all the practices are economically viable in all situations. However, depending on the initial energy efficiency of irrigation systems, several procedures are economically attractive and are currently underway.[50]

Six typical irrigation systems provide one meter depth of net irrigation annually and satisfy a peak water use rate of 8.4 millimeters (mm) per day. The six irrigation systems, with assumed irrigation efficiencies in parentheses, are hand-moved sprinkler (75%); two surface systems, (one 50% and one 70%); surface with an irrigation runoff recovery system (85%); trickle (90%) and low-energy precision application (90%). Pump efficiency is 50%, and pump-engine efficiency is 25%.[51]

It is evident that the total annual energy required for irrigation increases markedly with increased water lifts. This illustrates the importance of pumping lifts, including the operating pressure of pressurized systems, in irrigation energy considerations. Also, in a comparison of the surface systems, the curves dramatically show the importance of providing well-designed systems with high irrigation efficiency.[52]

Mechanization

United States agriculture has experienced profound changes since the early part of the twentieth century as tractor power and labor-saving machines have replaced draft animals and human power. These changes have contributed to the highly productive agriculture we know today, but they have made United States agriculture capital- and energy-intensive, while significantly reducing human labor input and drudgery. Considering the finite supply of diesel fuel and other petroleum-based fuels and chemicals, several questions arise. Is the current highly mechanized United States agriculture sustainable? Should energy-intensive machines be introduced in developing countries? The answers to these questions are complex and beyond the scope of this chapter, but farmers should adopt energy-conserving practices whenever possible.

In terms of direct fuel use (excluding chemicals), reduced tillage can save up to thirty-seven liters per hectare and cut farm fuel costs by 40%.[53] Plow-

---

50. Ibid.
51. E. T. Smerdon and E. A. Hiler, "Energy in Irrigation in Developing Countries," in *Proceedings of the International Conference on Water and Water Policy in World Food Supplies*, Texas A. & M. University, May 26–30, 1985.
52. Ibid.
53. Federal Energy Agency/United States Dept. of Agriculture, *A Guide to Energy Savings for the Field Crop Producer* (Washington, D.C.: U.S. Government Printing Office, 1977).

ing, disking, and/or harrowing a field before planting and crop cultivation are often unnecessary with reduced tillage practices. Potential benefits of reduced tillage include erosion reduction, less soil compaction, improved weed control, increased soil moisture storage, labor and cost savings, and more double-cropping opportunities. Stout lists many possibilities for saving fuel in field work and also provides a thorough discussion of achieving optimal tractor performance by minimizing energy wastes and reducing fuel costs.[54]

One way to save fuel in tractor operation is shifting up and throttling back. Engines operate most efficiently within a relatively narrow range of engine loads and speeds. When tractors are operated at part load, a common practice on farms, significant fuel savings can be achieved by the practice of shifting up and throttling back.

Livestock Production

Figure 5.6 shows a breakdown of consumption of nondietary energy in United States livestock production operations in 1974. For the future, solar heating is being tested for livestock production. Attempts are being made to use as much of the animal's own heat as possible. Generally, swine and poultry houses are being insulated well in both walls and ceilings to maximize animal heat utilization. Swine and poultry numbers per unit of space are increasing markedly. The confinement rearing and high- density production associated with use of animal heat conserve land for crop production. Along with confinement rearing and high-density production, producers are improving structural ventilation for livestock so the ventilation automatically changes with the weather.[55]

Further information on energy management for livestock production, dairy operations, and poultry and egg production is found in Stout.[56] The major energy inputs for United States livestock operations are feed handling (29%), feed processing and distribution (21%), farm travel (20%), and waste disposal (11%). However, this distribution can vary widely depending on the type of livestock operation.[57] The energy consumption for poultry and egg production consists mainly of electrical energy utilization for mechanical ventilation, lighting, egg cooling, and materials handling.[58] Mechanical ventilation accounts for nearly two thirds (64%) of the electrical energy requirement on an annual basis for egg production. The energy re-

54. Stout, *Energy Use and Management*.
55. Ibid.
56. Ibid.
57. FEA, *Energy Use in the Food System*.
58. FEA, *Guide to Energy Savings*.

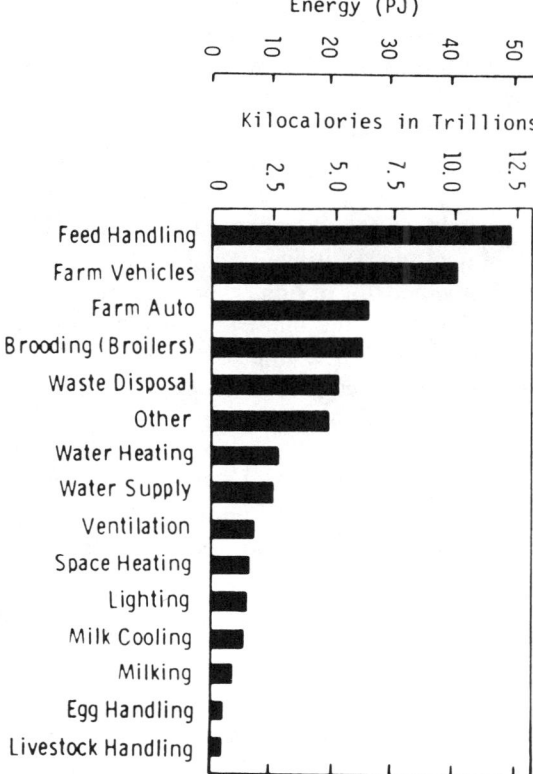

Figure 5.6. Consumption of nondietary energy in United States livestock production in 1974. (Source: United States Department of Agriculture and Federal Energy Administration, *Energy and U.S. Agriculture*, 1974 data base, Vol. I, FEA/D-76/459, Washington, D.C.: U.S. Government Printing Office, 1976.)

quired to grow a 20-week-old pullet is considerably more than for maintaining a laying hen for a year. Seventy-one percent of the energy for pullet growing is for brooding.[59]

Grain Drying

Drying grain with heated air requires large amounts of energy. Energy is used as a heat source to evaporate moisture and to run conveyors and fans.

---

59. M. L. Esmay et al., *Energy Management for Egg Production* (East Lansing, Mich.: Michigan State University, 1981).

By far, the greatest amount of energy is used to evaporate moisture. While the most positive way to reduce the energy needed to dry grain is to reduce the amount of moisture to be evaporated, other practices in planting, harvesting, handling, drying, and storing grain can also reduce the total energy requirements for a given system.[60]

A grain-handling system moves grain from harvest to utilization and includes machinery, equipment, and structures. An efficient system should: (1) provide the capacity to complete harvest in a reasonable amount of time; (2) maintain grain quality; (3) utilize available fuels efficiently; (4) provide some flexibility in methods of drying, handling, and cooling grain; and (5) utilize harvesting, handling, and marketing methods and techniques for reducing total energy needs. For example, one grain drying option is drying in the field. Harvesting below 30% moisture not only decreases the moisture to be removed and energy requirements, but also helps to maintain corn quality. The amount of water to be removed and the approximate heat required for drying is shown in Table 5.3. Drying from 26% moisture to 16% moisture, rather than from 30% initial moisture, reduces the energy requirements by 9 megajoules (MJ)/kg of corn, or about 32%.

Food Processing

During the last several decades, introduction of mechanization to achieve high processing capacity has caused the food industry to depend more heavily on energy derived from non-renewable resources. Energy, obtained mostly from petroleum products, has played a key role in processes, including those that involve heat transfer, mass transfer, and mechanical processing and handling. However, recently, the food industry has made considerable progress in improving energy conversion efficiencies and in instituting energy conserving policies.[61]

A report by FEA indicated that of total United States energy consumption, 2.9% is used for farm production, 4.8% for food processing, 4.3% for in-home food preparation, 2.8% for out-of-home food preparation, 0.5% for wholesale food trade, and 0.8% for retail trade.[62] These data include direct and indirect energy consumption and transportation costs for individual components.[63]

60. Stout, *Energy Use and Management in Agriculture*.
61. R. P. Singh, ed., *Energy in Food Processing* (Amsterdam: Elsevier Science Publishers, 1986).
62. FEA, *Enegy Use in the Food System*.
63. Ibid.

While farm production consumes only 18% of the energy expended by the food system, the remaining 82% is spent on processing, marketing, and preparation. The food processing (29%), in-home preparation (29%) and out-of-home preparation (17%) components of the food system are the leading energy consumers. These data emphasize the importance of various components in terms of the energy consumption and the opportunities for energy conservation in all components of the food system.[64]

In aggregate, the United States food and kindred products industry is the sixth largest energy consuming industry. The food industry consumed about 8% of the total energy used by all manufacturing industries. The ten food industries in order of their total energy consumption are: wet corn milling, meat packing, beet sugar, malt beverages, soybean oil milling, canned fruits and vegetables, bread and cake, fluid milk, food preparation, and corn-sugar refining. These ten food industries use more than half of the total energy consumed by all industries within the food and kindred products industrial group. In terms of energy intensities, beet sugar and wet corn milling clearly top the list, along with distilled spirits.[65] The energy to remove moisture is shown in Table 5.3.

Table 5.3. Amount of moisture removed and approximate energy required to dry 1 kg of corn

Initial moisture content (%)	Final moisture content							
	12%		14%		16%		18%	
	kg water	MJ[a]	kg water	MJ	kg water	MJ	kg water	MJ
30	6.5	35	5.8	31	5.1	27	4.4	24
26	4.8	25	4.1	22	3.4	18	2.8	15
122	3.3	17	2.6	14	1.9	10	1.2	6
18	1.9	10	1.2	7	0.6	3	1.2	6
16	1.0	5	0.4	2	0.6	3	1.2	6

(Source: Based on partial data from R. L. Maddex and F.W. Bakker-Arkema, *Reducing Energy Requirements for Harvesting, Drying, and Storing Grain*, East Lansing: Michigan State University, 1978. [Bull. no. E 1168])

[a]A range of 3–7 MJ may be required to evaporate 1 kg of water. For these calculations, 5.3 MJ/kg was used.

The United States food system is critically dependent on adequate supplies of energy. Because food is a perishable commodity, timely availability of energy is necessary to carry out the various processing operations. For example, many vegetable and fruit canning plants operate for only a few

64. Singh, *Energy in Food Processing*.
65. Ibid.

weeks of the year, and a critical shortage of fuel during the processing season would have a major impact on this industry. In addition, the key link provided by transportation between the producer and consumer would be seriously affected if a shortage of petroleum occurred.[66]

Energy in Food Processing is a compilation of research data and results by authors who have headed major studies on energy use in food processing in their respective countries.[67] Five major sections provide a comprehensive treatment of such topics as methods used in energy accounting, measurement of energy, and energy analysis. Quantitative data, much of which is presented from an international perspective, is included on energy consumption in a variety of food industries such as blanching, freezing, canning, irradiation, evaporation, membrane processing, and dairy and catering establishments.

D. Renewable Energy

In 1981, a task force was assembled by the USDA Science and Education Administration and charged with the responsibility of preparing a statement of the organization's energy capabilities and opportunities. The Science and Education Administration conducts and supports research to solve priority food, agriculture, family living, and community problems facing rural America. The task force report essentially was a plan for research and education programs on energy and agriculture.[68]

Solar Heating

The sun converts mass into energy at the rate of millions of tonnes per second. Each year, the solar radiation passing through the earth's atmosphere is about 700×10^{12} megawatt-hours (MWh). This is 13,000 times the current world energy use,[69] but it represents only 0.5×10^{-9} of the total energy radiated by the sun.[70]

---

66. R. L. Maddex and F. W. Bakker-Arkema, *Reducing Energy Requirements for Harvesting, Drying, and Storing Grain* (East Lansing, Mich.: Michigan State University, 1978 [Bulletin no. E 1168]).

67. Ibid.

68. USDA, *Energy Capabilities and Opportunities: A USDA Science and Education Administration Task Force Report* (Washington, D.C.: U.S. Government Printing Office, 1981).

69. D. S. Halacy, Jr., "Solar Energy and the Biosphere," in *Solar Energy Technology Handbook, Part A: Engineering Fundamentals,* ed. W. C. Dickinson and P. N. Cheremisinoff (New York: Marcell Dekker, 1980).

70. A. Androsky, *Uses of the Sun in the Service of Man* (El Segundo, Calif.: Aerospace Corp., 1973 [Report no. ATR-74-(9470)]).

About 30% of the radiation reaching the earth's atmosphere is reflected back into space. About 47% is absorbed as heat by the atmosphere, land, and water. But about 30% of the solar radiation supplied to the ground is reradiated into the atmosphere. Evaporation and precipitation use 23% of the atmospheric solar radiation.[71]

The simplest and most widely used method of obtaining solar energy is the flat plate collector, a sheet of blackened material positioned so the sun's rays shine on it. Only black absorbs all wavelengths of visible light; other colors reflect certain wavelengths. Heat loss, and the heat transfer rate to the water or air, are important considerations in flat plate collector designs. As the difference in temperature between the black plate and the outside air increases, heat lost through conduction, convection, and radiation also increases, reducing collector efficiency.

Water heating is one of the few commercially successful solar applications. An estimated several million solar water heaters are currently in use,[72] with more than 2 million units in Japan alone.[73] Like other solar energy devices, the units vary in efficiency, capacity, and costs, but all water heaters use flat plat collectors to capture solar heat. The type of storage tanks for heating water, insulation, materials, and design chosen depends on need and cost. The layer of water must be shallow, preferably 5–10 centimeters. In sunny, warm weather, the water temperature rises to about 50 degrees C in three to four hours. Efficiency ratings vary depending on design but are good even for simple constructions. Efficiencies of 50–70% are not unusual.[74]

The most energy-efficient solar drying of crops can occur in the field during the stages of crop maturity. To avoid serious field losses from unfavorable weather, farmers harvest high-moisture-content grains and then dry them to safe moisture levels before storage in portable solar collectors designed for grain and crop drying.[75] A solar collector incorporated into a building for heating is shown in Figure 5.7.

Solar Dryers: Their Role in Post-Harvest Processing[76] provides a comprehensive reference for the basic theories of drying, an outline of the prin-

71. Stout, *Handbook*.
72. Androsky, *Uses of the Sun in the Service of Man*.
73. Ken-ichi Kimura, "Present Technologies of Solar Heating, Cooling and Hot Water Supply in Japan," *Architectural Science Review* 19 (2) (1976).
74. R. N. Morse et al., "Solar Energy as a Major Primary Energy Source," in *Proceedings of the Symposium on Realistic Prospects for Solar Power in Australia*, International Solar Energy Society, 1973.
75. W. D. Peterson, "Agricultural Solar Systems Performance: Illinois Solar Demonstration Program," in *Solar and Wind Systems Workshop Proceedings*, Great Plains Agricultural Council, University of Nebraska, Oct., 1983, pp. 17–28.
76. B. Brenndorfer, et al., *Solar Dryers: Their Role in Post-Harvest Processing* (London: The Commonwealth Science Council, Commonwealth Secretariat, 1985).

Figure 5.7. Flat plate solar collector incorporated into the roof of a metal building. Air is drawn into the building between the glazing and the absorber plate for use in grain drying or space heating. (Source: DOE 007 003 001).

cipal operating features of solar dryers, and a detailed account of the construction and operation of the three most popular types of solar dryers. Information is also included on the theory and design of flat-plate solar collectors, which are an integral part of many dryers. Another useful reference for solar energy is *Handbook of Energy for World Agriculture*, which devotes a chapter to solar energy collection, storage, and applications.[77]

Numerous conferences have been conducted to present technical information relating to agricultural applications of solar energy.[78] These conference

77. Stout, *Handbook*.
78. See, for example: (a) Great Plains Agricultural Council, *Report of the Proceedings of the Solar and Wind Systems Workshop*, Lincoln, Nebr., Oct. 1983 (Great Plains Agricultural Council, 1983 [Great Plains Agricultural Council Pub. no. 108]). (b) J. G. Hartsock, ed., *Report of the Proceedings of the Solar Grain Drying Conference*, West Lafayette, Ind., May 1978. (c) H. H. Kluter and L. H. Soderholm, eds., *Report of the Proceedings of the Wind Energy Application in Agriculture Conference*, Ames, Iowa, May 1979. (d) G. C. Shove, ed., *Report of the Proceedings of the Solar Grain Drying Conference*, Urbana-Champaign, Ill., Jan. 1977. (e) USDA, *Report of the Proceedings of the Solar Grain Drying Conference*, Raleigh, N.C., June 1977. (f) USDA,

proceedings contain a wealth of detailed information about solar applications.

A series of on-farm demonstration projects was implemented by the USDA in cooperation with the U.S. Department of Energy. The objective was to demonstrate the technical and economic feasibility of using solar technology for drying crops and grains and heating livestock shelters. Conference discussions indicated that solar drying offers an alternative operating scheme for farmers if seasonal demands, emergencies, or political decisions cause conventional fuel shortages. In addition, careful planning, construction, management, and maintenance can make solar heating for livestock housing a significant economic alternative to conventional housing.[79]

Wind Energy

Wind energy, a form of solar energy, can be used to power mechanical devices such as water pumps, to provide heat or thermal energy, and to generate electricity. Variations in incident solar radiation in relation to time and location cause temperature differences in the atmosphere. The temperature differences lead to variations in atmospheric pressure, resulting in air movement from high to low pressure regions. Air movements which dominate large areas and are relatively constant in direction are called prevailing or planetary winds. Local topography, such as hills or bodies of water, can also cause local winds. The best wind power applications take advantage of both prevailing and local winds.[80]

Perhaps the most popular mechanical application for wind power is water pumping, which has been successfully applied for hundreds of years. Many types of systems have been designed for this purpose, but the farm, or multi-bladed windmill, and the sailwing are the most common today. A wind-assisted diesel irrigation pump is operated in the United States at Bushland, Texas. The power output of the diesel engine decreases as the wind speed increases, while maintaining constant pump power.[81]

There are two broad classifications of wind-electric systems: isolated and utility interconnected. Isolated wind-electric systems stand alone as com-

---

Report of the Proceedings of the Solar and Biomass Workshop, Atlanta, USDA and S & E Southern Agricultural Energy Workshop, April 1982. (g) USDA, *Application of Solar Energy to Agricultural Production Processes* (Washington, D.C.: U.S. Government Printing Office, 1985).
79. USDA, *Report of the Proceedings of the Solar and Biomass Workshop*.
80. Stout, *Handbook*.
81. R. N. Clark, *Co-Generation Using Wind and Diesel for Irrigation Pumping*, St. Joseph, Mich.: American Society of Agricultural Engineers, 1984 [*ASAE Paper* no. 84-2603]).

pletely independent power sources. Onsite battery banks store electrical energy for use during low wind periods. For example, the power produced by the rotor is fed into an alternator that generates AC, which is then rectified to DC. A regulator always controls the battery charging DC voltage; energy stored in the batteries is accessed as needed. If the battery storage system is full, the electrical power is diverted to some other useful application.[82] Utility interconnected wind systems correspond with the utility grid, producing electricity identical in frequency and voltage to utility supplied electricity, i.e. they are synchronized with utility lines. Systems utilize a) synchronous invertor, b) rectifier and synchronous invertor, and c) induction generator.[83]

Stout devotes a chapter to wind energy, in which wind energy conversion systems, wind characteristics, and wind energy feasibility studies are all discussed.[84] Current and future wind energy applications in agriculture are addressed by Kluter and Soderholm.[85]

Biomass Fuels

Biomass is defined as all organic matter except fossil fuels; that is, all crop and forest materials, animal products, microbial cell mass, residues, and byproducts that are renewable on a year-to-year basis. Biomass serves as food, feed, fiber, bedding, structural material, soil organic matter, and fuel.[86]

As shown in Figure 5.8, biomass may also be converted to liquid or gaseous fuels, which may be stored and subsequently burned to produce heat, or may be used to fuel internal-combustion engines (spark-ignition or diesel) to produce mechanical work. Many processes or technologies exist for converting biomass to a more useful form for fuel or industrial feedstocks. They are classified as wet or dry processes. Figure 5.9 shows some options for converting biomass to heat energy or to liquid or gaseous fuels. The dry processes shown include direct combustion, gasification, methanol production, and oil extraction. The wet processes include anaerobic digestion and ethanol fermentation.[87]

Efforts to assess the potential of biomass fuels include the U.S. Department of Energy's *Domestic Policy Review of Solar Energy*,[88] the Office of

82. W. T. Rose et al., *Survey of Commercially Available Wind-Electric Systems* (East Lansing: Agricultural Engineering Dept., Michigan State University, 1980 [AEIS 427, File no. 18.8)].
83. Ibid.
84. Stout, *Handbook*.
85. Kluter, *Report of the Proceedings of the Wind Energy Application*.
86. E. A. Hiler and B. A. Stout, eds., *Biomass Energy: A Monograph* (College Station, Tex.: Texas A. & M. University Press, 1985).
87. Stout, *Energy Use and Management in Agriculture*.
88. U.S. Department of Energy, *Domestic Policy Review of Solar Energy* (U.S. Department of Energy, 1979).

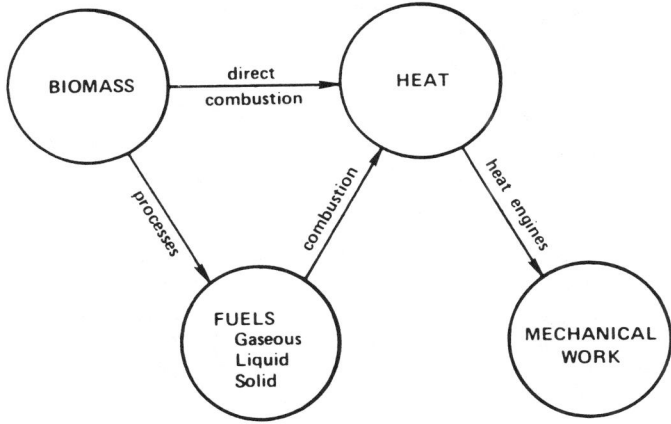

Figure 5.8. Biomass conversion to heat or mechanical work. (Source: B. A. Stout et al., *Energy Use and Management in Agriculture*, Boston: Breton, 1984.)

Technology Assessment, *Energy from Biological Processes*,[89] and the projections for the next two decades made by the DOE Energy Research Advisory Board's (ERAB) Biomass Panel.[90] The ERAB Biomass Panel estimates that about 635 million tonnes of dry biomass will be potentially available for fuel in the year 2000.

Ethanol: The ethanol production process is basically a biological process whereby yeast converts sugar to alcohol. It involves the grinding, mashing, cooking, and saccharification of grains, followed by fermentation and distillation. The prospects for using ethanol for fuel depend on the efficiencies of the various steps in the process, as well as on the cost of the feedstock used, balanced against the cost of gasoline and other conventional fuels. Ethanol is considered to be an excellent fuel for spark-ignition engines, but is generally unsuitable for diesel engines.[91]

Methane: Anaerobic digestion is a conversion process for wet biomass such as animal manure, municipal sewage, and certain industrial wastes. Through this process, complex organic matter is converted into methane and other gases. The byproduct effluent can be used as fertilizer or animal feed. Anaerobic digestion is a biological process carried out by living microorganisms. The feedstock organic matter is made into a slurry and then digested by bacteria.

89. U.S. Office of Technology Assessment, *Energy from Biological Processes* (Washington, D.C.: U.S. Government Printing Office, 1980).

90. U.S. Energy Research Advisory Board, *Biomass Energy: Report of the Energy Research Advisory Board, Panel on Biomass* (Washington, D.C.: U.S. Dept. of Energy, 1981 [Panel on Biomass chairman was David Pimentel]).

91. Stout, *Energy Use and Management in Agriculture*.

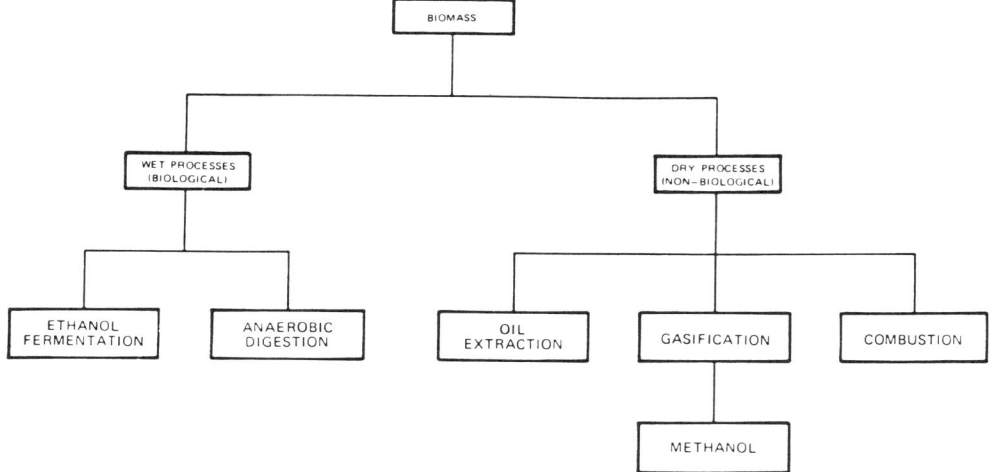

Figure 5.9. Options for converting biomass to heat energy or to liquid or gaseous fuels. (Source: B. A. Stout et al., *Energy Use and Management in Agriculture*, Boston: Breton, 1984.)

Plant Oils: Vegetable or plant oils are mixtures of triglycerides, i.e. glyceride esters of fatty acids. Reports of diesel engines operating on coconut oil, peanut oil, palm oil, and other vegetable oils or animal fats is not uncommon. Production of vegetable oil for diesel fuel is a relatively simple process which involves extracting the oil from the oilseed, filtering, degumming, and possibly reducing the viscosity of the liquid through transesterification. Currently, vegetable oil costs at least twice as much as diesel fuel. As long as petroleum-based diesel fuel is available, no great shift to vegetable oils is expected, but the technical prospects for this fuel are encouraging.[92] *Vegetable Oils Fuels: Proceedings of the International Conference on Plant and Vegetable Oils as Fuels* provides a valuable reference of research results on the future potential of this energy resource.[93]

Solid Fuels: Combustion is defined as the rapid chemical combination of oxygen with the combustible elements of a fuel. In agricultural crop residues such as corn stover or cobs, the combustible elements are carbon, hydrogen, and metallic elements such as potassium and sodium. The chemical products formed during the combustion process are carbon dioxide, water, and metallic oxides. The objective of good combustion is the release of

92. Stout, *Handbook*.
93. G. R. Quick, *Vegetable Oils Fuels: Proceedings of the International Conference on Plant and Vegetable Oils as Fuels*, Fargo, N.Dak., Aug. 1982 (St. Joseph, Mich.: American Society of Agricultural Engineers, 1982).

heat from the fuel while minimizing the losses from imperfect mixing and excess air. The actual combustion of a crop residue or wood occurs in three consecutive overlapping phases: (1) evaporation of moisture, (2) volatilization and burning of volatile matter, and (3) combustion of fixed carbon. Initially the moisture held by the residue is evaporated. Then heat is absorbed by the residue, raising its temperature and driving off the volatile gases. Hence, these gases are burned to sustain the combustion. When the majority of the volatiles are distilled out of the residue, the highly reactive surface of the remaining fixed carbon is burned in the presence of oxygen.[94] Direct combustion systems are generally classified as pile burners, semi-pile burners, suspension burners, and fluidized-bed combustors.[95]

Gasification (burning in a controlled atmosphere): the conversion of a solid or a liquid to a gas. If the oxygen supply is restricted to perhaps 40% of the stoichiometric requirements, incomplete combustion occurs, releasing combustible gases such as carbon monoxide, hydrogen, and methane. A solid residue or char remains. There are two main designs of producer gas generators: updraft and downdraft. In an updraft generator, the hot gases flow counter to the feedstock. A portion of the fuel stack is pyrolyzed (heated in the absence of air), and the resulting gas has a high tar content. In the downdraft system, the pyrolysis products are broken down while passing through the reaction zone before combining with the exiting gases. Downdraft generators, therefore, have the potential to eliminate tar from the gas and are probably better suited to use crop residues as a fuel source. Figure 5.10 shows the various zones in a downdraft gasifier and gives corresponding combustion and drying equations.[96] Gasifiers are classified as moving-bed, fluidized-bed, rotary-kiln, and multiple hearth types.[97] More research is needed on gasification to perfect it as a viable energy alternative.[98]

Other useful references for biomass fuels not cited in the previous text include *Biomass Energy Systems and the Environment*,[99] which provides basic information on biomass energy systems with emphasis on an examination of environmental and social issues related to producing energy from biomass. *Energy from Biological Processes*[100] presents analyses of promi-

94. Stout, *Energy Use and Management in Agriculture*.
95. Hiler, *Biomass Energy*.
96. Stout, *Energy Use and Management in Agriculture*.
97. Hiler, *Biomass Energy*.
98. Stout, *Handbook*.
99. H. M. Braunstein et al., *Biomass Energy Systems and the Environment* (New York: Pergamon Press, 1981).
100. U.S. Office of Technology Assessment, *Energy from Biological Processes* (Washington, D.C.: U.S. Office of Technology Assessment, U.S. Govt. Print. Off., 1980).

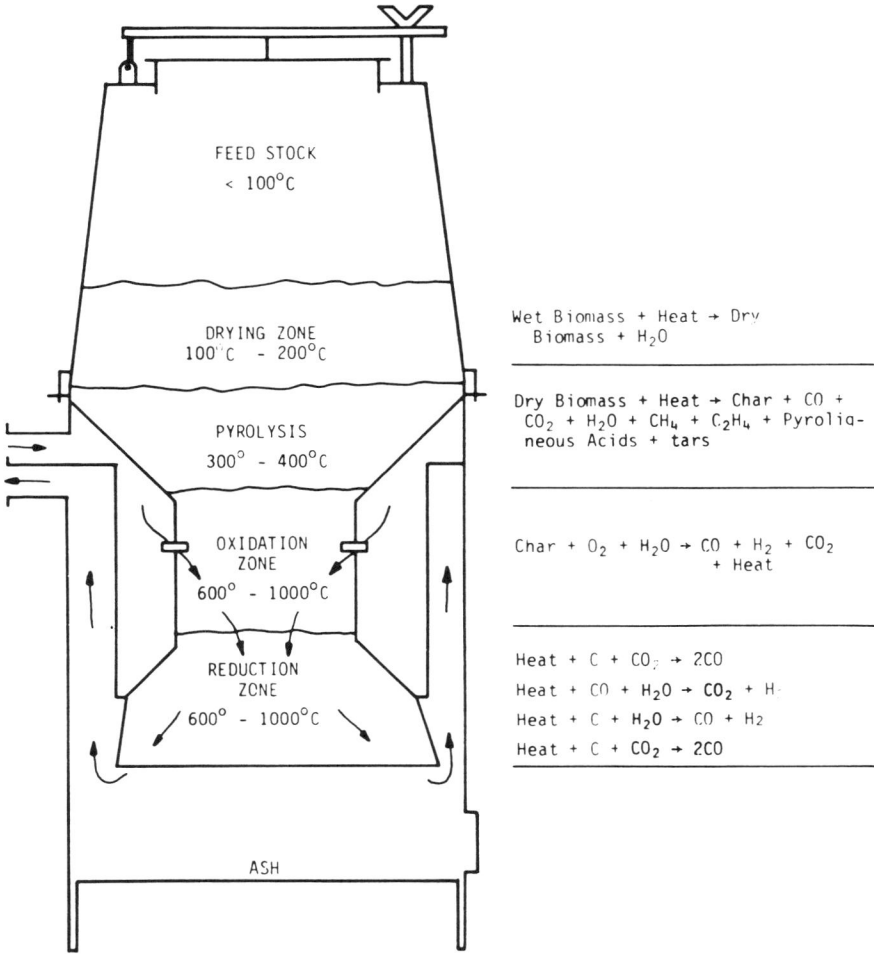

Figure 5.10. Reaction zones in a downdraft producer gas generator. (Source: T. B. Reed, *Problems and Opportunities for Solar Energy in Biomass, Pyrolysis, and Gasification,* Golden, Colorado: Solar Energy Research Institute, 1979.)

nent biomass issues, summaries of four biomass fuel cycles, a description of the role of biomass in two plausible energy futures, and discussions of policy options for promoting energy from biomass. An excellent general reference for renewable energy is *Renewable Energy Resources*[101] by John Twidell and Tony Weir.

101. J. W. Twidell and A. D. Weir, *Renewable Energy Resources* (London: E. & F. N. Spon Ltd., 1986).

E. International Dimension

Although numerous energy studies have been conducted in recent times, the amount of energy that is actually consumed in developing countries is difficult to determine. The developing countries where most of the world's population lives, consume only about 15% of the world's energy. As these countries develop, they will require more energy. One scenario portrays the energy share of the developing countries growing to 25% of the worldwide total by the year 2000. The International Institute for Applied Systems Analysis (IIASA) report, *Energy in a Finite World*, states "During this period (the next fifty years), the worldwide population will reach eight billion, and even with only modest economic growth and extensive conservation, global energy demand is likely to expand to three to four times today's level.[102]

Noncommercial energy constitutes significantly more than half the total energy used in many developing countries. Estimates indicate that in the 1970s, more than 50%, 60%, and 20% of all energy consumed in Africa, the Far East, and Latin America, respectively, came from noncommercial sources. For more than a third of the world's people, the real energy problem is a daily search to find the wood needed to cook dinner. Fuel wood is by far the most important form of noncommercial energy, accounting for 89% of the total energy used in Tanzania, 84% in Uganda, and 77% in Cameroon. Agricultural wastes, including animal dung, are also important noncommercial energy sources.[103]

In 1980, the World Bank initiated an Energy Sector Assessment Program to diagnose the most serious energy problems faced by individual developing countries.[104] More than fifty country assessments have been completed. Such assessments include:

(1) Analyzing the potential for changes in pricing, institutional arrangements, and other policies to encourage economically profitable production from indigenous sources, fuel substitution, and more efficient energy use;
(2) Assessing investment priorities in the energy sector; and
(3) Providing a framework for official and private assistance to this sector.

Rural development is affected by the severe problems faced by many developing countries that rely heavily on imported petroleum. As prices rise, governments face the necessity of using increasingly large sums of

---

102. International Institute for Applied Systems Analysis, *Energy in a Finite World: Executive Summary* (Laxenburg, Austria: IIASA, 1981).
103. Stout, *Handbook*.
104. World Bank, *The UNDP/World Bank Energy Sector Programmes* (Washington, D.C.: International Bank for Reconstruction and Development, 1984).

foreign exchange for imported oil, thus diverting resources that might be used for other development activities. In Kenya, for example, 85% of the commercial energy is derived from imported petroleum at a cost of 25% of the country's foreign exchange earnings. However, many developing countries have important characteristics that make increased use of renewable and nonconventional energy sources attractive, making them ideal for small-scale rural development programs.

D. E. Morrison expressed quality of life (QOL) as a function of energy use in Figure 5.11. At low levels of energy use (quadrant III), basic need satisfaction is linearly related to energy use. As the amount of energy increases (quadrant II), two paths may be hypothesized: Option A projects a linear relationship between QOL and energy use; whereas, Option B suggests an optimum QOL at a moderately high level of energy use, followed by a deterioration for excessively high energy-use rates. Energy-intensive industrialized countries like the United States are, no doubt, operating in quadrant II, and excessive energy use has often resulted in environmental deterioration and, subsequently, a lower quality of life. The developing countries, however, are unquestionably operating in quadrant III, at least as far as their rural populations are concerned. Thus a formula for agricultural and rural development is to supplement the meager work capacity of humans with as much energy as is efficient and cost-effective.[105]

Dozens of authors have attempted to address the international energy dimension. Only a few are mentioned here. Makhijani and Poole described life in energy-poor developing countries and provided a variety of examples to illustrate that people in developing countries are not receiving their share of energy.[106] The Food and Agriculture Organization of the United Nations (FAO) devoted a major section of one of its annual reports to an analysis of energy and agriculture.[107] *Energy and the Developing Countries* contains thirty-one chapters on various development issues related to energy.[108] Parikh's *Energy Systems and Development* presents current energy consumption statistics, projects energy requirements until the year 2000, and examines various supply options.[109] *Energy in the Developing World* focuses on the largest and most populous developing countries where populations total

---

105. Stout, *Handbook*.
106. A. Makhijani and A. Poole, *Energy and Agriculture in the Third World* (Cambridge, Mass.: Ballinger, 1975).
107. Food and Agriculture Organization, "Energy and Agriculture," Chapter 3 in *FAO Annual Report: The State of Food and Agriculture* (Rome; Food and Agriculture Organization, 1976), pp. 82–111.
108. P. Auer, *Energy and the Developing Countries* (Oxford: Pergamon, 1981).
109. J. K. Parikh, *Energy Systems and Development* (Bombay, India: Oxford University Press, 1980).

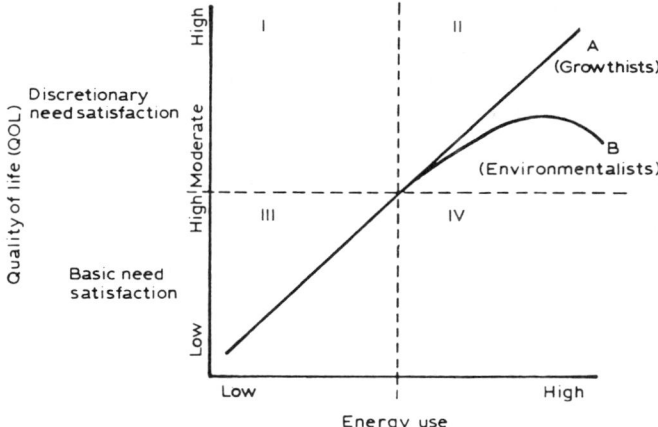

Figure 5.11. Conflicting growthist and environmentalist perspectives on the relationship of energy use and quality of life. (Source: D. E. Morrison, "Equity Impacts of Some Major Energy Alternatives," in *Energy Policy in the United States: Social and Behavioral Dimensions*, S. Warkov, ed., New York: Praeger, 1978, pp. 164–193.)

more than 2.2 billion, or nearly 75% of the Third World.[110] Readers who want to collect information directly from the source will find the United Nations Educational, Scientific, and Cultural Organization (UNESCO)/Solar Energy Research Institute (SERI) directory of information sources and research centers useful. It lists hundreds of information sources from around the world.[111] These books and references serve as examples of the large collection of internationally related literature that emerged in the last two decades, but they only scratch the surface of the information that is available.

F. Environmental Implications

In addition to an energy crisis that poses hazards for our economy, we also have an environmental crisis that poses hazards to our air, water, and food supply.[112] Energy, other natural resources, and the environment are

---

110. V. Smil and W. E. Knowland, eds., *Energy in the Developing World* (Oxford: Oxford University Press, 1980).
111. UNESCO, *International Directory of New and Renewable Energy Information Sources and Research Centers*, 2d ed. (Paris: UNESCO, jointly with SERI, USA, 1986).
112. J. A. Carver, "The Challenge of Energy Generation," in *Energy Needs and the Environment*, eds. R. L. Seale and R. A. Sierka (Tucson: University of Arizona Press, 1973).

interwoven in a complex fashion. When social and economic considerations are superimposed, the global interdependence becomes overwhelming. Disturbances introduced anywhere in the system may have immediate or delayed repercussions elsewhere.[113]

Environmental quality is not easy to quantify, but several indicators may be used to show trends. In 1950, worldwide fossil fuel combustion emitted 1.6 billion tonnes of carbon. By 1979, carbon emissions had climbed to more than 5 billion tonnes. Since 1979, carbon emissions have fallen off somewhat, perhaps due to reduced oil consumption and improved emission controls.[114] Future environmental quality will be influenced significantly by the level of world energy consumption. Acid-rain-forming sulfur emissions are projected to increase from 100 million tonnes in 1984 to 170 million tonnes in the year 2000. Reduced energy consumption and implementation of improved pollution control technology would limit sulfur emissions to a 20% increase by the year 2000 and provide comparable control of carbon dioxide emissions. Although the subject is complex and controversial, many scientists have strong convictions that dramatic change in the global climate will occur if carbon dioxide emissions reach 600 parts per million (ppm), from 343 ppm in 1980.[115]

Public concern for a quality environment and the preservation of natural resources has significantly increased, as has concern about the potential impact of agri-chemicals on human health. Water pollution control became an early national objective and its institutional structure has been expanded by successive amendments. Air quality and solid waste disposal are more recent sources of concern. The National Environmental Protection Act provides an umbrella-type statement of national policy, requiring that projects involving government participation or approval be subjected to careful environmental impact scrutiny. These measures have created constraints on the production, consumption, and transmission of energy resources and supplies.[116] An adequate energy supply is necessary for human welfare and economic development, but a clean and healthy environment is also necessary. A primary goal of the energy sector should be to ensure a balance between the adverse environmental effects and the human and economic benefits of energy development.[117]

---

113. Stout, *Handbook*.
114. L. R. Brown et al., *State of the World 1985* (Washington, D.C.: Worldwatch Institute and New York: W.W. Norton, 1985).
115. Stout, *Handbook*.
116. Carver, "Challenge of Energy Generation."
117. Stout, *Handbook*.

G. Energy Alternatives and Policies

No single fuel source is likely to meet future energy needs. Rather, a great diversity of options is desirable. The United States has large reserves of petroleum and natural gas, and even larger reserves of coal. Development of these domestic energy reserves, in order to reduce dependence on imported oil, has been a matter of high national priority. Engineers and scientists continue working on energy research programs to develop new and improved technology to effectively utilize domestic oil, gas, and coal, as well as nuclear power. Because fossil fuel supplies are finite, energy consumption at present levels must eventually depend on renewable or inexhaustible sources, such as solar energy or nuclear fusion. Numerous discussions and debates have occurred regarding the assumptions, strategies, policies, and timetable for shifting from finite to nearly infinite energy forms.[118]

Although developing countries strive to maintain a series of balances—between exports and imports, population and food supply, and resource development and utilization—the evidence suggests that they have real integration problems and that their schisms and internal conflicts hinder the development of rational energy strategies. A change in energy policy usually means treading on the interests of an economic or a political interest group; hence, energy questions are settled in the realm of politics rather than in a rational and objective context.[119]

Nevertheless, to feed the expanding world population adequately and to meet other social and economic goals, the amount of energy used per person per hectare in agricultural production, processing, and distribution should be significantly increased. Poor countries need as much energy as they can get, as cheaply as possible. Much research has been devoted to increasing the efficiency of agricultural production. For example, better water management; fertilizer use; and machinery design, selection, and, operation are constantly being sought. Changing consumer habits and improving food processing are effective ways to conserve energy. The great potential for increased photosynthetic efficiency must be studied by agronomists and biochemists. But these are long-term prospects. In the short-term, it is necessary to strive for increased efficiency in agricultural production. Energy conservation in agriculture often consists of better management of all necessary operations or elimination of unnecessary activities.[120] In chapter 9 of the *Handbook*, titled "Alternatives," Stout provides a detailed discussion of increasing the energy supply through such options as energy cropping, solar

118. Hiler, *Biomass Energy*.
119. Stout, *Energy for World Agriculture*.
120. Ibid.

energy alternatives, integrated energy systems, and priority allocation of scarce fuels. A series of workshops on energy conservation was conducted by the Energy Research and Development Administration (ERDA), Division of Industrial Energy Conservation. These workshops generated a wide range of energy conservation project ideas for the ERDA (now DOE) program.[121]

Energy problems in the rural areas of developing countries cannot be solved by energy conservation alone. Efficient energy management must be a primary method of decreasing energy demands. Energy management for subsistence farmers, who use little or no commercial energy, means efficiently applying their own manual efforts, or, for some, efficiently using animal power. Rural non-farmers, who use some commercial energy for lighting, heating, and cooking, could save the most energy through the development and use of improved cooking stoves. Energy is a relatively small consideration in the total agricultural production budget of commercial farmers. Therefore, many commercial farmers are not exceedingly concerned about energy as long as plentiful supplies are available when needed.[122]

The politics and economics of energy are much more volatile than the technical and engineering aspects. Energy policies are formulated in the political arena. Energy prices are influenced by many factors that have little to do with physical supply. The magnitude of energy resources is often a matter of opinion and controversy. Gloomy forecasts are available for those who want to believe them, and rosy predictions of future energy supplies are available for those who are more optimistic. Thus both energy policies and energy economics are quite unpredictable.[123]

One of the most difficult dimensions of energy policy is the uncertainty of future energy prices and supplies. We must make appropriate investments in new energy production as well as adjust consumption to the changing patterns of energy availability in a timely manner to avoid serious economic and social disruptions. Uncertainty makes timely investment and economic adjustment very difficult. We simply do not know, for example,

---

121. See: (a) *Report of the Proceedings of the Agriculture Processing Industry Workshop on Energy Conservation, Washington, D.C., March 1976* (Washington, D.C.: Washington Scientific Marketing, available from Springfield, Va.: National Technical Information Service, 1976). (b) *Report of the Proceedings of the Energy Research and Development Administration Workshop on Energy Conservation in Agricultural Production, Washington, D.C., July 1976* (Washington, D.C.: Washington Scientific Marketing, available from Springfield, Va.: National Technical Information Service, 1976). (c) *ERDA Programs and Objectives: Energy Conservation in Food Processing* (Washington, D.C.: Washington Scientific Marketing, available from Springfield, Va.: National Technical Information Service, 1977).
122. Stout, *Handbook*.
123. Stout, *Energy Use and Management in Agriculture*.

what technology will develop or what will influence the supply or price of energy in the future. Public and private planning and policy must be conducted in the context of this uncertainty.[124]

H. Summary

Energy to supplement our meager work capacity is a necessity, not a luxury. United States agriculture uses less than 3% of the total energy supply—but it is a vital 3%. Adequate food supplies and a high standard of living are dependent on that 3% of energy being available in the right form at the right time. Precious supplies of conventional fuels must be stretched through conservation and efficient management as we seek to maintain and improve our quality of life. In addition, new and renewable energy forms must be developed and utilized as they become economically viable to ensure a continued abundant energy supply.[125]

---

124. J. Shaffer, "Energy, Transportation and Food: Interdependent Policy Issues," Rochester Seminars, unpublished.

125. Stout, *Energy Use and Management in Agriculture*.

6. Literature in Instrumentation, Control, Electronics and Automation

G. C. ZOERB
Agricultural Engineering Department
University of Saskatchewan

A. Introduction and Background

Essentially all agricultural engineering research and development activities are based on the necessity to make fundamental measurements. Numerical data must be obtained in order to evaluate processes or systems. Unless one is able to state interrelationships of physical phenomena in terms of numerical values, it is often difficult if not impossible to understand these interrelationships. Instruments and/or instrumentation systems are important components of all research, development and testing programs.

The terms "instrumentation," "control," "electronics" and "automation" used in the title are interrelated. Instrumentation used to make a measurement may contain electronics in one or more parts of the system, usually as amplifiers. Automation is the application of a control system to perform one or more functions with little or no human input. All instrumentation and control systems must have sensors (detectors) or transducers. (A transducer converts energy from one form to another, for example, a thermocouple converts from thermal to electrical energy). Recently electronic devices through the use of integrated circuit (IC) chips have been used as sensors. Finally, the term "instrumentation" is usually taken in the general sense to include control, electronics and automation.

A complete instrumentation (or measurement) system can be classified or analyzed according to the functions of the various components. Such an analysis is presented in the initial chapters of *Measurement Systems: Application and Design.*[1] This book contains a good balance of theory and hardware and is an excellent text for a graduate level course in instrumentation.

---

1. Ernest O. Doebelin, *Measurement Systems: Application and Design*, 4th ed. (New York: McGraw-Hill Book Company, Inc., 1990).

The development and use of instrumentation in agricultural engineering has paralleled other advances in the application of engineering to the field of agriculture. Early papers found in the American Society of Agricultural Engineers (ASAE) publications used only elementary instruments, usually with an analog (meter) readout. One of the earliest books in agricultural engineering, *Farm Machinery and Farm Motors* by J. B. Davidson and L. W. Chase, was published in 1908 (one year after the founding of the ASAE in 1907 at Madison, Wisconsin).[2] It is interesting to note that the authors covered dynamometers, including "traction dynamometers." In addition to a spring-type direct-reading dynamometer, they have illustrated a German-built recording dynamometer. The chart (reel or spool) was ground driven. Another unit shown had the chart driven by clock-work. A planimeter was used to determine average draft for a test run. An engine indicator used to obtain PV diagrams for steam and gas engines was illustrated also.

Agricultural engineering subject areas evolved by the mid-fifties into four main classifications which comprised the technical divisions of ASAE. These were Power and Machinery, Soil and Water, Electric Power and Processing, and Farm Structures. While research, teaching and extension programs have been conducted within these areas, the topic of instrumentation has been characterized by its "cross-cutting" nature. In other words, basic instrumentation is required, particularly in research, in all the above areas of specialization. This need for instrumentation was expressed by Robert Trullinger, ASAE President, in his address, "Research for Tomorrow's Agriculture" at the June 1950 ASAE Meeting. He stated that "Engineers will need knowledge of advanced scientific instrumentation to solve future problems."

The *Agricultural Engineering Index (1907–1960)* by C. W. Hall shows that prior to 1950, only 36 papers are listed which deal with instrumentation.[3] Notable in terms of their later significance and widespread use were papers by S. C. Heth on the use of strain gages for farm equipment design in 1947 and by G. J. Bouyoucos and G. A. Crabb, Jr., on the measurement of soil moisture by the electrical resistance method in 1949. Slightly over four times as many papers under the headings of "Instruments, Instrumentation, Controls" are to be found from 1950 to 1960 as compared to those listed before 1950. Many of these papers dealt with the measurement of variables such as temperature, humidity, flow (air and water), pressure, moisture content, soil properties, noise, torque and even radiation. It is

---

2. J. B. Davidson and Leon W. Chase, *Farm Machinery and Farm Motors* (New York: Orange Judd Company, 1908).

3. Carl W. Hall, *Agricultural Engineering Index* (1907–1960) (St. Joseph, Mich.: American Society of Agricultural Engineers, 1961).

interesting to note a paper by M. N. Langley in 1960 on developments in sprinkler irrigation automation. A significant paper in this era in terms of initial developments in the assessment of quality of agricultural products is one by K. H. Norris in 1958 on measuring light transmittance properties of agricultural commodities.

The importance given to the measurement of animal and tractor power is illustrated by the fact a total of twenty-two papers were listed in the above *Index* between 1910 and 1951, under the heading of "dynamometer." Sixteen of these were before 1942. Only seven papers are listed in the *Index* under "Computers" by 1960, these being published between 1958 and 1960.

The literature in agricultural engineering appears to be expanding at an exponential rate. In order to obtain information in any subject area, it is important to have access to a comprehensive index. The four volumes of the *Agricultural Engineering Index* by Carl W. Hall, and co-authors Glenn E. Hall and James A. Basselman[4] are essential in any literature review. These indexes reference world-wide literature published in the discipline of agricultural engineering. In addition, technical papers presented at the ASAE meetings are listed. References are listed by subject matter. The annual *ASAE Comprehensive Index of Publications* also references most of the world-wide publications (plus ASAE papers presented). This annual index includes listings by subject matter and by author.

Abstracts are closely related to indexes and very useful for determination of technical content of a paper. Noteworthy are the *Agricultural and Horticultural Engineering Abstracts* which were published from 1951 to 1966. These quarterly abstracts were prepared by the Information Department of the National Institute of Agricultural Engineering (NIAE), and published by the Commonwealth Agricultural Bureaux (CAB) in England. Since 1976 the abstracts, which cover the world's scientific and technical literature, have been published monthly by CABI as *Agricultural Engineering Abstracts*. An author and subject index as well as a cumulated annual index is included. These abstracts provide one of the best sources of literature in English and other languages. Considerable applied work is being done throughout the world, certainly with people in Japan and Western Europe making valuable advancements. Morgan reported in the mid-fifties that about 4,500 items per year were listed. A useful reference in terms of

4. (a) Ibid. (b) Glenn E. Hall and Carl W. Hall, *Agricultural Engineering Index (1961–1970)* (St. Joseph, Mich.: American Society of Agricultural Engineers, 1972). (c) Carl W. Hall and James A. Basselman, *Agricultural Engineering Index (1971–80)* (St. Joseph, Mich.: American Society of Agricultural Engineers, 1982). (d) Carl W. Hall and James A. Basselman, *Agricultural Engineering Index (1981–85)* (St. Joseph, Mich.: American Society of Agricultural Engineers, 1987).

searching the literature is the book by Bryan A. Morgan, *Keyguide to Information Sources in Agricultural Engineering.*[5] Morgan, head of the library at Silsoe College in England, also provides an extensive list of computerized data bases. During a one-year period (1982–83) at NIAE of 128 searches using 293 databases, 64% were provided by AGRICOLA (*Bibliography of Agriculture*) and CAB's abstract services. The third most popular data base was COMPENDEX (*Engineering Index*).

Since research papers involving instrumentation, control, electronics and automation may be found in any of the discipline areas of agricultural engineering, the above indexes, abstracts and data bases are especially useful. For example, if one wishes to search the literature on internal combustion engines, it is easy to locate references under this topic in a subject matter listing. The application of instrumentation, control, electronics, and automation has taken place in practically all the subject matter areas of agricultural engineering. The indexes provide cross-reference listings, so relevant papers can be located with relative ease. Of course, most agricultural engineers are involved in the use of instrumentation with agricultural equipment or processes as applied to crop and livestock production.

B. Some Reference Books in Instrumentation for Agricultural Engineers

In the ASAE, the increasing importance of instrumentation to agricultural engineers was recognized by the formation of a Committee on Instrumentation and Controls in 1950. At the 1951 meeting the committee decided to publish a monthly feature called "Instrument News" in *Agricultural Engineering*. The first article (on thermistors) was in the August 1952 issue.

During the fifties, several agricultural engineering departments in the United States introduced a course in instrumentation in their graduate curricula. An initial problem was that of obtaining a suitable textbook. Even in the other engineering disciplines, there were few instrumentation textbooks. One of the first books to appear was D. P. Eckman's *Industrial Instrumentation* in 1950.[6] This was an excellent book at that time in terms of measurement of variables such as temperature, pressure, flow, level, and other mechanical measurements.

---

5. Bryan A. Morgan, *Keyguide to Information Sources in Agricultural Engineering* (London: Mansell Publishing Ltd., 1985).
6. Donald P. Eckman, *Industrial Instrumentation* (New York: John Wiley and Sons, Inc., 1950).

The shortage of textbooks in the area of instrumentation and the fact that no books were available which were directed to agricultural engineering, led the Instrumentation and Controls Committee to take its own action. Members of the committee agreed to develop one or more topics which were later presented as papers on programs sponsored by the Committee at ASAE meetings. These papers, when reviewed, were incorporated as chapters in the textbook, *Instrumentation and Measurement for the Environmental Sciences*.[7] The second edition, edited by Bailey W. Mitchell, was published in 1983. This book with fifteen chapters covers four main subdivisions: System Analysis, Primary Sensing Elements, Signal Conditioning and Coupling, Data Output and Controlling.[8]

No doubt the need for textbooks in the general area of instrumentation was felt in other disciplines as well. Consequently, by the time the ASAE text was published in 1975, many other books had appeared. A few of these which provided excellent references to agricultural engineers are mentioned below. Other instrumentation books are listed in the additional references at the end of the chapter.

The Strain Gage Primer by Perry and Lissner was initially published in 1955 when many agricultural engineers in industry and universities were just beginning to learn how to use this revolutionary device, both for strain measurement and as the sensing element in many transducers. A second edition appeared in 1962.[9]

Two handbooks provided valuable and extensive reference material during this period. The *Process Instruments and Controls Handbook*, edited by D. M. Considine, was published in 1957 with second and third editions in 1974 and 1985.[10] H. P. Kallen's *Handbook of Instrumentation and Controls* was available in 1961.[11]

The book *Basic Electrical Measurements* by M. B. Stout provides a fundamental background in theory of electrical measurements.[12] Emphasis is placed on galvanometers, potentiometer and bridge circuit analysis, shielding, magnetic measurements. Chapters on experiments and statistical anal-

---

7. Zachary A. Henry, ed., *Instrumentation and Measurement for the Environmental Sciences* (St. Joseph, Mich.: American Society of Agricultural Engineers, 1975).

8. Bailey W. Mitchell, ed., *Instrumentation and Measurement for the Environmental Sciences*, 2d ed. (St. Joseph, Mich.: American Society of Agricultural Engineers, 1983).

9. C. C. Perry and H. R. Lissner, *The Strain Gage Primer*, 2d ed. (New York: McGraw-Hill Book Company, Inc., 1962).

10. D. M. Considine, ed., *Process Instruments and Controls Handbook*, 3d ed. (New York: McGraw-Hill Book Company, Inc., 1985).

11. H. P. Kallen, *Handbook of Instrumentation and Controls* (New York: McGraw-Hill Book Company, Inc., 1961).

12. M. B. Stout, *Basic Electrical Measurements*, 2d ed. (Englewood Cliffs, N.J.: Prentice-Hall, Inc., 1960).

ysis and on electrical indicating instruments are pertinent for all agricultural engineers.

A book by N. H. Cook and E. Rabinowicz, *Physical Measurement and Analysis*, covers conventional measurement topics of temperature, force, torque, flow, pressure, but also provides an excellent chapter at that time on radiotracer techniques.[13]

The text *Experimental Stress Analysis* by Dally and Riley gives comprehensive coverage of electrical resistance strain gage circuitry and recording equipment for strain measurement.[14] Stress analysis by the brittle coating method, photoelasticity and photoelastic coatings are covered also. This book, like *The Strain Gage Primer*, is useful for those in the power and machinery area.

A popular reference book has been J. P. Holman's *Experimental Methods for Engineers*.[15] In addition to the usual variables covered in this type of instrumentation book, there is a chapter on air pollution sampling and measurement, as well as a portion of a chapter dealing with nuclear radiation measurements.

Instrumentation in Scientific Research by Kurt S. Lion provides an excellent treatise on transducers.[16] It is still considered by many to be one of the best introductory books available on the topic of transducers.

At many universities, electrical engineering departments have provided one or more graduate classes in electronics for non-electrical engineering students. An excellent book for this purpose has been *Basic Electronics for Scientists* by J. J. Brophy.[17]

One of the most used reference books by agricultural engineers has been *Mechanical Measurements* by Beckwith, Buck and Marangoni.[18] This book presents an excellent coverage of the fundamentals of technical measurement in Part I and of applied mechanical measurements in Part II. A chapter on the application of digital techniques to mechanical measurements has been added in the third edition. This is one of the few textbooks which presents a chapter on acoustic measurements.

13. Nathan H. Cook and Ernest Rabinowicz, *Physical Measurement and Analyses* (Reading, Mass.: Addison-Wesley Publishing Company, Inc., 1963).
14. James W. Dally and William F. Riley, *Experimental Stress Analysis*, 3d ed. (New York: McGraw-Hill Book Company, Inc., 1991).
15. J. P. Holman, *Experimental Methods for Engineers*, 5th ed. (New York: McGraw-Hill Book Company, Inc., 1989).
16. Kurt S. Lion, *Instrumentation in Scientific Research* (New York: McGraw-Hill Book Company, Inc., 1959).
17. James J. Brophy, *Basic Electronics for Scientists*, 5th ed. (New York: McGraw-Hill Book Company, Inc., 1990).
18. Thomas G. Beckwith, N. Lewis Buck, and Roy D. Marangoni, *Mechanical Measurements*, 3d ed. (Reading, Mass.: Addison-Wesley Publishing Company, 1982).

In recent years, books have been published which deal with the application of instrumentation to agriculture or agricultural engineering problems. One example is *Electricity and Electronics for Agriculture* by Allen F. Butchbaker.[19] R. J. Gustafson, Chairman of the Agricultural Engineering Department, the Ohio State University, has written a textbook, *Fundamentals of Electricity for Agriculture*.[20] This book, in addition to covering fundamentals of electricity with applications to farmsteads, provides an introduction to electrical controls and solid state electronics. The second edition was published by the ASAE in 1988. S. W. R. Cox at the National Institute of Agricultural Engineering (NIAE) has published three related books which provide numerous examples of electronic (or instrumentation) applications in the production of field crops, protected crops (vegetables and flowers) and livestock. The first book, entitled *Instrumentation in Agriculture*, with D. E. Filby as coauthor, was published in 1972.[21] In 1982, the book *Microelectronics in Agriculture and Horticulture: Electronics and Computers in Farming* was published.[22] The third book, *Farm Electronics*, appeared in 1988.[23] An excellent reference for those working in the electric power and processing or food engineering areas is *Near-Infrared Technology in the Agricultural and Food Industries* by P. Williams and K. H. Norris.[24]

Agricultural engineers, perhaps more than engineers in other disciplines, have worked with other groups, especially disciplines within agriculture and veterinary medicine as well as with electrical engineers on interdisciplinary research projects. Often the agricultural engineer is the "middle person" in a research project involving the use of electronics or control systems in solving a problem in a biological or environmental system. Several books are available which deal with instrumentation outside the usual applications in engineering. For example, *Principles of Applied Medical Instrumentation* by L. A. Geddes and L. E. Baker presents an extensive coverage of bioinstrumentation used to measure physiological phenomena.[25] Many types of

---

19. Allen F. Butchbaker, *Electricity and Electronics for Agriculture* (Ames, Iowa: Iowa State University Press, 1977).
20. Robert J. Gustafson, *Fundamentals of Electricity for Agriculture*, 2d ed. (St. Joseph, Mich.: American Society of Agricultural Engineers, 1988).
21. S. W. R. Cox and D. E. Filby, *Instrumentation in Agriculture* (New York: Hafner Publishing Co., 1972).
22. S. W. R. Cox, *Microelectronics in Agriculture and Horticulture: Electronics and Computers in Farming* (Totowa, N.J.: Allenheld, Osmun, 1982).
23. S. W. R. Cox, *Farm Electronics* (Oxford and Boston: BSP Professional Books, 1988).
24. P. Williams and K. H. Norris, *Near-Infrared Technology in the Agricultural and Food Industries* (St. Paul, Minn.: American Association of Cereal Chemists, Inc., 1987).
25. L. A. Geddes and L. E. Baker, *Principles of Applied Biomedical Instrumentation*, 3d ed. (New York: John Wiley and Sons, Inc., 1989).

transducers (resistive, inductive, photoelectric, piezoelectric, thermoelectric, chemical) are described with applications to physiological measurements. Chapters on electrodes for bioelectric events, detection of physiological events by impedance and the bioelectric events comprise about half of the 600 pages.

A source of information often overlooked is that available from transducer and instrumentation manufacturers. Perhaps this occurs because such information is not normally found in the literature indexes or abstracts. These company handbooks, technical notes and catalogs are extremely valuable to anyone who is "building" a measurement device or system. Certainly, more specific performance data and operating details are given than can be found in any textbook. Furthermore, many agricultural engineers and others, simply *select* instruments, transducers or other components for a particular application where such information is a necessity. Hopefully, the person making such a selection has some background training in the instrumentation field. As an example, publications such as a manufacturer's analog-digital conversion handbook and an advanced TTL logic databook may be found on the bookshelf of a modern-day agricultural engineer. Or an electronics company many publish a more general book or bulletin, such as *Understanding Microprocessors*.

Most companies are eager to furnish product information. Some specific examples are listed here. Omega Engineering, Inc.[26] provides a series of instrumentation publications or manuals. Among the more prominent of these are: *Pressure and Strain Measurement Handbook and Encyclopedia*; *Temperature Measurement Handbook and Encyclopedia*; and *Data Acquisition and Computer Interface Handbook and Encyclopedia*. The Eaton Corporation provides a *Lebow Loadcell and Torque Sensor Handbook*, which includes transducer design fundamentals and product listings. Another example is the *Solid State Sensor Handbook: Pressure Sensors and Accelerometers* published by SenSym, Inc.

One of the additions to the technical divisions of the ASAE in the 1980s was the Food and Process Engineering Institute. In 1961 the ASAE Committee, Physical Properties of Agricultural Products was formed. The activities of this Committee have played a part in development of the processing and food engineering area. Throughout the years, the Committee has sponsored technical programs at the ASAE meetings dealing with mechanical, rheological, electrical, thermal and optical properties of agricultural products and foodstuffs, as well as particle statistics. A knowledge of the physical properties of agricultural products is essential in the design of machines

---

26. Mention of commercial companies or products does not constitute endorsement.

and structures for production, handling and processing of food and fiber. An important component in research in this area has been the measurement techniques used to determine physical properties. In many cases, no standard measurement procedure has been adopted. A review paper by G. Zoerb on instrumentation and measurement techniques for determination of physical properties of agricultural products was published in the 1967 *ASAE Transactions*. In 1973, Finney produced a publication with numerous references, *Measurement Techniques for Quality Control of Agricultural Products*.[27] The three main measurement techniques covered are light transmittance and reflectance, X-radiation, and sonics and ultrasonics.

No doubt the most comprehensive publications available dealing with properties of agricultural products are N. N. Mohsenin's three books: *Physical Properties of Plant and Animal Materials*, *Thermal Properties of Foods and Agricultural Materials*, and *Electromagnetic Radiation Properties of Foods and Agricultural Products*. The first book provides a comprehensive study of plant and animal materials.[28] Topics covered are structure and chemical composition, water retention, physical characteristics, contact stresses, mechanical damage, aero-and hydrodynamic characteristics, friction and rheological properties. Three chapters are devoted to rheology— concepts, properties and measurements. The second book covers heat transfer, specific heat, thermal conductivity, thermal diffusivity, unit surface conduction, applications of thermal properties in cooling, freezing, heating and heat treatment of foods and agricultural materials.[29] The third book covers electromagnetic radiation properties, with at least two-thirds of the book devoted to applications.[30]

A topic often missed in instrumentation textbooks is sound or acoustic measurement. However, agricultural engineers have been concerned with this topic. In the 1961–70 *Agricultural Engineering Index*, twenty-five papers on noise measurement and control are listed. For the 1971–80 period, the *Index* lists sixty-eight papers, but only nine for the 1981–85 period. In 1971, measurement of tractor and tractor cab noise was instituted as part of the Nebraska Tractor Test. Noise is defined as unwanted sound. With the sound level data available, tractor manufacturers concentrated on making

---

27. Essex E. Finney, Jr., *Measurement Techniques for Quality Control of Agricultural Products* (St. Joseph, Mich.: American Society of Agricultural Engineers, 1973).

28. Nuri N. Mohsenin, *Physical Properties of Plant and Animal Materials*, 2d ed. (New York: Gordon and Breach Science Publishers, Inc., 1986).

29. Nuri N. Mohsenin, *Thermal Properties of Foods and Agricultural Materials* (New York: Gordon and Breach Science Publishers, Inc., 1980).

30. Nuri N. Mohsenin, *Electromagnetic Radiation Properties of Foods and Agricultural Products* (New York: Gordon and Breach Science Publishers, Inc., 1984).

reductions in these noise levels. In a few years levels near the operator's ears dropped typically from 92 to 97 dB(A) to around 85 dB(A).

In the United States, the Noise Control Act was passed in 1972. This act empowered the U.S. Environmental Protection Agency (EPA) to set limits on the amount of noise permitted from trucks, buses and trains. The agency also is allowed to regulate sources of noise of newly manufactured products such as jackhammers, motorcycles and snowmobiles. EPA can require labeling of noise levels on household appliances. *The Noise Handbook*, edited by W. Tempest, covers all aspects of noise measurement, control and legislation.[31] An excellent reference book is the *Handbook of Noise Measurement* by Peterson and Gross.[32] For anyone interested in the effects of noise on hearing impairment, work performance, physiological stress and other factors, Kryter's book *The Effects of Noise on Man* is an excellent reference.[33] There is no doubt that as concerns on environmental pollution increase, noise measurement and control will be a greater consideration in the agricultural engineering workplace.

C. Interaction with Other Organizations

The American Society of Agricultural Engineers has had interaction with other organizations throughout the years, particularly since 1950. In the early fifties the ASAE had representatives working in cooperation with a dozen organizations. Significant in terms of relevance to the instrumentation (measurement) area was the American Society of Testing Materials, the National Farm Electrification Conference, and the American Standards Association. The ASAE was admitted to constituent membership in the Engineers Joint Council (EJC) in 1961. This organization was formed by the founder engineering societies after World War II. Among the constituent societies were the American Society of Civil Engineers, the American Society of Mechanical Engineers, and the American Society for Engineering Education. The discipline of agricultural engineering has drawn on these organizations and societies for technical support, although the extent is difficult to pinpoint.

The ASAE's association with the International Congress of Agricultural Engineering (CIGR, for Commission Internationale du Génie Rural) has provided contact with agricultural engineers worldwide. This association

---

31. W. Tempest, ed., *The Noise Handbook* (New York: Academic Press, 1985).
32. Arnold P. G. Peterson and Erwin E. Gross, Jr., *Handbook of Noise Measurement*, 9th ed. (West Concord, Mass.: Genrad, Inc., 1980).
33. Karl D. Kryter, *Effects of Noise on Man*, 2d ed. (New York: Academic Press, 1985).

has been strengthened in recent years. The first meeting of the CIGR to be held in North America was at Michigan State University in 1979.

The ASAE and Canadian Society of Agricultural Engineering (CSAE) have worked together since its inception in 1959, and particularly since 1986 when CSAE became affiliated with ASAE. Instrumentation was strengthened by these two organizations working so closely together.

D. Evolution of Instrumentation, Control, Electronics, and Automation

As mentioned earlier, instrumentation, control, electronics and automation are interrelated. As a result, these topic areas have developed almost in parallel. Many measurement devices and instrumentation systems have been developed first for industrial and/or military use. Later these found application in the agricultural industry. In terms of chronological development the sensor or transducer was first. Instruments were developed which used signal conditioning to transform the transducer output into an analog readout either by a meter or via a chart recording. These instruments contained amplification, so electronics (vacuum tubes, then transistors) were employed. As applications were perceived, open loop and then closed loop control systems were applied to agriculture. In the seventies and eighties much emphasis was placed on the development of data acquisition techniques with the use of data loggers connected to computers through interfaces. Since the digital computer can accept data only in digital format, analog signals from transducers must be converted with an A/D converter. These units are now available as inexpensive integrated circuit (IC) chips. Likewise D/A converters must be used at computer outputs for data examination. The analog meter readings in early measurement had to be manually recorded. The data then were often inserted into a mathematical expression to obtain the result using a slide rule. Modern instrumentation conditions the output signal from the transducer or from several transducers through multiplexing, converts to digital form for input to a computer which performs the calculations and prints out the answer. Or the output may be a graphical representation of the relationships under study.

The tremendous advances in electronics in the last twenty years have led to solid state integrated circuit devices used both as sensors and in instrumentation systems including computers. The popular junction semiconductor temperature sensor is produced as an IC chip, about 12 mm by 1.5 mm in size. It contains eleven transistors, six resistors and a capacitor. This sensor is available in cans, miniature flat packages, chip form and stainless steel probes. Large scale integration (LSI) refers to an integrated circuit with between 100 and 1,000 gates per chip. Very large-scale integration devices (VLSI) with over 1,000 gates have been developed. The heart of the modern microcomputer is the LSI IC chip, the microprocessor unit (MPU), also called the central processing unit (CPU). An excellent periodical publication with a relatively high technical level is *Computers and Electronics in Agriculture*, J. R. Lambert, editor-in-chief,[34] published quarterly since 1985.

It is interesting to note the wide variety of topics with which agricultural engineers become associated. This can be illustrated if we look at any section of the *Agricultural Engineering Index*. As an example, let us examine the listing under "Control, Controlled, Controller" for the 1971–80 period. The following are some of the areas covered: digital temperature recorder, load management, IC circuits for control and monitoring, Boolean algebra for switching and control, control by passive solar design, control for lawn and garden tractors, implement coupling and control, multiple water bath temperature control, digital control of agricultural processes, solid-state relays for control, control of tractors on sloping ground, control for draught force sensing, control unit for milking systems, controlled atmosphere (CA) for cabbage, controlled environment (several applications), controller for livestock feeding system, trickle irrigation control, control of a furrow irrigation valve, controllers for agricultural processing and production systems, controls for automated plant juice concentrate production, controls for grain drying, electronic controls—key to production efficiency on the farm, trends in robot controls.

Following the development of mechanization the next logical step is automation in various degrees. Beginning about 1965 research on the application of automatic controls to agriculture began to appear. The justification for automatic control has been to increase output per man hour, improve machine efficiency, provide better quality control of the agricultural product and relieve operators of much drudgery. In North America and in Europe, a large number of research papers have been published, particularly between 1965 and 1985, dealing with automatic control applied to both field and farmstead operations.

Early farmstead applications dealt with automatic or semi-automatic feed processing and handling systems. In dairying, the use of a transponder attached to a cow's neck permitted automatic identification, feed dispensing, combined with transducers to measure milk yield and to detect the onset of health problems such as mastitis. Other examples were automatic livestock

---

34. *Computers and Electronics in Agriculture, An International Journal* (Amsterdam, Netherlands: Elsevier Science Publishers, 1985 [published quarterly]).

waterers and solid-state temperature controllers for animal and poultry ventilation systems. In field applications, moisture sensors and timer-controlled structures (gates) have been incorporated into automatic surface irrigation systems. In harvest operations, several researchers in North American and in Europe have developed automatic feed-rate control systems for combines to reduce losses and to optimize throughput.

One of the most fascinating applications of automatic control has been in tractor guidance. The principal objective has been to relieve the operator of the continuous and tiresome job of steering between ends of a field. In addition to the reduction of operator fatigue, less overlap or miss should occur. Roy E. Young has compiled and edited a monograph, *Automatic Guidance of Farm Vehicles*.[35] The publication contains a review paper by Gerhard Jahns (Institut für landtechnische Grundlagenforschung (Institute for Basic Research), Bundesforschungsanstalt für Landwirtschaft (Agricultural Research Center), Braunschweig-Völkenrode, West Germany) on automatic steering systems for farm vehicles. The monograph has a bibliography of 233 items as well as a list of ninety-five patents dealing with vehicle guidance. In a good portion of the research on tractor guidance, the system was designed to retain the operator on the tractor because of variable field conditions and obstacles, such as wet spots and trees. The guidance system was intended to serve as a steering assist. Guidance devices such as furrows, standing crops or buried wires have been used. Fully automatic spatial positioning systems have not been developed to date.

In typical farm operations (except for gantry controlled traffic applications), it appears that agricultural engineers and others have concluded that an operator must be present on the tractor or self-propelled machine. Consequently, emphasis has shifted somewhat towards monitoring operations. The first type of monitor used on farm tractors was simply a dial gauge to indicate engine functions such as oil pressure, coolant temperature and fuel tank level. Present commercially available tractor monitors indicate drive wheel slip, ground speed, area covered per hour, and total area covered. In the near future tractor monitors will also provide information in terms of efficiency of fuel use (litres per hectare or gallons per acre). Also, a continuous readout showing the combination of engine speed (throttle setting) and transmission gear selection will be provided for best fuel efficiency.

Monitors have also become commonplace for other field equipment, particularly for grain seeding equipment, corn planters, crop sprayers and combines. The combine grain loss monitor has been accepted as an important

---

35. Roy E. Young, *Automatic Guidance of Farm Vehicles: A Monograph* (Auburn, Ala.: Auburn University Agricultural Experiment Station, 1976).

unit to assist the operator in maximizing throughput while minimizing grain losses. In some respects, moisture testers for grain and forage and electronic livestock scales can be considered as monitors. A significant factor in the adoption of monitors and agricultural control systems during the past twenty years has been the increased degree of user confidence in electronic instrumentation.

The ASAE literature shows a number of symposia dealing with electronics and automatic control applications in the agricultural industry. In 1974, the International Federation of Automatic Control (IFAC) held a symposium on *Automatic Control for Agriculture* at the University of Saskatchewan, Saskatoon, Canada.[36] This has been the only IFAC symposium to date pertaining to agriculture. A total of thirty-six papers were presented in the following areas: Tractor design and guidance systems, instrumentation, automation of irrigation and spraying, automation in agricultural production, automation in dairy and livestock production, and harvest automation.

The ASAE National Conference on *Agricultural Electronics—1983 and Beyond*[37] has provided the most comprehensive state-of-the-art presentation to date of instrumentation, control, electronics and automation applications to agriculture. The ninety-eight papers presented cover applications to field equipment, irrigation and drainage, controlled environments, livestock production systems, materials and processing, and special topics.

Considering the increased use of robots in industry, it is natural that agricultural applications would be investigated. Tasks in production agriculture, however, seldom permit the repetitive, same-path operation characteristics of industrial use. In 1983 the first International Conference on Robotics and Intelligent Machines in Agriculture was held in Tampa, Florida. The objectives of the conference were to "reviews the current research related to this topic, to project future applications in agriculture and to stimulate new research."[38] The papers presented cover a wide range of applications and potential applications. The Australian experimental sheep-shearing robot is a unique example.

---

36. Preprints of the *IFAC Symposium on Automatic Control for Agriculture*, sponsored by the Associate Committee on Automatic Control and the National Research Council of Canada, Saskatoon, Sask., Canada, June 1974.

37. *Agricultural Electronics—1983 and Beyond*, Vol. I. *Field Equipment, Irrigation and Drainage*, Vol. II. *Controlled Environments, Livestock Production Systems, Materials Handling and Processing*; Proceedings of National Conference, Dec. 11–13, 1983, Chicago (St. Joseph, Mich.: American Society of Agricultural Engineers, 1984 [*ASAE* Publ. 8-84]).

38. *Robotics and Intelligent Machines in Agriculture*, Proceedings of the 1st International Conference, Tampa, Fla., 1983 (St. Joseph, Mich.: American Society of Agricultural Engineers, 1984 [*ASAE* Publication 4-84]).

The Royal Agricultural Society of England sponsored an international symposium on *Farm Electronics and Computing*[39] in October 1985 at Stratford-upon-Avon, England. Of the thirty-two main papers and eleven poster papers, most deal with computer applications in agriculture.

Two additional important conferences and expositions were Agri-Mation I in 1985 and Agri-Mation II in 1986.[40] These were cosponsored by the ASAE with the Society of Manufacturing Engineers and held in Chicago. These conferences stressed the use of sensors and automated systems, including robotics, in field and farmstead applications.

The previous discussion, to some extent has, shown the chronological evolution and application of instrumentation, control electronics, and automation in agriculture with significant publications highlighted. Table 6.1 presents the relative emphasis of these interrelated topics and others included in this general area. The table attempts to illustrate the change in topic emphasis during the time period from 1961 to 1988. Although many of the subject listings appear in one or more of the headings given, the trends are still evident. Paper numbers have increased rapidly. Different terminology has been adopted, for example, note the decrease in the use of "measurement" and the increase in the use of "instrumentation." The large number of computer-oriented papers is noteworthy. The table also shows when topics began to reach the agricultural engineering literature. In certain periods, the numbers are distorted because of special conferences and symposia. For example, the numbers for electronics and robotics in 1981–85 are relatively high due to the Robotics Conference in Florida and the 1983 Agricultural Electronics Conference in Chicago. Other related headings could be used in this table as well, such as modeling and simulation.

The topic of remote sensing in Table 6.1 deserves consideration. Relatively few papers which deal with remote sensing are included. This fact is surprising considering the numerous applications and potential applications to agriculture. A comprehensive paper on the role of remote sensing in agriculture was presented at the 1978 American Society for Engineering Education (ASEE) Conference by David T. Higgins, professor of civil engineering at Washington State University. A basis for his paper was his participation in the 1977 National Air and Space Agency-ASEE summer sys-

---

39. *Farm Electronics and Computing International Symposium*, Oct. 1985, Stratford-upon-Avon, Warwickshire, Proceedings (Stoneleigh, Kenilworth, Warwickshire, Eng.: Royal Agricultural Society of England, National Agricultural Centre, 1985 [Monograph Series no. 4]).

40. (a) *Agri-Mation I. Proceedings of the Agri-Mation I Conference and Exposition*, Feb. 25–28, Chicago, Ill. (St. Joseph, Mich.: American Society of Agricultural Engineers, 1985 [*ASAE* Publication 01-85]). (b) *Agri-Mation II: Proceedings of the Agri-Mation II Conference and Exposition*, Mar. 3–5, Chicago, Ill. (St. Joseph, Mich.: American Society of Agricultural Engineers, 1986 [*ASAE* Publication 01-86]).

Table 6.1. Frequencies of listings of published and presented papers in the *Agricultural Engineering Indexes* by topic

	1961–70	1970–80	1981–85	1986	1987	1988
Artificial Intelligence	—	—	—	9	16	5
Automation	29	29	44	27	19	20
Computers	46	240	360	87	112	85
Control and Controllers	70[a]	47	108[a]	38[b]	39[b]	50[b]
Electronics	2[c]	30	232	37	55	32
Expert Systems	—	—	12	18	40	29
Instrumentation	12	93	87	27	36	21
Machine Vision	—	—	—	18	10	25
Measurement and Measuring	32	10	3	4	1	11
Microcomputers and Microprocessors	—	60	240	41	51	32
Monitors	1	18	55	23	22	17
Remote Sensing	3	8	—	5	3	9
Robotics	—	1	46	13	14	—

[a]Also controlled atmosphere.
[b]Includes microcomputer and microprocessor controllers.
[c]2 papers on electrostatics—charging of chemical sprays, dusts.

tem design study on the role of aerospace technology in agriculture. A long list of potential applications of remote sensing in agriculture is given in the paper. These include weather forecasting, soil surveys, water resource surveys, crop vigor, soil fertility, soil salinity, insect and disease infestations in crops, health of animals and many more.

At a recent ASAE national meeting, the topic of spatial farming created much interest. This concept requires an accurate grid map of soil types and fertility levels. These levels vary from point to point in any field. With this information and a means of accurately determining implement location, inputs such as seed, fertilizer and chemicals can be applied according to yield potential in various parts of the field. Measurement techniques for yield mapping are being developed. Savings in inputs plus overall yield increases have been shown to occur as a result of this "prescription" farming technique. Remote sensing is bound to play an important role in this technology. A very comprehensive reference is *The Manual of Remote Sensing*.[41] Remote sensing is an area in which agricultural engineers should be more heavily involved.

The computer-based area of "expert systems" is another topic which has been of considerable recent interest to agricultural engineers. The above

41. R. G. Reeves, ed., *The Manual of Remote Sensing*, Vol. I. *Theory, Instruments and Techniques*, Vol. II. *Interpretation and Applications* (Falls Church, Va.: American Society of Photogrammetry, 1975).

table shows no listings prior to 1980. The ASAE's *Agricultural Engineering* has produced a number of review or state-of-the-art articles on new technologies in the past three or four years. For example, one such paper on expert systems appears in the Jan/Feb 1986 issue. In this paper, the authors state: "Expert systems are computer programs designed to emulate the logic and reasoning processes human experts would use to solve a problem in their field of expertise."[42]

Other topics covered recently in *Agricultural Engineering* deal with biosensors, robotics, computer-aided design, information technologies, artificial intelligence-based sensing, computer-aided control, smart sensing and machine vision. In many of these emerging technology papers, the authors emphasize the importance of suitable sensors. The need to develop sensors designed especially for agriculture has been demonstrated by Vanden Berg in the ASAE publication *Agricultural Sensors*.[43]

E. Summary

The use of instrumentation, control, electronics and automation in the discipline of agricultural engineering has occurred in parallel with engineering developments in agriculture. Early instrumentation was simple, analog in nature. Modern systems are much more sophisticated, usually digital, and microcomputer-based for measurement and control. The application of instrumentation (control, electronics, automation) to agricultural crop and livestock production systems, has been very important in providing for improvements in machine and process efficiency, improved product quality and in the reduction of operator drudgery.

Agricultural engineers have contributed significantly to the research and to the literature in the application of instrumentation in agriculture. The use of instrumentation, control, electronics and automation has occurred in all of the agricultural engineering technical divisions. Publication sources such as the ASAE's *Agricultural Engineering* and *Transactions* are most important in keeping abreast with technical developments and applications. Books and symposia are, of course, important sources of subject matter theory and design as well as for instrumentation applications for the agricultural industry. New technology in the area of instrumentation is emerging rapidly. More will be heard in the near future about robotics, artificial intelligence,

---

42. Larry T. Huggins, John R. Barret, and Don D. Jones, "Expert Systems, Concepts and Opportunities," *Agricultural Engineering* (Jan./Feb. 1986).

43. Glen E. Vanden Berg, *Agricultural Sensors* (St. Joseph, Mich.: American Society of Agricultural Engineers, 1988 [*ASAE* Publication 09-88]).

biosensors, smart sensing and machine vision. It is an exciting time to be alive.

Bibliography

Ahmed, M., "A Plant Analogue Sensor for Irrigation Scheduling," Thesis, University of Newcastle-upon-Tyne, U.K., 1987.
Allocca, John A., and Allen Stuart, *Transducers: Theory and Applications* (Reston, Va.: Reston Publishing Company, 1984).
Amlaner, C. J., and D. W. Macdonald, eds., *A Handbook on Biotelemetry and Radio Tracking* (Oxford: Pergamon Press, 1980).
Arthur, Kenneth, *Transducer Measurements* (Beaverton, Oreg.: Tektronix, Inc., 1970).
Barney, George C., *Intelligent Instrumentation: Microprocessor Applications in Measurement and Control* (Englewood Cliffs, N.J.: Prentice-Hall Publishing Company, 1985).
Barrett, E. C., and L. F. Curtis, *Introduction to Environmental Remote Sensing*, 2d ed. (London and New York: Chapman and Hall, 1982).
Barrett, J. R., and D. D. Jones, *Knowledge Engineering in Agriculture* (St. Joseph, Mich.: American Society of Agricultural Engineers, 1989).
Bauer, M. E., *Remote Sensing of Agricultural Crops and Soils* (W. Lafayette, Ind.: Purdue University Lab for Application of Remote Sensing, 1982).
Bell, David A., *Electronic Devices and Circuits*, 3d ed. (Englewood Cliffs, N.J.: Prentice-Hall Publishing Company, 1986).
Brindley, Keith, *Sensors and Transducers* (London: Heinemann Professional Publishing, 1988).
Brown, Paul B., Gunter N. Franz, and Howard Moraff, *Electronics for the Modern Scientist* (Amsterdam and New York: Elsevier, 1982).
Buban, P., Sr., et al., *Understanding Electricity and Electronics Technology*, 5th ed. (New York: McGraw Hill Book Company, Inc., 1986).
Chironis, Nicholas P., ed., *Machine Devices and Instrumentation* (New York: McGraw Hill Book Company, Inc., 1966).
Clark, John H., *Agricultural Materials Handling Manual* (Ottawa: Information Services, Agriculture Canada, 1979 [Earlier ed., Canada Committee on Agricultural Engineering of Canadian Agriculture Services Coordinating Committee, Canada Dept. of Agriculture, 1962]).
Colclaser, Roy A., D. A. Neamen, and C. F. Hawkins, *Electronic Circuit Analysis: Basic Principles* (New York: John Wiley and Sons, 1984).
Colijn, Hendrik, *Weighing and Proportioning of Bulk Solids* (Bay Village, Ohio: Trans. Tech. Publications, 1975).
Considine, Douglas M., and Glenn D. Considine, *Standard Handbook of Industrial Automation* (New York: Chapman and Hall, 1986).
Cunningham, John E., and Delton T. Horn, *Handbook of Remote Control and Automation Techniques*, 2d ed. (Blue Ridge Summit, Pa.: TAB Books, Inc., 1984).
Curran, P. J., *Principles of Remote Sensing* (New York: John Wiley and Sons, Inc., 1986).
Dally, James W., William F. Riley, and Kenneth G. McConnell, *Instrumentation for Engineering Measurements* (New York: John Wiley and Sons, 1984).
Davies, Gomer L., *Magnetic Tape Instrumentation* (New York: McGraw-Hill Book Company, 1961).

Diefenderfer, A. James, *Principles of Electronics Instrumentation*, 2d ed. (Philadelphia: W.B. Saunders Company, 1979).
Doebelin, Ernest O., *Control System Principles and Design* (New York: John Wiley and Sons, 1985).
Dove, Richard C., and Paul H. Adams, *Experimental Stress Analysis and Motion Measurements: Theory, Instruments and Circuits, Techniques* (Columbus, Ohio: Charles E. Merrill Books, 1964).
Goldsmid, H. J., *Applications of Thermoelectricity* (New York: John Wiley and Sons, 1960).
Hall, Carl W., *Errors in Experimentation* (Champaign, Ill.: Matrix Publishers, 1977).
Hassall, J. R., and K. Zaveri, *Acoustic Noise Measurements* (Naerum, Denmark: Bruel and Kjaer, DK-2850, 1979).
Hollier, R. H., ed., *Automated Guided Vehicle Systems* (Kempston, Bedford, U.K.: IFS Publications, 1987).
Klemm, W. R., ed., *Applied Electronics for Veterinary Medicine and Animal Physiology* (Springfield, Ill.: Charles C. Thomas, 1976).
Krutz, Ronald L., *Interfacing Techniques in Digital Design with Emphasis on Microprocessors* (New York: John Wiley and Sons, 1988).
Kumar, M., "Spectral Reflectance and Light Interception by Crop Canopies," Thesis, Nottingham University, 1981.
Marshall, B., and F. I. Woodward, eds., *Instrumentation for Environmental Physiology* (Cambridge and New York: Cambridge University Press, 1985 [*Society for Experimental Biology Seminar Series* no. 22]).
Møller, A., *Applications of Microprocessors within Agriculture*, Translation from Danish of *Anvendelse af Mikroprocessorer i Landbruget II* (Hundested, Denmark: Agric. Contact, 1985).
Moverley, J., *Microcomputers in Agriculture* (London: Collins, 1986).
Mulders, M. A., *Remote Sensing in Soil Science* (Amsterdam and New York: Elsevier Science Publishers, 1987).
Norton, Harry N., *Biomedical Sensors, Fundamentals and Applications* (Park Ridge, N.J.: Noyes Publications, 1982).
Parker, Sybil P., ed., *Dictionary of Electronics and Computer Technology* (New York: McGraw-Hill Book Company, 1984).
Pfeffer, Phillip E., and W. W. Gerasimowicz, eds., *Nuclear Magnetic Resonance in Agriculture* (Boca Raton, Fla.: CRC Press, 1989).
Pratt, William K., *Bibliography on Digital Image Processing and Related Topics* (Los Angeles, Calif.: University of Southern California, Electronic Sciences Laboratory, 1973).
Prensky, S. D., *Electronic Instrumentation* (Englewood Cliffs, N.J.: Prentice Hall, 1982).
Price, William J., *Nuclear Radiation Detection*, 2d ed. (New York: McGraw Hill, 1964).
Rasmussen, William O., et al., *Computer Applications in Agriculture* (Boulder, Colo.: Westview Press, 1985).
Sanglerat, G., *The Penetrometer and Soil Exploration* (Amsterdam: Elsevier, 1979).
Scheingold, Daniel H., ed., *Transducer Interfacing Handbook* (Norwood, Mass.: Analog Devices, Inc., 1981).
Schenck, Hilbert, *Theories of Engineering Experimentation*, 3d ed. (Washington, D.C.: Hemisphere Publishing, 1979).
Schweppe, J. L., et al., *Methods for the Dynamic Calibration of Pressure Transducers* (Washington, D.C.: U.S. Government Printing Office, 1963 [*National Bureau of Standards Monograph* no. 67]).

Serridge, Mark, and Torben R. Licht, *Piezoelectric Accelerometer and Vibration Preamplifier Handbook* (Naerum, Denmark: Brel and Kjaer, DK-2850, 1986).
Sharland, Ian, *Woods Practical Guide to Noise Control* (Woods of Colchester, U.K.: Woods of Colchester, Ltd., 1972).
Sistler, Fred, *The Farm Computer* (Reston, Va.: Reston Publishing, 1984).
Smith, Robert A., ed., *Very High Resolution Spectroscopy* (New York: Academic Press, 1976).
Stevens, G. N., *Equipment Testing and Evaluation* (Silsoe, Bedford, Eng.: National Institute of Agricultural Engineering, 1982).
Sydenham, Peter H., ed., *Handbook of Measurement Science*, Vol. I. *Theoretical Fundamentals*, Vol. II. *Practical Fundamentals* (New York: John Wiley and Sons, 1982, 1983).
Vose, P. B., *Introduction to Nuclear Techniques in Agronomy and Plant Biology* (Oxford and New York: Pergamon Press, 1980).
Wells., David, ed., *Guide to CPS Positioning* (Fredericton, New Brunswick, Canada: Canadian GPS Associates, 1987).
Wilson, J. A., *Control Electronics with an Introduciton to Robotics* (Chicago, Ill.: Science Research Associates Inc., College Division, 1985).
Window, A. L., and G. S. Holister, eds., *Strain Gauge Technology* (London and Englewood Cliffs, N.J.: Applied Science Publishers, 1982).
———. *Microprocessor Basics* (Eleven articles reprinted from Machine Design) (Cleveland, Ohio: Penton Publishing Co., 1980).

Conferences, Symposia and Workshops

Balasubrahmanian, A., ed., *Computer Applications in Food Production and Agricultural Engineering—II*, Proceedings of the 2d IFIP TC5 Working Conference on Food Production and Agricultural Engineering, New Delhi, Mar. 1984 (New Delhi, India: Defence Res. and Development Org., 1985).
Beecher, G. R., ed., *Research Instrumentation for the 21st Century*, Symposium, May 1986 (Beltsville, Md.: Beltsville Agricultural Research Center, Dordrecht and Boston: Nijhoff, 1987).
Dodd, Vincent A., and Patrick M. Grace, eds., *Proceedings of the Eleventh International Congress on Agricultural Engineering*, Dublin, Ireland, Sept. 1989 (Rotterdam, Netherlands: A. A. Balkema, 1989).
Fraysse, Georges, ed., *Remote Sensing Application in Agriculture and Hydrology*, Proceedings of a Seminar, Joint Research Centre of the Commission of the European Communities, Ispra (Varese), Italy, Nov.–Dec. 1977 (Rotterdam, Netherlands: A. A. Balkema and Salem, N.H.: Distributed in the United States and Canada by MBS, 1980).
Warnecke, H. J., ed., *Agricultural Robots*, Agrotique 86 Association pour le Developpement de l'Electronque dans le Sud-Oest Bodeaux, France, 1986 (French with English summary).
———. *Automation in Dairying*, Proceedings of the Symposium, Wageningen, Netherlands, Apr. 1983.
———. *Automated Guided Vehicle Systems: Proceedings of the 2d International Conference on AGVS 7 16th PIA Conference*, Stuttgart, June, 1983 (Amsterdam: Elsevier Science Pub., 1984).
———. *Knowledge Based Systems in Agriculture: Prospects for Application*, Proceedings of the 2d International DLG-Congress, Frankfurt, June, 1988 (Frankfurt am Main: Deutsche Landwirtschafts-Gesellschaft (DLG), 1988).

———. *Livestock Environment III*, Proceedings of the 3d International Livestock Environment Symposium, Toronto, Apr. 1988 (St. Joseph, Mich.: American Society of Agricultural Engineers, 1988).

———. *Microelectronics in Agriculture*, International DLG-Congress for Computer Technology, Hanover, May 1986, (Frankfurt am Main: Deutsche Landwirtschafts-Gesellschaft, 1986).

———. *National Conference and Exhibition on Robotics*, Melbourne, Aug., 1984 (Barton ACT, 1984).

———. *Solid-State Transducers 83*, 2d International Conference on Solid-State Sensors and Actuators, Delft University of Technology, May–June 1983 (Delft University of Technology, 1983).

———. *Third International Conference on Automotive Electronics, Institution of Electrical Engineers,* London, Oct. 1981 (London: Automobile Div. of Instit. of Mech. Engineers, Mechanical Engineering Publications Ltd., 1981).

———. *Vision '85 Conference Proceedings*, Detroit, Mar. 1985 (Dearborn, Mich.: Machine Vision Association of SME, 1985).

———. *Workshop on Remote Sensing of Snow and Soil Moisture by Nuclear Techniques*, Voss, Norway, Apr. 1979 (Oslo, Norway: Norwegian National Committee on Hydrology, 1979).

7. The Current Core Literature of Food Engineering

R. Paul Singh
Department of Agricultural Engineering
University of California, Davis

The United States food industry emerged from World War II as a highly competitive group of companies involved in the production, processing, and distribution of foods. The accelerating pace of developments during the last four decades has been greatly influenced by the changing consumer preference for foods and the expanding nutritional knowledge. Engineering applications in food processing have continued to increase due to the increased reliance on large-scale equipment to handle, process, and distribute foods. Within this context, food engineering has emerged as a requisite profession to implement the technological advances made in food processing.

The successes of the food processing industry in the industrialized nations are in large part due to the major investments made in the past on research and development. Both food scientists and technologists have made major contributions in shaping the growth of this industry. Nicholas Appert is credited with one of the earliest applications of food technology, the discovery of the canning process. His monograph on canning, written in French in 1810, was translated into English and published in 1811.[1]

In the United States, the practitioners of food technology found a professional home in 1939 when the Institute of Food Technologists was formed. The food technologist was envisioned to use chemistry and engineering to solve problems in food processing. In the words of Dr. Samuel Cate Prescott, Dean of Science at the Massachusetts Institute of Technology, "from . . . our extended work in the bacteriological aspects of food preservation there has grown . . . a continually broadening and healthy interest leading to numerous courses of instruction and to research in what we call Food

---

1. Nicolas Appert, *L'Art de Conserver, Pendant Plusieurs Années, Toutes les Animales et Végétales* (Paris: Paris, 1810 [English translation by N. Appert, *The Art of Preserving All Kinds of Animal and Vegetable Substances* (London: Black, Perry and Kingsbury, 1811]).

Technology, but which might perhaps with equal correctness be named Food Engineering."[2]

According to Parker et al.,[3] three major developments since the 1900s had a marked influence on the transformation of the food industry in the United States. First, the increasing role of food processing as a vital link between the farmer and the consumer; second, the increased recognition that any nutritional or other quality degradation that may occur during food processing must be minimized; and third, the realization that in many instances the same operation can be commonly used in processing of different foods, thus allowing for diversification in food manufacturing. Food engineering took its roots as a consequence of this third development. The need for efficient use of raw materials and manual labor, reduction in the wastage of energy, developments of new machines, improvement in plan organization, and design of new processes are some of the areas where food engineers began to play an important role. Realizing the need for teaching food engineering in the food science curriculum, Parker et al. wrote one of the first textbooks on this subject, entitled *Elements of Food Engineering*.[4] The contents of this book emphasized the role of engineering principles in processing of different food commodities.

Food engineering is a quantitative discipline based on the "engineering sciences." Its key foundations are in physics, mathematics, transport phenomena, thermodynamics, reaction kinetics, machine design, and food sciences. During the last three decades, process engineering topics have dominated the contents of several books written in the general area of food engineering. In many regards, these books contain material that has a distinct similarity to topics commonly found in the field of chemical engineering. In fact, certain books written for chemical engineers have included examples pertaining to foods, and they have found use in teaching courses taught in food engineering curriculum, for example, *Transport Process and Unit Operations* by Geankoplis.[5]

In the early 1960s, a major contribution to the food engineering field was made by Charm who wrote the treatise *Fundamentals of Food Engineering*.[6] This book, now in its third edition, has been translated into several languages. It was one of the first attempts to relate basic engineering principles

2. M. E. Parker, E. H. Harvey, and E. S. Stateler, *Elements of Food Engineering* (New York: Reinhold Publishing Corp., 1952–54).
3. Ibid.
4. Ibid.
5. C. J. Geankoplis, *Transport Processes and Unit Operations*, 2d ed. (Boston: Allyn and Bacon Inc., 1983).
6. S. E. Charm, *Fundamentals of Food Engineering* (Westport, Conn.: AVI Pub. Co., Inc., 1963; 3d ed., 1978).

to the fundamentals of food processing. The author used the deductive approach to several physical problems found in the food industry. The advantages of the analytical methods, leading often to a reduction in the experimental work and providing a guide to direct research, are illustrated in several examples.

Following Charm's book, several authors have elucidated the use of engineering principles in the design of processes and systems for the food industry. The first edition of *Food Process Engineering* by Heldman[7] and its second edition[8] presented the analytical and numerical techniques useful in designing mechanical and thermal unit operations important in food processing. A unit operation is defined as the method by which an intentional or controllable change of form or place of a food material or ingredient is effected. The authors presented methods commonly used for engineering analysis in allied fields, such as chemical and mechanical engineering, with applications specific to food processing.

Leninger and Beverloo wrote *Food Process Engineering*[9] using their lecture material in teaching engineering aspects of food technology at the Agricultural University of Wageningen, Netherlands. This textbook contains a descriptive presentation of principles important in designing mechanical operations including handling and mixing of materials, size reduction and mechanical separations. In addition, a variety of physical methods of food preservation are presented in a quantitative manner. A more rigorous treatment of food engineering topics was provided by Loncin and Merson in *Food Engineering: Principles and Selected Applications.*[10]

Food engineering academic programs have grown in several United States universities. By the late 1980s, there were three United States universities offering academic programs designated as "food engineering": Purdue University, University of Massachusetts and Michigan State University. In Canada, the University of Guelph provides a similar program. According to Steffe, many agricultural engineering departments around the country offer food engineering "options" such as the University of California at Davis, the University of Florida, the University of Kentucky, the University of Missouri, Ohio State University, the University of Tennessee and Texas A & M University.[11] A minor in food engineering is offered by the Univer-

---

7. D. R. Heldman, *Food Process Engineering* (Westport, Conn.: AVI Publ. Co., Inc., 1974).

8. D. R. Heldman and R. P. Singh, *Food Process Engineering* (Westport, Conn.: AVI Pub. Co., Inc., 1981).

9. H. A. Leninger and W. A. Beverloo, *Food Process Engineering* (Dordrecht, Netherlands: D. Riedel Publ. Co., 1975).

10. M. Loncin and R. L. Merson, *Food Engineering: Principles and Selected Applications* (Orlando, Fla.: Academic Press, 1979).

11. J. F. Steffe, "Chairman's Column," *Food Engineering Division Newsletter* 14 (2) (1990).

sity of Illinois for students enrolled in any field of engineering. In addition, several agricultural engineering departments offer courses in food engineering: Colorado State University, Cornell University, University of Georgia, Iowa State University, Kansas State University, University of Maine, University of Maryland, University of Minnesota, Mississippi State University, University of Nebraska, Oregon State University, Pennsylvania State University, Rutgers University, Washington State University, and University of Wisconsin. A number of books mentioned in the preceding paragraphs are regularly used in teaching food engineering courses at these universities.

Food engineering is also an important component of the food science curriculum as recommended by the Institute of Food Technologists. A number of educational institutions offer food engineering courses to students enrolled in food science and related majors. Several books have been written to introduce food engineering principles to a food science student. The contents of these books build on information given in the introductory courses in physics and mathematics. One of the earlier books, with a delineation of engineering principles for food science students, was *Unit Operations in Food Engineering* by Earle.[12] Harper wrote *Elements of Food Engineering*, now in its second edition.[13] Other books written with a similar focus include *Fundamentals of Food Process Engineering* by Toledo,[14] *Food Engineering Fundamentals* by Batty and Folkman,[15] and *Introduction to Food Engineering* by Singh and Heldman.[16] The books contain quantitative descriptions of a number of important unit operations commonly found in the food industry. The mathematical treatment in these books is usually at a level appropriate for students with non-engineering backgrounds.

A more descriptive treatment of unit operations commonly found in the food industry was provided in *Food Engineering Operations* by Brennan et al.[17] This book is based on lectures given on a course in food engineering at the National College of Food Technology (University of Reading, England). While most of the operations used in food processing have common applications in other fields, some are unique, such as individual quick freezing. These and other operations are described with illustrations of com-

12. R. L. Earle, *Unit Operations in Food Processing*, 2d ed. (Oxford, Eng.: Pergamon Press, 1983).
13. E. Watson and J. C. Harper, *Elements of Food Engineering*, 2d ed. (New York: Van Nostrand Reinhold, 1988).
14. R. T. Toledo, *Fundamentals of Food Process Engineering* (Westport, Conn.: AVI, 1980; 2d ed., 1991).
15. J. C. Batty and S. L. Folkman, *Food Engineering Fundamentals* (New York: Wiley and Sons, 1983).
16. R. P. Singh and D. R. Heldman, *Introduction to Food Engineering* (Orlando, Fla.: Academic Press, 1984).
17. J. G. Brennan, J. R. Butters, N. D. Cowell, and A. E. V. Lilly, *Food Engineering Operations* (London: Applied Science Pub., 1969).

monly used equipment. Another descriptive treatment of food processing operations in terms of engineering applications has been presented in *Food Engineering Systems,* in two volumes, by Farrall. Many unique engineering aspects that are important in processing of different food commodities are described. Examples of food products discussed include manufacture of cheese, potato products, egg processing, fruit and vegetable processing, candy and confectionery manufacturing. These volumes contain valuable information on plant operation, plant design, and specific requirements in processing selected commodities.[18]

Bowen is credited with writing one of the earliest books on dairy engineering.[19] Farrall, with several contributors, wrote *Engineering for Dairy and Food Products*[20] based partially on his earlier volume *Dairy Engineering.*[21] This was a pioneering effort to apply engineering know-how to food processing. This book was one of the most widely used food engineering textbooks in the 1960s in teaching food engineering in several institutions around the world. With a similar emphasis on the dairy industry, Kessler wrote *Food Engineering and Dairy Technology.*[22] This book was written for food technologists and mostly uses a descriptive approach. Some relevant mathematical expressions that are useful in the analysis of engineering operations are included. While most of the illustrations are from the dairy industry, the principles discussed in the book are common to processing of other foods. Another comprehensive treatment of technological issues relevant to the dairy industry is *Dairy Technology and Engineering* by Harper and Hall.[23] This book capitalized on the true multi-disciplinary nature of the dairy food industry. The engineering and technological issues are considered in relation to principles of microbiology, physical chemistry, and biochemistry.

The progress made by the food industry in producing safe, wholesome food products is in part due to the increased emphasis given to proper sanitation practices. The task of improving design of food processing equipment that allows for easier cleaning has been within the realm of food engineers. In *Engineering for Food Safety and Sanitation*, Imholte[24] presented an over-

18. A. W. Farrall, ed., *Food Engineering Systems* (Westport, Conn.: AVI Pub., 1976–1979).
19. J. T. Bowen, *Dairy Engineering* (New York: Wiley and Sons, 1925).
20. A. W. Farrall, *Engineering for Dairy and Food Products* (New York: J. Wiley & Sons, 1963).
21. A. W. Farrall, *Dairy Engineering* (New York: Wiley and Sons, 1942).
22. H. G. Kessler, *Food Engineering and Dairy Technology* (Freising, Germany: V.A. Kessler Publ., 1981).
23. W. J. Harper and C. W. Hall, *Dairy Technology and Engineering* (Westport, Conn.: AVI Pub. Co., Inc., 1976).
24. T. J. Imholte, *Engineering for Food Safety and Sanitation* (Crystal, Minn.: Technical Institute of Food Safety, 1984).

view of modern food sanitation for practicing engineers. In addition, several standards written for specific segments of the food industry such as the 3–A Dairy Standards, Baking Industry Sanitation Standards Committee Bakery Standards, and the select group of United States Department of Agriculture standards for the meat industry are excellent sources for information relevant to sanitation.

In the production of antibiotics, vitamins, steroid hormones, and foods, many of the manufacturing operations have a common basis. This fact is further explored in two volumes entitled *Biochemical and Biological Engineering Science* by Blakebrough.[25] Several issues associated with the production, isolation and processing of biological materials are delineated. These volumes were one of the first comprehensive treatments linking food engineering with biological engineering sciences. Food engineers, who have been trained in the fundamentals of biology and engineering, now find an important role in the emerging fields of biotechnology.

Processing of agricultural materials starts after the crops are harvested. Food engineers get involved with several farm operations that are concerned with increasing the value of the commodity and reducing food losses. The engineering aspects of on-farm processing were first dealt with in a quantitative manner by Henderson and Perry in 1955 in their classic book *Agricultural Process Engineering*, which is now in its third edition.[26] Both mechanical and thermal unit operations are described with sufficient information to design processing systems. Drying is an important unit operation that has wide applications in cereal grain processing. This was further explored by Brooker et al. in *Drying Cereal Grains*.[27] They introduced a new approach based on computer-aided modeling for the design and analysis of drying systems. Simulation techniques suggested in this pioneering book are now widely used in research and development work related to grain drying.

One of the important challenges to a food engineer in the design and analysis of food processing systems is obtaining suitable information on physical and thermal properties of foods. The availability of reliable data on food properties is often scanty. Recently, a number of coordinated efforts, both at national and international levels, have addressed these issues. In Europe, a COST project (European Cooperation in the Field of Scientific

---

25. N. Blakebrough, *Biochemical and Biological Engineering Science* (London: Academic Press, 1967–68).

26. S. M. Henderson and R. L. Perry, *Agricultural Process Engineering*, 3d ed. (New York: AVI Pub., 1976 [*Ferguson Foundation Agricultural Engineering Series*]).

27. D. B. Brooker, F. W. Bakker-Arkema, and C. W. Hall, *Drying Cereal Grains* (Westport, Conn.: AVI Publ., Inc., 1974).

and Technical Research) resulted in publication of two books that provide data and details on methods for measuring properties: *Physical Properties of Foods*, *I*[28] and *II*[29] edited by Jowitt et al. Under the auspices of the American Society of Agricultural Engineers, a compilation of comprehensive data on food properties obtained by researchers from various agricultural experiment stations has appeared in *Physical and Chemical Properties of Foods* by Okos.[30] Additional books with a comprehensive treatment of food properties include *Engineering Properties of Foods* by Rao and Rizvi,[31] *Thermal Properties of Foods and Agricultural Materials* by Mohsenin,[32] and *Electromagnetic Radiation Properties of Foods and Agricultural Products* by Mohsenin.[33] Research activities aimed at obtaining physical properties of foods by scientists in North America and Europe are reviewed in *Food Properties and Computer-Aided Engineering of Food Processing Systems*, by Singh and Medina.[34] This book is a result of an Advanced Research Workshop held on this subject under the sponsorship of NATO.

A set of five volumes have appeared in England since 1981 titled *Developments in Food Preservation*, edited by Thorne.[35] These volumes contain exhaustive reviews of a number of unit operations important to the food industry. Several chapters are an excellent resource for both quantitative and descriptive material relevant to food preservation processes. Other recent books address areas of special interest in food manufacturing, such as *Energy in Food Processing* by Singh[36] and *Heat Transfer and Food Products* by Hallstrom et al.[37]

As the food engineering field has expanded its domain to include preprocessing, sorting, grading, processing, packaging and distribution of foods, the terminology used in various steps has also increased. The definitions and descriptions of important terms common to food engineering were

28. R. Jowitt, F. Escher, B. Hallstrom, A.F. Th. Meffert, W.E.L. Spiess, and G. Voss, *Physical Properties of Foods*, I (London: Elsevier Applied Science, 1983).
29. R. Jowitt, F. Escher, M. Kent, B. McKenna and M. Roques, *Physical Properties of Foods*, II (London: Elsevier Applied Science, 1987).
30. M. A. Okos, *Physical and Chemical Properties of Foods* (St. Joseph, Mich.: American Society of Agricultural Engineers, 1986).
31. M. A. Rao and S. S. H. Rizvi, *Engineering Properties of Foods* (New York: Marcel Dekker, Inc., 1986).
32. N. N. Mohsenin, *Thermal Properties of Foods and Agricultural Materials* (Reading, Mass.: Gordon & Breach Science Pub., 1980).
33. N. N. Mohsenin, *Electromagnetic Radiation Properties of Foods and Agricultural Products* (New York: Gordon Breach Science Pub., 1984).
34. R. P. Singh and A. Medina, *Food Properties and Computer-Aided Engineering of Food Processing Systems* (Dordrecht, Netherlands: Kluwer Academic Publishing, 1989).
35. S. Thorne, ed., *Developments in Food Preservation* (London: Elsevier Science Pub., 1981–89).
36. R. P. Singh, *Energy in Food Processing* (Amsterdam: Elsevier Science Publishers, 1986).
37. B. Hallstrom, C. Skjoldebrand, and C. Tragardh, *Heat Transfer and Food Products* Elsevier Applied Science, 1988).

incorporated in *Encyclopedia of Food Engineering* by Hall et al. in 1970, now in its second edition.[38]

Technical articles pertaining to food engineering have appeared mostly in *Journal of Food Science* published by the Institute of Food Technologists, and *Transactions of the American Society of Agricultural Engineers*. Two journals that specifically address food engineering topics are *Journal of Food Process Engineering* and *Journal of Food Engineering*. Other English-language journals containing articles on food engineering topics include, *Bioprocess Engineering* (Springer-Verlag), *Biotechnology Progress* (American Institute of Chemical Engineers, American Chemical Society), *Canadian Journal of Food Science and Technology* (Canadian Institute of Food Science and Technology), *Food Biotechnology* (Marcel Dekker), *International Journal of Food Science and Technology* (Institute of Food Science and Technology, U.K.), *International Journal of Refrigeration* (Butterworth, U.K.), *Journal of Agricultural Engineering Research* (The British Society for Research in Agricultural Engineering and Academic Press), *Journal of Food Processing and Preservation* (Food and Nutrition Press), *Journal of Microwave Power* (International Microwave Power Institute), and *Lebensmittel-Wissenschaft und Technologie* (Swiss Society of Food Science and Technology).

An important source for technical information relevant to food engineering is the proceedings of the International Conferences on Engineering and Food. These proceedings contain a wealth of information on current research being conducted in laboratories around the world. Notable publications include *Food Processing Engineering*, in two volumes, by Linko et al.,[39] *Engineering and Food*, Volumes 1 and 2 by McKenna,[40] and *Food Engineering and Process Applications* also in two volumes, by LeMaguer and Jelen.[41]

With a broader application of emerging technologies in the food industry, new publications have appeared that describe technical innovations and the engineering basis for using such technologies. Examples of such books are *Ultrafiltration Handbook* by Cheryan,[42] *Microwave Processing and Engi-*

---

38. C. W. Hall, A. W. Farrall, and A. L. Rippen, *Encyclopedia of Food Engineering*, 2d ed. (Westport, Conn.: AVI Pub., Inc., 1986).
39. P. Linko, Y. Makki, J. Olkku, and J. Larinkari, *Food Process Engineering* (London: Elsevier Applied Science, 1980).
40. B. McKenna, *Engineering and Food Industry* (London: Elsevier Applied Science, 1984).
41. M. LeMaguer and P. Jelen, *Food Engineering and Process Applications* (London: Elsevier Applied Science, 1986).
42. M. Cheryan, *Ultrafiltration Handbook* (Lancaster, Pa.: Technomic Pub. Co., 1986).

neering by Decarau and Peterson,[43] *Food Extrusion* by Harper,[44] *Concentration and Drying of Foods* by McCarthy,[45] *Preconcentration and Drying of Food Materials* by Bruin,[46] and *Process Engineering in Food Industry*, by Field and Howell.[47] Future titles in the food engineering area are expected to focus on new technologies and modifications of existing techniques that result in improvements in the processing efficiency.

It is evident from the preceding paragraphs that most of the authors of food engineering textbooks are from North America and Europe. These books are commonly used across the Atlantic in teaching as well as for reference purposes by practicing engineers. These books have also found use in several developing countries in Asia and Africa. In most cases, books published in the Western nations are too expensive for widespread use in the developing countries. Libraries in these countries are a major source of books for food engineering students and practitioners. In some cases, local publishers have made arrangements to print low cost editions; however there is a need to expand such efforts.

While several publishing companies have been involved in printing books on food engineering topics, a notable supporter of food engineering books has been the AVI Company. The founder of the AVI Company, Dr. Donald Tressler, was himself a prolific writer of books on food processing. The AVI Company was recently purchased by Van Nostrand Reinhold Company, Inc.

As the food engineering profession continues to evolve, the demand for literature on food engineering topics will continue to grow. It is expected that in the near future the number of food engineering curricula offered by major universities in North America and Europe will increase. This is expected in response to the rapid changes taking place in the food industry. Similar increases are also expected in several developing countries where food processing is receiving increased attention in attempts to reduce food losses. These trends are in consonance with the recent actions of several major publishing companies who have initiated special book and monograph series to focus on issues important to food engineering.

43. R. V. Decarau and R. A. Peterson, *Microwave Processing and Engineering* (Chichester, Eng.: Ellis Horwood Ltd., 1986).
44. J. M. Harper, *Extrusion of Foods*, Boca Raton, Fla.: CRC Press, 1981).
45. D. MacCarthy, *Concentration and Drying of Foods* (London: Elsevier Applied Science, 1986).
46. S. Bruin, *Preconcentration and Drying of Food Materials* (London: Elsevier Applied Science, 1988).
47. R. W. Field and J. A. Howell, *Process Engineering in Food Industry* (London: Elsevier Applied Science, 1989).

8. Development Forest Engineering and Its Literature

J. L. FRIDLEY
Department of Forest Products and Engineering, University of Washington

J. A. MILES
Department of Agricultural Engineering, University of California, Davis

F. E. GREULICH
Department of Forest Products and Engineering, University of Washington

A. First Academic Programs

The Royal Indian Engineering College at Cooper's Hill (Great Britain) was among the first colleges to blend engineering with forestry when in 1884 it established a school of forestry.[1] Headed by William Schlich, the school taught its students technical material in preparation for employment in the Colony of India. Forestry program graduates would aid in the harvesting of India's forests. *Schlich's Manual of Forestry* became the definitive English language forestry textbook of that era.[2] The volume on forest utilization written by Karl Gayer, Professor of Forestry at the University of Munich, was the recognized standard on forest utilization in Germany. Within this volume the section on wood transportation describes road, rail, cable and water based transportation systems. There are good structural illustrations but no numerical analysis is offered. The academic program at Cooper's Hill was moved to Oxford in 1905, marking a transition in becoming what we now view as modern forestry instead of an area of applied engineering.

The first American Forest Engineering degree was awarded at Cornell University in 1902, where Bernard E. Fernow, a German educated forest

1. B. E. Fernow, *A Brief History of Forestry* (Toronto: University Press, 1913), p. 377–78.
2. William Schlich, *Schlich's Manual of Forestry*, 2d ed. (London: Bradbury, Agnew & Co., Ltd., 1908 [Vol. V, "Forest Utilization," English translation by W. R. Fisher of *Die Forstbenutzung* by Karl Gayer]).

engineer, led a short lived program (1896–1903). Only seventeen students graduated with this degree, which Fernow described as follows: "This degree, it is believed, expresses more adequately . . . the fact that not a science but an art of technical character has been studied as a profession; it is a title indicating practical rather than literary attainments and describes the work for which the student has prepared, namely, the application of technical scientific knowledge to a business and in a productive industry."[3] Fernow clearly identified the engineering aspects of forestry; a profession whose mission reached beyond the logging operation to address such issues as regeneration and protection of the soil and water resource.[4]

In the southeastern United States, the Biltmore Forest School operated from 1898 to 1913 under the direction of another German forestry school graduate, Carl A. Schenck. The Biltmore School did not have a forest engineering curriculum but did emphasize the practical aspects of forestry, including substantial attention given to logging.[5] Schenck's text, *Logging and Lumbering or Forest Utilization; A Textbook for Forest Schools*, was apparently published in 1912, although the book does not in fact give a publication date.[6] Part I dealt with logging operations while Part II discussed the manufacture of wood products. Part I consists of three chapters and is an excellent descriptive treatment of logging engineering practice during that era. The first chapter covers camps, duration of employment, enumeration, and animals, The discussion is non-technical and is a generally thorough introduction to the topic. The second chapter describes the tools used (including discussion on setting and filing saw teeth), tree felling and bucking. Chapter 3, transportation, is quite detailed and covers land transportation (without vehicles), water transportation, vehicles, roads, loading, cable logging, and choosing between transportation systems. The section on roads includes equations for grade, curve layout, cuts and fills, etc., along with prediction equations for the rolling resistance of wheels.

An early textbook by Ralph Clement Bryant, a forest engineering graduate of the program at Cornell, is similar in topical content to Schenck's book.[7] Bryant does not present any engineering analysis of harvesting activity but, somewhat unique for that period, he does give an extensive bibli-

3. R. S. Hosmer, *Forestry at Cornell* (Ithaca, N.Y., Dec. 1950).
4. Bernard E. Fernow, "The Forester, an Engineer," *Journal of the Western Society of Engineers* 6 (5) (1901): 402–420.
5. J. O'Hearne, "How Shall We Teach Logging Engineering?" 5th Annual Session, *Pacific Logging Congress*, 1913.
6. C. A. Schenck, *Logging and Lumbering or Forest Utilization: A Textbook for Forest Schools* (Darmstadt, Germany: L. C. Wittich, 1912).
7. Ralph C. Bryant, *Logging: The Principles and General Methods of Operation in the United States* (New York: John Wiley & Sons, Inc., 1913).

ography. Bryant's text was the first widely distributed textbook on logging in North America. This text was followed with a thoroughly revised second edition in 1923.

B. *The Emergence of Logging Engineering*

The need for engineering skills in North American logging and forest management continued to grow during the first decade of this century, especially in the west. The U.S. Forest Service responded to this need for engineers when, in 1905, F. G. Plummer was transferred from the Geological Survey to the Forest Service. Plummer was the first engineer in what was to become, in 1907, the U.S. National Forest System. By 1910, activity on the National Forest had produced 320 miles of road, 2,225 miles of trail, 1,888 miles of telephone lines, 464 cabins and barns, and fifty-one corrals.[8] Industry was similarly responding. Turn-of-the-century loggers in the Pacific Northwest well understood the value of a capable engineer.

Western loggers were contending with steep rugged terrain, big trees, and logging technology that was based on cable systems and railroads. They needed persons who could survey railroad lines and property boundaries as well as oversee road layout and construction. Many civil and mechanical engineers had the necessary educational background, but two problems prevented these engineers from being employed by the logging firms. First, young mechanical and civil engineering graduates were not compatible with the loggers. The Pacific Northwest was extremely rural, logging enterprises operated out of camps, and the climate was, true to the northwest's reputation, conducive to wet socks and webbed feet for those who undertook an engineering career in the outdoors. Formally educated engineers were finding preferable employment in more urban settings at more attractive salaries. Further, the logging business did not know what to do with engineers. It was not appropriate to turn an engineer into an overpaid logger; yet with no existing career track for engineers in the business, there did not exist means for an engineer to become sufficiently familiar with the operations of the business.

The second problem impeding the employment of engineers in the logging industry was the poor performance of engineers who had been hired by the industry. When civil and mechanical engineers were employed by the logging industry the results were frequently disastrous. E. T. Clark de-

---

8. United States Forest Service, *History of Engineering in the Forest Service, 1905–1969* (Washington, D.C.: USFS Engineering Staff, 1972).

scribes a case where a logging outfit engaged a civil engineering firm to locate some boundary lines. The surveyors, not understanding the nature of logging, located the section corners but, to the loggers' dismay (and too late discovery), did not blaze the boundary lines for the fallers. Clark also told of a logging company that hired a "gang of civil engineers" to survey a few miles of rail line. The construction crew discovered the newly surveyed line would have required excavation at a cost not justifiable by the timber to be extracted. The company then called in their own timber cruiser (Note: probably a college educated forester) to "spot in a road that could be built without bankrupting the company."[9]

In August of 1908, George Cornwall, editor of a trade magazine called *The Timberman*, and Edward English, an influential logging firm owner from Mt. Vernon, Washington, visited in the Dillar Hotel in Seattle. At that meeting Mr. Cornwall proposed his ideas for a meeting of what was to become the Pacific Logging Congress. The congress would be a "friendly powwow of other loggers for an exchange of ideas (pertaining to logging)." It is evident that Cornwall understood the nature of the men operating logging operations and the importance of their perceptions, thus he allowed that "the idea of this congress was therefore a mutual and simultaneous inspiration."[10]

Cornwall subsequently met with Dean Frank G. Miller and Professor Hugo Winkenwerder of the University of Washington to discuss the upcoming Pacific Logging Congress to be held on the campus of the University of Washington in Seattle. Miller and Winkenwerder had recognized the growing need for engineering talents in the logging camps and were interested in establishing a program in logging engineering. But, they too recognized that the success of their endeavor would depend on acceptance by the loggers, and that the easiest way to attain acceptance was to allow the idea to come from the loggers themselves. George Cornwall, for his part, needed little convincing and took it upon himself to become a champion for the effort to establish a "new" profession. So, the profession and academic field of "logging engineering" became an important component of the Congress' mission.

The first Pacific Logging Congress (PLC) was held July 19–21, 1909, in the Hoo-Hoo House at the Alaska Yukon Pacific Exposition in Seattle. At that first PLC, George Cornwall stated:

---

9. E. T. Clark, "Logging Engineering Should Be Recognized by Institutions of Learning," 4th Annual Session, *Pacific Logging Congress*, 1912.
10. G. M. Cornwall, "Development of the Logging Industry of the Pacific Coast States," 1st Annual Session, Pacific Logging Congress, *The Timberman* 10 (1909): 32.

Logging is an engineering science and as such it must be considered in the future to a greater extent than it has in the past. The country is doing bigger things in every department of human activity, and the logging business is no exception to the rule. It takes close application and a high grade of engineering skill to be able to lay out the proper location of roads, which will intersect and draw to one common point the greatest amount of timber in any one tract. The grasp of this one problem is the deciding factor in determining the ability of the engineer, which often can be realized only after the tract is well opened up. There is a growing field on the Pacific Coast for young men with a knowledge of engineering, both civil and mechanical, who will devote their time to a study of the Pacific Coast logging requirements, with a view of being able to present in an intelligent and practical manner a working plan for opening up and logging a tract of timber, This is practically an unoccupied field, and one of the underlaying motives which dominated the congress.[11]

Although Cornwall may well have been the one to coin the term logging engineering, Frank Lamb is one of the earliest to use the term in publication. In a paper presented to the first PLC, Lamb of Lamb Timber Company in Hoquium, Washington, discussed some of the subjects that compose logging engineering and suggested: "I hope that I have briefly outlined a few of the subjects comprised under the general term logging engineering, and while it would not make us more valuable men or more successful in our business, yet I think that if we practical men were to call ourselves logging engineers instead of simply loggers or boss loggers it might give us greater pride in our profession. I use the term profession advisedly, because I think the act of drawing logs out of the woods to the markets of the world is fully as elevating, fully as useful an occupation as is the drawing of useless teeth out of another man's head, and if one is a profession, so should the other be."[12]

The following year, 1910, a short course in logging engineering was taught at the University of Washington by Professor W. T. Andrews.[13] One year later Elias T. Clark was hired to take charge of the forest engineering program at the University of Washington.[14] A major strength of that program was in the extensive use of a capstone field exercise that is still the trademark of the program today.[15]

11. Ibid.
12. Frank Lamb, "Logging Engineering Requires Skill and Experience for Success," 1st Annual Session, Pacific Logging Congress, *The Timberman* 10 (1909): 32.
13. W. T. Andrews, "Introduction to the Practical Teaching of Logging Engineering and Lumber Manufacture at the University of Washington," *Forest Club Annual* (University of Washington) 4 (3) (1925):34–37.
14. Clark, "Logging Engineering . . . "
15. J. L. Fridley and P. Schiess, "A Successful Senior Forest Engineering Capstone Design Course" (1989 [*ASAE Paper* no. 89-5510]).

Oregon State University (then Oregon State Agricultural College) established a department of logging engineering in 1913. A well-respected logger from industry, J. P. Van Orsdel was hired as the program's first professor of logging engineering and the first logging engineer graduated from this program in 1915.[16] The new curriculum was outlined to the PLC as follows:

> In the student's freshman year he is taught, aside from (citizenship, executive training, military training) trigonometry, analysis, general forestry, elementary mensuration, plane surveying, general chemistry, and wood work. In the second year, engineering physics, blacksmithing, tool making and tempering, machine shop, practice, mechanical drawing, topographic surveying, railroad surveying and dendrology and mensuration. In the third year this is followed up by advanced mensuration, forest appraisals and reports, log scaling, logging railroads, logging machine design, elements of steam engineering and steam laboratory, mechanism, lumber rates and tariffs. The senior year is devoted entirely to specialized work and the following ground is covered: topographic logging plans, logging devices and equipment, logging methods, timber technology and testing, and lumber manufacture.[17]

By 1920 logging engineering programs had been established at the University of California (Berkeley), Oregon State University, the University of Washington, the University of Idaho, and the University of British Columbia. These early curricula, like the one described by Van Orsdel, stressed traditional forestry, logging planning and setting layout, surveying (land and railroad), topographic maps, and steam engines. However, by 1920, the very nature of logging engineering was beginning to change.

C. Re-emergence of Forest Engineering

Two forces were acting to change the fundamental nature of the problem addressed by logging engineering. First, technology was changing. Advances in the internal combustion engine and manufacturing processes during World War I were enabling the development of tractors and motor trucks suitable for logging. The result was a change in logging methods that reduced dependence on railroads. Second, concern for the forest resource

---

16. W. A. Davies, "Western Logging Engineering Schools—Oregon State College," 42nd Annual Session, Pacific Logging Congress, *Loggers Handbook* 11 (1951): 87–89.

17. J.P. Van Orsdel, "Presentation of Logging Engineering Curriculum at the Oregon State Agricultural College," 8th Annual Session, Pacific Logging Congress, 1916.

was building and with that concern came increased interest in regeneration and selective logging.

In 1919 the Oregon Engineers Registration Law was passed and logging engineering was included as one of the branches.[18] The passing of legislation that provided for professional licensing of logging engineers was an acknowledgement of the importance of engineering to the protection of forest resources.

The primary emphasis of the early logging engineering programs was however directed at the problem of economic development of a timber resource located on difficult terrain. The requisite system of railroads and cable yarders represented a substantial capital investment. Poor harvest design, resulting in high logging costs, were of constant concern. The preparation of boundary and topographic maps, the development of a rail and cable transport system and the actual railroad survey, design and location called for the skills of an engineer. That forestry knowledge was also required in equal measure was not as clear. Indeed it was not until the early 1920s when public concern about sustained forest yield became a political issue that forestry skills were accorded significant recognition in the conduct of harvesting operations. In 1922 George Cornwall, writing for the industry, observed: "From now forward the growing of timber will become a recognized and essential part in logging. A good fundamental knowledge of forestry will be helpful, in fact necessary, in conducting logging operations in the future; where the question of how best to remove the present crop with a view of providing for a continuous future supply will be regarded as a test of efficiency."[19]

In 1924 the director of the newly formed Pacific Northwest Forest Experiment Station of the U.S. Forest Service, Thornton T. Munger, called logging without forest replacement "industrial suicide."[20] It was during this period that the term forest engineer first appeared in the Pacific Northwest. As noted by Cornwall in the same paper, it was felt that the name of "logging engineer" should be widened to "forest engineer" to reflect adequately the scope of these new responsibilities. As previously noted the term "forest engineer" had already been introduced by Fernow whose earlier definition of the scope of the forest engineer's activity is consistent with the ideas advocated by Cornwall and Munger.[21]

---

18. Davies, "Western Logging Engineering Schools."
19. G. M. Cornwall, "The Profession of Logging Engineering," *Forest Club Quarterly* (University of Washington) 10 (1922): 17–19.
20. T. T. Munger, "Objectives of the New Federal Forest Experiment Station," 15th Annual Session, Pacific Logging Congress, 1924. p. 6–7.
21. Fernow, "The Forester, an Engineer."

The confluence of advancing technology and increasing concern about the forest resource had then forced a reconsideration of the role of the logging engineer. If logging (forest) engineers should once again enjoy a high profile in corporate operations, it was thought that it would be because of the broader issues of forest resources management and the ability of forest engineers to address those issues with a uniquely appropriate set of skills. But during the economic depression of the 1930s and the Second World War, the interest of the forest industry focused on short term economic efficiency. Logging time and cost studies were increasingly applied, and the forest engineer began to use many of the techniques popularized by industrial engineers. Interest in and development of the broader role of the forest engineer in forest resource management seemed to have waned.

The CCC program (Civilian Conservation Corps) initiated in the early 1930s supplied over 250,000 young men to do conservation work. Hundreds of engineers were employed to design and supervise the construction of roads, trails, bridges, etc. While we usually do not think of these CCC related activities as forest engineering per se, this activity provided much of the infrastructure which has been essential to efficient forest transportation, and was a catalyst in stimulating the U.S. Forest Service to publish their own *Engineering Field Tables* in 1935.[22] This handbook concentrated on practical surveying, earthwork, road drainage and surfaces, and concrete and timber construction.

The 1940s brought many advances to steep terrain harvesting technology, which had been the impetus for logging engineering in the Pacific Northwest. The appearance of track mounted steel towers, wide use of rubber and track mounted cable loaders and the wide acceptance of the power chain saw in felling and bucking operations were some of the more significant advances. Along with the improvements in technology came greater interest in forestry as a component of logging engineering. During the 1940s the logging engineering curricula began to show changes reflecting this new emphasis. The program of Oregon State University was renamed to forest engineering,[23] and the program at the University of British Columbia was changed to add more English, technical forestry and forest products in place of the applied engineering courses.[24]

---

22. United States Forest Service, *Engineering Field Tables* (Washington, D.C.: U.S. Government Printing Office, 1935).
23. Davies, "Western Logging Engineering Schools."
24. L. Besley, "Western Logging Schools—University of British Columbia," 42nd Annual Session, Pacific Logging Congress, *Loggers Handbook* 11 (1951): 79–84.

In Washington State, a tax law designed to encourage forestry on private land was passed in 1941.[25] The first forest practices act for the State was passed in 1945 and was directed at achievement and maintenance of adequate regeneration on cut-over land.[26] Further significant forest practices legislation would not be seen again until early in the 1970s. In 1949 logging engineering was granted recognition as a distinct branch of engineering, for purposes of professional licensing, by the Washington State Legislature.[27]

D. Professional Society Activities

The previously mentioned Pacific Logging Congress was, until about 1930, an organization that functioned much like today's technical societies. The annual meetings consisted of presented papers, formal discussion, field trips and a business meeting. The presented papers were very often of high caliber and some of them remain landmark papers in forest engineering research. The fundamental nature of the PLC began to change during the thirties. This change manifests itself as the presentations change from the technical to the business side of logging.

The Canadian Institute of Forestry was founded in 1908 as the Canadian Society of Forest Engineers with the participation of Fernow. The current name, adopted in 1950, more accurately reflects the members' preponderant professional interest in forestry rather than engineering. This professional society continues to publish the *Forestry Chronicle* which contains only occasional articles of minor engineering content.

The Society of American Foresters (SAF) is yet another professional forestry association. Lacking a traditional interest in the engineering aspects of forestry it provides only limited support to forest engineering activities. In spite of this limitation it has historically been widely subscribed to as a professional organization by American forest engineers. Its publications such as *Forest Science* and the regional applied forestry journals (*Western, Southern and Northeastern Journals of Applied Forestry*) provide an important outlet for forest engineering articles.

The Forest Products Research Society is an organization of researchers with a common interest of solid wood products. Through such publications as the *Forest Products Journal* and the *Timber Harvesting and Merchandis-*

---

25. D. H. Basinger, "The Status of Forest Taxation in the State of Washington," *Forest Club Quarterly* (University of Washington) 15 (2) (1941): 17–20.

26. Lloyd T. Webster, "Washington's Forest Practices Act," *Forest Club Quarterly* (University of Washington) 19 (1–3) (1945–46): 5–7.

27. G. D. Markworth, "Western Logging Engineering Schools—University of Washington," 42nd Annual Session, Pacific Logging Congress, *Loggers' Handbook*, 11 (1951): 85–87.

ing Newsletter the FPRS has been active in the publication of forest engineering research.

In the later 1960s and early seventies, protection of the public resources adversely affected by forest harvesting operations became a front page political issue. The increased public awareness sparked interest in forest engineering among other disciplines. The American Society of Agricultural Engineers (ASAE) held two forest engineering conferences in 1968 and 1969. The interest of the ASAE serves to illustrate that (1) the public concern for the forest resources was sparking interests of professionals outside of forestry and (2) the broader scope of forest engineers (outlined by Fernow in 1901 and further discussed by Cornwall and others during the 1920s) was becoming recognized. B. Y. Richardson wrote, in the Foreword to the proceedings of the first ASAE sponsored forest engineering conference, "Good engineering is also required in site preparation, regeneration, cultural and protective functions. These needs take the form of design, development and testing of machines for precise planting, seeding, fertilizer application, nursery operations as well as insect, disease and fire control."[28]

The second ASAE sponsored forest engineering conference, held in 1969, is significant because it was the first conference held since the early Pacific Logging Conferences that was directed at teaching and curricula in forest engineering. (The PLC had, as previously mentioned, evolved so as to place dominant emphasis on the business and occupation of logging as opposed to the profession and discipline of forest engineering.) S. J. Coughran noted in the opening remarks of the conference that" it was quite evident that the subject matter to be explored in this conference is extremely controversial."[29] The controversy he refers to was one of determining whether forest engineers are or should be foresters or engineers. George Cornwall's notion of a distinct profession of forest engineering had perhaps become forgotten.

In 1981, ASAE sponsored a third forest engineering conference, the Forest Regeneration Symposium. This conference was held in Raleigh, North Carolina, and published a proceedings under the same title.[30] This conference identifies Forest Engineering as a profession which serves all the as-

---

28. B. Y. Richardson, "Foreword," *Proceedings of the Forest Engineering Conference*, American Society of Agricultural Engineers, Michigan State University, East Lansing, Sept. 25–27, 1968.

29. S. J. Coughran, "Opening Remarks," *Proceedings of the Forest Engineering Conference on Education*, American Society of Agricultural Engineers, Chicago, Dec. 8–9, 1969.

30. American Society of Agricultural Engineers, *Forest Regeneration*, Proceedings of a Symposium on Engineering Systems for Forest Regeneration, Mar. 2–6, Raleigh, N.C. (St. Joseph, Mich.: American Society of Agricultural Engineers, 1981 [*ASAE* Publication no. 10-81]).

pects of forestry, where most previous works concentrated on the removal of timber and the associated transportation systems.

The ASAE's technical journal, *Transactions of the ASAE*, has served as an important outlet for the more engineering oriented research papers since the 1968 and 1969 conferences.

By the end of the 1970s, most of the western states had toughened and enlarged the scope of their forest practice legislation. Companies engaged in the harvest of a very valuable timber resource were operating on difficult terrain under restrictive forest practice acts. Forest engineers were again in high demand. The forest engineering programs in the Northwest were strong and numerous and others had materialized throughout the country. Some of the newer programs had affiliation with Agricultural Engineering Departments. In 1979 a forest engineering conference, independent of any existing organization, was held in Corvallis, Oregon. This conference marked the formation of the Council on Forest Engineering (COFE), a proximate professional organization for persons interested in forest engineering. At that first meeting of COFE it was decided that no affiliation should be sought with either the ASAE or the Society of American Foresters. The "controversy" of the 1969 meeting was still a concern. By the beginning of the eighties, however, even the forest engineering profession was impacted by the industry-wide recession. Academic concerns were replaced by institutional concerns as employment opportunities and student enrollment declined.

E. Development of the Current Literature

The previously mentioned early texts by Schlich, Schenck and Bryant were followed by J. P. Stewart's 1927 *Manual of Forest Engineering and Extraction*.[31] At the time, Stewart was a lecturer in Forest Engineering at the University of Edinburgh. His examples draw from extensive experience in North America, Africa and India where he had served as an advisor to various forest managers. His manual provides practical solutions to a variety of forest engineering problems ranging from protection from wild animals and malaria to a variety of logging and transportation schemes. Surveying, sleds, petrol and steam tractors, wire rope operations, slides and flumes, road, railways, trestles, water transport, permanent buildings and timber conversion and seasoning are included. Published with his manual

---

31. J. P. Steward, *Manual of Forest Engineering and Extraction* (London: Chapman and Hall, 1927).

are twenty-four pages of advertisements for goods commonly needed in a logging camp.

Some of the books by Brown were written as forest engineering texts. These books are generally descriptive of logging practices and contain but little engineering analysis.[32] One point of significance in the 1936 volume is the inclusion of silvicultural considerations within the chapters on logging methods. Only a loose tie is made between forestry and engineering.

In 1942 Professor Donald Matthews published his book *Cost Control in the Logging Industry*.[33] This text reflects the concern for economic efficiency prevalent during the 1930s. Among the topics covered are the economic location of roads and landings, economic service standards for roads and the selection of logging equipment by economic criteria. Despite the voluminous research that has been inspired at least in part by this book, it remains the only English language text written specifically on the topic of forest engineering economics. A major weakness of this poorly referenced text is its lack of a bibliography.

In 1947, the American Pulpwood Association initiated the *Technical Release* series oriented toward solving problems and presenting innovative ideas for loggers. Each monthly mailing to the Association membership contained several "Releases," each devoted to a single topic. In many cases these were written by the logger who actually developed the problem solution. Approximately 100 of these articles are published each year. While they normally are not written or reviewed by professional engineers, many do address the art and occasionally the science of forest engineering. These articles, although not of the technical quality exhibited in the early PLC papers, probably served to fill some of the void created as the PLC departed from its strong technical beginnings and became more of a social and trade organization.

A second contribution from the University of Edinburgh appeared in 1951 in the form of *Forest Engineering Roads and Bridges*.[34] James L. Harrison had been Forest Officer in India and following his retirement from foreign service he lectured on forest engineering and utilization at Edinburgh. This text discusses road reconnaissance and location, drainage structures, quarrying, retaining walls and river crossings. Harrison points out the essential need for a transportation network regardless of the particular log-

---

32. (a) Nelson C. Brown, *Logging—Principles and Practices in the United States and Canada* (New York: John Wiley, 1934). (b) Nelson C. Brown, *Logging—Transportation: The Principles and Methods of Log Transportation in the United States and Canada* (New York: John Wiley, 1936). (c) Nelson C. Brown, *Logging: The Principles and Methods of Harvesting Timber in the United States and Canada* (New York: John Wiley, 1949).
33. D. M. Matthews, *Cost Control in the Logging Industry*. (New York: McGraw-Hill, 1942).
34. J. L. Harrison, *Forest Engineering Roads and Bridges* (Edinburgh: Oliver and Boyd, 1951).

ging system to be used. This textbook is one of the first to offer rigorous engineering analysis directed at bridge and retaining wall structures typical of forestry operations. This book represents an engineering version of the type of text written by Schenck or Bryant.

With the exception of the text by Harrison, the previously mentioned books dealt with forest engineering topics in a descriptive fashion, lacking engineering analysis and synthesis. Filling this void are a number of manuals and handbooks.

Logging and forest road construction appeared as chapter topics in the *Forest Handbook for British Columbia* in 1953.[35] The original *Handbook* was primarily written by students at the University of British Columbia, under the direction of John (Jack) Walters. Walters was later to become internationally recognized for his work in developing a mechanical tree planter and serving as one of the first chairmen of the Forest Machine Committee of the American Society of Agricultural Engineers. It is interesting that this handbook had two sections devoted to engineering topics. The first, simply titled "Engineering," dealt with planning issues such as surveying and setting layout. The second section was titled "Logging Safety," and consisted of seven pages of direct quotations from the new British Columbia Safety code, implemented by the Workman's Compensation Board on September 1, 1950. This handbook is currently in its fourth edition.

In 1955 the Society of American Foresters published its first handbook, the *Forestry Handbook*.[36] A committee led by A. M. Koroleff of the Pulp and Paper Research Institute of Canada prepared the chapter on logging. The chapters on road engineering and surveying were prepared by a committee chaired by Anthony P. Dean of the U.S. Forest Service. A total of 167 pages of this handbook are dedicated to forest engineering, which is indicative of the need for published materials at the time. Although not explicitly an engineering handbook the *Forestry Handbook* is a common reference book in the libraries of most forest engineers.

A *Forester's Engineering Handbook* was written by Eric R. Huggard, a lecturer in forest engineering at the University College of North Wales, in 1958.[37] It covered the familiar topics of surveying, road and bridge design and construction, use of explosives, and timber extraction.

The United Nations through the Food and Agriculture Organization has also published a variety of manuals starting with one that gives a good

---

35. J. Walters, *Forest Handbook for British Columbia*, 1st ed. (Vancouver: University of British Columbia, 1953).

36. American Society of Foresters, *Forestry Handbook*, ed. by R. D. Forbes (New York: Ronald Press, 1955).

37. E. R. Huggard, *Foresters' Engineering Handbook* (Bangor: University College of North Wales, 1958).

blend of practical application and theory, *Tractors for Logging*.[38] In 1958, the United Nations Economic Commission for Europe (ECE) and the Food and Agriculture Organization of the United Nations (FAO) formed a joint FAO/ECE Committee on Forest Working Techniques and Training of Forest Workers. An International Training Course on Mechanized Forest Operations was held in Sweden in 1959, and a lengthy publication of this work was published by the U.N. in 1960.[39] Topics included equipment analysis, work-study, transportation systems, road standards, detailed descriptions of logging systems, and even human physiological requirements related to woods operations. Many of the topics, including some of the same diagrams, still appear in the current research literature. Several additional manuals of interest have been published by FAO.[40]

A short time later, J. Kenneth Pearce of the University of Washington published the first of several versions of the *Forest Engineering Handbook*.[41] Pearce is a registered civil engineer but his handbook is written for practitioners who may not have had formal engineering training. It also filled an important gap in the literature because it was specifically written for use in western North America where European literature had not gained wide acceptance.

During this same period in Canada, Lussier published a textbook dealing with the application of management science techniques to forest engineering problems. This book has stimulated numerous research papers and remains an excellent source of material for both teaching and research.[42]

Another publication which has been reprinted in several versions and has served as source material for several texts is the *Skyline Tension and Deflection Handbook*, by Hilton Lysons and Charles Mann.[43] The authors presented tables and graphical techniques to aid in the design of skyline yarder settings. Presented in 1967, these techniques became the standard yarder

38. United Nations, *Tractors for Logging* (Rome: FAO, 1956 [*FAO Forestry Development* Paper no. 1]).
39. United Nations, *ECE/FAO Joint Committee of Forest Working Techniques and Training of Forest Workers* (Geneva, 1960).
40. (a) United Nations, *Logging and Log Transport in Tropical High Forest* (Rome: FAO, 1974 [*FAO Forestry Development Paper* no. 18]). (b) United Nations, *Harvesting Man-Made Forests in Developing Countries* (Rome: FAO, 1976). (c) United Nations, *Planning Forest Roads and Harvesting Systems* (Rome: FAO, 1977 [*FAO Forestry Paper* no. 2]). (d) United Nations, *Assessment of Logging Costs from Forest Inventories in the Tropics* (Rome: 1978 [*FAO Forestry Papers* no. 10]).
41. J. K. Pearce, *Forest Engineering Handbook* (Department of the Interior, Oregon State Office, 1961).
42. L. J. Lussier, *Planning and Control of Logging Operations* (Québec, Canada: Forest Research Foundation, Université Laval, 1961).
43. H. Lysons and C. Mann, *Skyline Tension and Deflection Handbook* (Washington, D.C.: U.S. Government Printing Office, 1967 [*U.S. Forest Service* PNW 39]).

setting design guide until they were replaced by computer techniques in the early 1980s.[44]

Several descriptive manuals of logging operations were published by the U.S. Forest Service during the 1970s.[45] Norman Sears was responsible for initiating a continuing series of publications known as *Engineering Field Notes*.[46] While this series is intended as a U.S. Forest Service internal communication network to provide guidance on engineering methods, information exchange, continuing training and awareness of new developments and technical literature, the *Notes* are widely distributed and commonly used by the forest engineering community.

It was not until 1972 when J. Kenneth Pearce and George Stenzel published *Logging and Pulpwood Production* that a textbook was written in the United States as a replacement for the texts of the 1910–40 era.[47] This text addresses the same familiar topics as earlier published handbooks, but with much more attention to referenced research and some attempts to expose basic principles as well as problem solutions. By contrast, Steve Conway's books, *Timber Cutting Practices*[48] and *Logging Practices*,[49] contain detailed descriptions of logging practices, but are of limited scholastic value due to scant reference to research papers and the absence of any discussion of fundamental engineering principles.

In 1974, the Woodlands Research Division of the Pulp and Paper Research Institute of Canada and the Logging Development Program of the Canadian Forestry Service were merged to form the Forest Engineering Research Institute of Canada (FERIC). With a 1990 budget in excess of seven million dollars and a Canadian staff of 84 people, this cooperative government-industry alliance is the largest and most prolific source of forest engineering literature. Major activity areas are harvesting, secondary transportation, silvicultural operations and woodlot technology. They have staff support in the areas of design engineering, instrumentation and computers, and library functions. Their *Log Bridge Construction Handbook* is indica-

---

44. Virgil W. Binkley and John Sessions, *Chain and Board Handbook for Skyline Tension and Deflection* (USDA FS PNW Region, 1979 [GPO 799-549]).

45. (a) Fred C. Simmons, *Handbook for Eastern Timber Harvesting* (Broomall, Pa: USDA FS NE Area State & Private Forestry, 1979). (b) Keith L. McGonagill, *Logging Systems Guide* (1978 [USDA FS Alaska Region, Div. of Timber Management, Series no. R10-21]). (c) Donald D. Studier and Virgil W. Binkley, *Cable Logging Systems* (Corvallis, Oreg.: O.S.U. Book Stores, Inc., 1975).

46. United States Forest Service, *Engineering Field Notes* (Washington. D.C.: U.S. Department of Agriculture, 1969+).

47. J. K. Pearce and G. Stenzel, *Logging and Pulpwood Production* (New York: Ronald Press, 1972).

48. S. Conway, *Timber Cutting Practices* (San Francisco: Miller Freeman Publishers, 1968).

49. S. Conway, *Logging Practices* (San Francisco: Miller Freeman Publishers, 1976).

tive of their orientation toward the forestry construction practitioner rather than the design engineer.[50] Although it lacks much needed engineering analysis this manual is an excellent handbook for field design of log stringer bridges.

The following two manuals are characteristic of the vast quantity of published material from a variety of sources. The manual *Trucks and Trailers and Their Application to Logging Operations* by J. McNally represents a good blend of analytical and descriptive material.[51] The *Manual for Roads and Transportation,* most recently revised by David Holmes in 1978 and brought out in two volumes, is an excellent textbook for students.[52] It employs a good blend of numerical analysis with the practical and descriptive. These two manuals and many others of similar high quality have been published but are not widely distributed.

A number of forest engineering texts and references have been generated by European authors during the 1980s. Ivar Samset from the Norwegian Forest Research Institute produced a very complete cable logging text written in Norwegian. Fortunately, *Winch and Cable Systems* was translated into English in 1985.[53]

In 1981, the Skogsarbeten organization in Sweden produced three volumes on forest machinery systems. Unfortunately, the *Terrangmaskinen* series 32 was only available in Swedish until 1989, when the first volume was translated into English.[54] The excellent diagrams used in these books provide a valuable resource even if the reader does not understand the written text.

Tree Harvesting Techniques was written by K. A. F. Staaf from the Swedish University of Agriculture at Uppsala and N. A. Wiksten.[55] This text appears to rely heavily on earlier Swedish works and is clearly a descendent of the FAO/ECE work listed above. The two volumes on *Operational Efficiency in Forestry* edited by U. Sunburg from the Swedish University of Agricultural Sciences at Farpenberg and C. R. Silversides from

---

50. Michael M. Nagy, J. T. Trebett, G. V. Wellburn, and L. E. Gower, *Log Bridge Construction Handbook 1980.* (Vancouver, British Colombia, 1980 [*Engineering Research Institute of Canada Handbook* no. 3]).

51. J. A. McNally, *Trucks and Trailers and Their Application to Logging Operations* (Fredericton, New Brunswick: University of New Brunswick, Department of Forest Engineering, 1975).

52. David C. Holmes, *Manual for Roads and Transportation,* Rev. ed. (Burnaby, British Colombia: British Columbia Institute of Technology, 1978).

53. I. Samset, *Winch and Cable Systems* (Translated from Norwegian) (Dordrecht: Martinus Nijhoff/Dr. W. Junk Publishers, 1985).

54. (a) C-E. Malmberg, *Terrangmaskinen* (Stockholm, Sweden: Skogsarbeten, 1981). (b) C-E. Malmberg, *The Off-Road Vehicle* (Translation of *Terrangmaskinen,* vol. 3) (Montreal: Joint Textbook Committee of the Paper Industry, 1989).

55. K. A. G. Staaf and N. A. Wiksten, *Tree Harvesting Techniques* (Dordrecht: Martinus Nijhoff/Dr. W. Junk Publishers, 1984).

the Canadian Forestry Service also seem to have grown from the FAO/ECE roots.[56] This is the first book to seriously address such things as ergonomics of forest operations, problem analysis, energy analysis and the interaction between the stand, the prescription and the machine. Measurements and logging systems are confined to four pages.

Another step in the development of the forest engineering literature occurred in late 1989, when *The Journal of Forest Engineering* began to be published under the sponsorship of the Forest Engineering Department at the University of New Brunswick. Representatives of twelve countries sit on the editorial board, so it is clear that an international scope is intended.

F. Challenges for the Future

The development of the forest engineering discipline and its literature has been influenced by Western society's deep concern for the world's forests and by a pragmatic need for wood based products. It is reasonably clear from the literature however that the discipline has spent the last eighty to 100 years responding to the change in technology and the shifting emphasis on environmental concerns, as opposed to building a foundation from which tomorrow's new technology and solutions will arise. The result is a discipline with a body of literature that has never developed a cohesive framework of information that can serve to increase the awareness of a novice or enhance the analytical and design capabilities of advanced students. A perusal of the forest engineering literature would lead one to conclude that forest engineering design, the area of synthesizing new solutions, is not dependent upon analysis but rather upon an orally transmitted collection of field procedures.

The forest engineering profession now faces a serious challenge. Scholars must address the need for a literature that will serve to fully describe and define forest engineering. This new literature must not be merely descriptive nor handbook presentation of known solutions of limited analytical value. This new literature must present a synthesis of the dispersedly published engineering analysis that has been directed at forestry problems.

Practitioners are also facing serious challenges that call for innovative solutions not to be found in the handbooks. Low elevation second-growth stands in the Pacific Northwest are being harvested and are the focus of intensive management activity. These stands, as compared to the old-

---

56. U. Sunburg and C. R. Silversides, eds., *Operational Efficiency in Forestry* (Dordrecht: Martinus Nijhoff/Dr. W. Junk Publishers, 1988).

growth stands of former years, have more homogeneous timber located on gentler terrain. Road location and logging are not as challenging in this regard. The use of computers has greatly reduced office engineering time, thus a given quantity of design activity can be accomplished with fewer engineer hours. Easier conditions for roading and harvesting have reduced the obvious financial benefit associated with careful planning and engineering design.

If planning and conducting the roading and harvesting operation have been made easier by the terrain and technology, in at least two aspects it has become, and will become, much more difficult. First, the large and highly valuable logs of the old-growth forest have been replaced with small diameter lower value logs. This change in log size and value has made log handling critical to the profitability of a logging enterprise. The homogeneous nature of the timber resource better lends itself to mechanized harvesting and handling operations than did the old growth timber. Successful mechanized logging operations are highly engineered systems. Second, increased recognition and legislation for the protection of the public resources of air, water, fish and wildlife have placed major constraints on timber harvesting and other forest management activities. The resources belonging to society at large cannot be dismissed as illegitimate or ephemeral concerns. It is here that the forest engineer can make a substantial contribution to the forestry industry and to society. The engineering design of forest roads, harvest systems, or other forest management operations is the key to the integration of the many constraints currently placed on forest management and utilization. To the extent that engineering skills and accountability can contribute to the identification and implementation of environmentally acceptable, ecologically desirable and financial attractive management and harvesting activities forest engineers must be involved.

The future of the profession depends on practicing forest engineers and educational institutions cooperating to redefine the areas of technology or bodies of knowledge that constitute forest engineering. The continued development of forest engineering should address the technology and problems of today and anticipate those of tomorrow. It is also the time to examine the profession in a new way, not engineering with some forestry, not foresters with some engineering, not even a hybrid engineer-forester, but as a distinct profession and academic discipline.

9. Aquacultural Engineering

JAW-KAI WANG
Department of Agricultural Engineering
University of Hawaii

A. Prologue

"Products and processes are the successful result of a chain of events that starts with the discovery of new truths about the natural world. Abstract knowledge first has to be 'translated' into technology before it can be applied to the manufacture of marketable goods. This vital link, the Board (Engineering Research Board, National Research Council) explained, is provided by engineering research. . . . The annual world market for biotechnology has been estimated to reach $100 billion by the year 2000. To reach this level, however, will require new techniques for large-scale culture of plant and animal."[1]

Aquacultural engineering is the application of science and technology to the production of aquatic animals and plants. It is primarily concerned with the efficient use of the inputs required in the production of aquatic products and the economic efficiency of the production system. It is a branch of the agricultural engineering profession.

Modern aquaculture is committed in large part to the production of aquatic animals and plants for profit. Aquaculture must compete with fisheries and agriculture as the supplier of food, fertilizer, and industrial and pharmaceutical raw materials. Therefore, the overall efficiency of an aquaculture production system is always paramount in the minds of aquacultural engineers.

One of the challenges encountered in the development of aquacultural engineering is the development of mutual understanding and acceptance be-

---

1. National Research Council, *News Report* (Washington, D.C.) 27 (8) (Aug.–Sept. 1987).

tween agricultural engineers and aquaculturists. It is a challenge similar to that the agricultural engineers themselves experienced not too long ago.

The challenge can be illustrated by the following example:

The Bioengineering Symposium for Fish Culture conference, sponsored by the Fish Culture Section of the American Fisheries Society and the Northeast Society of Conservation Engineers, was held in Traverse City, Michigan, in October 1979. It seems the sponsors and the participants at this meeting accepted the premise that developments in fish culture should primarily be the responsibility of either fish culturists or research biologists, and after the cultural practices have been established, that the engineers should have the responsibility for the design of the systems that put the findings of the biologists to profitable practices. In the paper entitled "Symposium Perspective," which set the tone of the conference, Roger Burrows stated: "Most developments in fish culture are derived from either fish culturists or research biologists. The practical application is dependent on the ability of the engineers. The field of engineering requires capability in the mechanical, structural, hydraulics, electrical, and chemical disciplines. Unfortunately most engineering firms do not have all these capabilities confined to one person or a few persons who are willing to work together and listen to the biologists and fish culturists." This viewpoint is commonly accepted and widely held by biologists and aquatic farmers. The basic problem with such mistaken beliefs is the implied assertion that if one is able to find a jack-of-all-trades engineer who has a nice personality, one's problems are over. This is simply not true. The practice of engineering is an activity that has its roots in science and technology, and is constrained by economics. Aquacultural engineers cannot design aquaculture systems without data and experience (training), which can only be provided by aquacultural engineering research and the operation of pilot facilities.

The typical biologist will ask how to raise a fish; the engineer will ask how to design and operate a production facility in the most efficient and profitable manner. The uniformed biologist thinks that a pond is needed to hold and raise the animal or plant, but the aquacultural engineer needs to know the effects of shape, size, and construction details on pond productivity and maintenance costs, and the ease with which the animal or plant can be harvested. It is essential for the engineer to know the system's response to changes in facility design and operation so that each dollar of investment can be justified. Without that knowledge the aquacultural engineer cannot produce the best "compromise" design and will invariably either over- or under-design the system.

The difficulty in designing an efficient production system for aquatic animals and plants lies primarily in knowing what to design. In other words, existing literature does not provide engineers with a firm guideline on how to determine the functional requirements of such systems. How does one determine the size of a shrimp pond and whether it should be round, square or rectangular. Should the pond be built to last three or ten, or twenty years?

Once the functional requirements of the pond are properly specified by an aquacultural engineer, a civil engineer can finish the design and a contractor can complete the construction. If a jack-of-all-trades engineer is all that is required, engineering would not have been diversified into the specialties that one sees today.

Proceedings of the Bio-Engineering Symposium for Fish Culture (1981) provides an excellent example of the gap between aquacultural engineers and aquatic biologists, a gap yet to be bridged. The way to bridge the gap is to start the development of production research with joint efforts, to avoid less useful after-the-fact dialogue.

B. Introduction

Aquacultural engineering is an emergent technology.

The establishment of the Aquacultural Engineering Committee by the American Society of Agricultural Engineers (ASAE) in the 1970s and the formation of the Aquacultural Engineering Subject Group within the Institution of Chemical Engineers (IChemE) in 1985 are, in themselves, important events in the application of engineering technologies and sciences to the production of aquatic animals and plants. Many papers on aquacultural engineering can be found in the *Transactions of the ASAE*. In addition, in recent years, many papers related to aquacultural engineering have been published in the *Journal of the World Aquaculture Society*.

Aquacultural Engineering started publication in January 1982 with quarterly issues published every year. Its refereed articles cover the entire spectrum of aquacultural engineering, and, in a sense, by the publication of a wide range of articles, the journal is helping to define the meaning of aquacultural engineering.

There are very few existing publications that can be considered true aquacultural engineering literature, but many are related to the subject. The most difficult task in doing this review, therefore, was in defining and choosing the publications to be included. It has also become obvious in reviewing the literature that it is much easier to find a publication that tells you what to do once you have determined the type of system required, than to find a publication that will lead you logically to a choice of a system. In other words, the practice of aquacultural engineering profession today is much more of art than science.

The publications reviewed have been divided into six major categories. General Literature, Aquarium and Closed Systems, Production Facility Design, Effluent Management, Feasibility Studies, and Conference Proceedings. It was difficult to categorize certain publications, however some classification was needed and a few arbitrary choices were made.

Aquariums is devoted to the display of aquatic animals and plants and not to commercial production. However, it is an important subject in aquacultural engineering and many of the practices in aquarium design are applicable to small household aquariums and other closed systems. "Closed Systems" is a subject that is gaining importance and is combined with aquarium systems because the two share many common concerns in design and operation.

Production Facility Design is divided into four subcategories: Freshwater Systems, Seawater Systems, Cage Culture Systems, and Integrated Systems. The increasing world population and the intensifying demand for a better material life have led to an increasing societal emphasis on environmental quality, which in turn is increasing the demand to improve the efficiency of resource utilization in aquatic production. As the cost of effluent management will frequently determine whether or not an aquaculture enterprise will be profitable, it has become increasingly important for an aquaculture production system to reduce the need for waste management by eliminating waste in the system.

Effluent Management covers aquaculture effluent and water quality management. As aquaculture production gains significance, it is important to realize that aquaculture effluent is different from municipal waste. It is even different from most agriculture discharge because aquaculture operations add no toxic material to the water used. In fact, most aquaculture effluent is created in the effort to maintain water quality in the production system.

C. Literature Review

General Literature

This section covers publications which are not easily fitted into any of the other categories, and which cover a broad range of topics.

Aquacultural Engineering is the first major book on this subject and an excellent representative of general literature in this field.[2] The book covers

---

2. Frederick W. Wheaton, *Aquacultural Engineering* (New York: John Wiley & Sons, 1977).

the basic engineering principles, as well as the physical and chemical sciences, which are useful in the design of aquacultural production systems. What it does not cover is how this information should be applied to the design or operation of an aquacultural system. In other words, the author assumes that the reader knows what the "functional design" of an aquaculture system should be and has designed the book to supply the reader with the necessary tools to design such a system. The extensive reference list is still useful in 1991 and the book is an excellent base from which a first year graduate course in aquacultural engineering can be developed.

Instrumentation in Aquaculture is a collection of papers presented at the 1989 World Aquaculture Society Meeting.[3] The papers give a state-of-the-art review of the subject and each article contains an excellent reference list. Instrumentation is a critical problem facing the aquaculture industry as it seeks to reduce labor while increasing production intensity. Sensors for continuous water quality monitoring were primarily developed for the wastewater industry and are not, in general, suitable for aquaculture use. The existing market for specialized instrumentation, however, has been insufficient to generate manufacturers' interest in developing specialized equipment for aquaculture. The *Proceedings* gives a good review of the instrumentation that is available and what can be done with them. The papers on process control are interesting.

Crustacean and Mollusk Aquaculture in the United States is a good, quick reference book containing information on some of the working aquaculture production systems in the United States.[4]

Literature for United States Aquaculture 1970–1982 is an excellent compilation of aquaculture literature published in periodicals, including the *Journal of Aquaculture Engineering* during 1970–82.[5] Most aquacultural engineers should find it a useful addition to their everyday library. The book contains 228 pages and can be obtained from the National Agricultural Library free of charge. The literature is divided into eleven categories and an author index. The sections on Production and Management Systems, Harvesting and Handling, and Marketing and Economics are particularly useful to engineers.

Selected Aquaculture Publications is a compilation of serials, newsletters, conference and symposium proceedings, and bibliographies. It is not

---

3. James A. Wyban and Ellen Antil, eds., *Instrumentation in Aquaculture*, Proceedings of a Special Session at the 1989 Annual Meeting of the World Aquaculture Society (Honolulu, Hawaii: The Oceanic Institute, 1989).

4. Jay V. Huner and E. Evan Brown, eds., *Crustacean and Mollusk Aquaculture in the United States* (Westport, Conn.: AVI Publication Co., 1985).

5. John B. Forbes and Charles N. Bebee, *Literature for United States Aquaculture 1970–1982* (Beltsville, Md.: National Agricultural Library, 1983).

an aquacultural engineering publication, but aquacultural engineers may find it useful, as well as Coche's *A List of Selected FAO Publications Related to Aquaculture, 1966–1989.*[6]

Aquarium and Closed Systems

Fish and Invertebrate Culture: Water Management in Closed Systems is the first book devoted to closed aquatic production system design.[7] It is probably still the best first book for an aquacultural engineer who is interested in closed systems design. In semi-closed and open systems, water is continually discarded and replenished from a natural source and effluent must be discharged. The closed system offers both economy in water usage and excellent environmental control. That is not to say, however, that economically competitive closed systems can be developed today for commercial production. The book covers the important subjects of biological, mechanical, and chemical filtration concisely and quite well. Toxicity, disease prevention and disinfection are also described. A list of cited literature is included.

Anyone designing an aquarium will find *Aquarium Design Criteria* useful.[8] The pamphlet gives good coverage of aquarium project development and the articles were written by experts in their fields. Except for the chapters on filtering and materials, one is likely to be surprised by how little progress has been made in aquarium design during the last twenty years.

Sea-Water Systems for Experimental Aquariums is a collection of papers.[9] Long-term recycle systems are described along with the changes that were observed in the systems over time. Use of artificial seawater is compared to use of natural seawater as a culture media. Aeration and removal of supersaturated gases from seawater is described.

Solution Equilibria is a rather detailed description of the measurement, use and interpretation of the thermodynamics of the solution state.[10] The measurement of the constants which describe ionic equilibria in water, the

6. André G. Coche, *Selected Aquaculture Publications: Serials, Meeting Proceedings, and Bibliographies/Directories/Glossaries* (Rome: Food and Agriculture Organization, 1987). Coche has also compiled *A List of Selected FAO Publications Related to Aquaculture . . .* in two revisions the most recent being for 1966–89 (Rome: FAO, Feb. 1990 [*FAO Fisheries Circular* no. 744, Rev. 2; FIRI/C744]). Each revision supersedes the earlier compilation.

7. Stephen H. Spotte, *Fish and Invertebrate Culture: Water Management in Closed Systems* (New York: John Wiley and Sons, 1979).

8. William Hagan, ed., *Aquarium Design Criteria* (National Fisheries Center and Aquarium, U.S. Dept. of the Interior. Drum and Croaker Pub., 1970).

9. J. R. Clark and R. L. Clark, eds., *Sea-Water Systems for Experimental Aquariums* (Washington, D.C., 1964 [*U.S. Fish and Wildlife Service Research Report* no. 63]).

10. F. R. Hartley, C. Burgess, and R. Alcock, *Solution Equilibria* (New York: John Wiley & Sons, 1980).

buffering capability of seawater, ionic exchange between the various important nutrients and metabolites, soluble and solid ion-exchange, and adsorptive materials are detailed.

Aquarium Systems describes a variety of practical suggestions and concerns for those who intend to culture freshwater or marine organisms.[11] The section on the chemical and physical treatment of the seawater in recirculating systems is quite detailed.

Marine Aquariums in the Research Laboratory is a short description of the use of submerged biological filtration as a means of maintaining marine systems.[12] The aerobic bioconversion of toxic metabolites to CO_2 and NO_3 is discussed. A list of cited literature is included.

D. Production Facility Design

Freshwater Systems

Ponds—Planning, Design, Construction was published by the Soil Conservation Service of the United States Department of Agriculture for use by ranchers and farmers to design and build ponds for livestock and irrigation water.[13] It is an excellent pamphlet, very well written and covering all the basics one needs to know to design small freshwater ponds, whether for aquatic production or for the originally intended purposes. The section on waterproof linings, though slightly outdated by newly available materials, is very useful as it is a subject not commonly covered by other books. The sections on maintenance of the pond, installing vegetation, and protecting the pond are also useful as they at least call attention to areas of which one needs to be mindful. Some of the FAO/UNDP sponsored publications go into detail on engineering design but pay little or no attention to maintenance.

Inland Aquaculture Engineering is a collection of lectures presented at the Aquaculture Development and Coordination Programme (ADCP) Inter-regional Training Course in Inland Aquaculture Engineering.[14] The publication covers site selection, maintenance and planning. It is aimed primarily

---

11. A. D. Hawkins, *Aquarium Systems* (New York: Academic Press, 1981).
12. J. M. King and S. Spotte, *Marine Aquariums in the Research Laboratory* (Placitas, N. Mex.: Aquarium Systems, Inc., 1974).
13. United States Department of Agriculture, Soil Conservation Service, *Ponds—Planning, Design, Construction* (Washington, D.C.: U.S. Govt. Print. Off., 1982 [*USDA Agriculture Handbook* no. 590]).
14. Aquaculture Development and Coordination Programme, *Inland Aquaculture Engineering: Lectures Presented at the ADCP Inter-regional Training Course in Inland Aquaculture Engineering*, Budapest, June 6–Sept. 3, 1983 (Rome: United Nations Development Programme and FAO, 1984 [ADCP/REP/84/21]).

at project managers to lead them step-by-step in producing project documentation useful for bank financing review. Mastery of the subjects covered by the publication and careful adherence to its instructions do not, however, guarantee a good design. The book provides good coverage of engineering subjects useful to aquaculture production systems, but it does not provide the essential knowledge that would enable one to judge the goodness of a design. It is a collection of useful information and a useful handbook.

Seawater Systems

In *Design and Operating Guide for Aquaculture Seawater Systems*, Huguenin and Colt present a well-balanced approach to seawater aquaculture systems design.[15] One of the fundamental differences between the freshwater and the seawater or brackish water systems is the problem with biofouling. The book has a well-written but brief section on biofouling which compares the different methods of biofouling control. On the other hand, the chapter on solid removal is too brief to be useful. One of the most useful features of the book is its appendices on selected bibliography. The references are divided in the following appendices, such as flow-through seawater system, reuse seawater system, biofouling, etc. The author(s) and source for each of these references is listed and a brief comment is given. It is an interesting approach and certainly makes the reference list much more useful than the simple list that one generally finds at the end of a book. The most recent comprehensive book is *Coastal Aquaculture Engineering*.[16]

Engineering Aspects of Brackish Water Aquaculture in the South China Sea Region gives a summary of the status of brackish aquaculture in Indonesia, Malaysia, Philippines, Singapore and Thailand.[17] If you know what you want in terms of ponds and other facilities, these publications are designed to lead you to documents that can be used by contract officers and contractors.

Report of Consultation/Seminar on Coastal Fishpond Engineering, publication of *Inland Aquacultural Engineering* (Aquaculture Development and Coordination Programme, 1982), is a collection of reports concentrating on

---

15. John E. Huguenin and John Colt, *Design and Operating Guide for Aquaculture Seawater Systems* (Amsterdam: Elsevier Scientific Publication Co., 1989).
16. A. N. Bose et al., *Coastal Aquaculture Engineering* (London: Edward Arnold, 1991).
17. South China Sea Fisheries Development and Coordinating Programme, *Engineering Aspect of Brackish Water Aquaculture in the South China Sea Region* (South China Sea Fisheries Development Coordinating Programme, 1975 [SCS/75/WP/16]).

the development of brackish fish and shrimp production in the Southeast Pacific coastal region and includes the swamp land.[18]

Comparative Study of Tidal and Pumped Water Supply for Brackishwater Aquaculture Ponds in Malaysia is a useful and interesting publication.[19] Although a small volume, it carefully documents comparative costs and management factors involved in a tidal water exchange aquaculture farm production system and a pumped water system. The two farms studied are located adjacent to the Brackishwater Aquaculture Research Station at Gelan Patah in Johor Bharu, Malaysia, but the conclusion of the study—that a pump-operated system is more economical than a tidal system—should have wide application elsewhere. It is indeed refreshing to read a publication that indicates what system to design rather than merely how to design it.

Guide to Design and Construction of Coastal Aquaculture Ponds is well written and reflects the leading position of Japan in the development of coastal brackish aquaculture.[20] It is written for fishery specialists. Chapters 3 and 6 offer useful discussions on water management, flood control, floating breakwater, and aeration for marine ranching.

Fishpond Engineering: A Technical Manual for Small- and Medium-Scale Coastal Fishfarms in Southeast Asia is the result of a regional consultation/seminar held in Surabaya, Indonesia, August 4–12, 1982.[21] The seminar was jointly sponsored by the FAO/UNDP and the Agency for Agricultural Research and Development of the Directorate General Fisheries of Indonesia. As the title implies, the pamphlet covers the necessary subjects from site selection to survey, layout, and construction, and includes a chapter on feasibility study. It is intended to be used as a guide in project preparation.

Recent Advances in Aquaculture is a helpful review of mangrove swamps and aquaculture.[22] Mangrove swamps dominate sheltered coastlines throughout the tropics and there is increasing pressure for their development.

---

18. South China Sea Fisheries Development and Coordinating Programme, *Report of Consultation/Seminar on Coastal Fishpond Engineering* (South China Sea Fisheries Development Coordinating Programme, 1982 [SCS/GEN/82/42]).

19. Robert H. Gedney, Yung C. Shang, and Harry Cook, *Comparative Study of Tidal and Pumped Water Supply for Brackishwater Aquaculture Ponds in Malaysia* (Rome: Food and Agriculture Organization and the United Nations Development Programme, 1983 [SCS/83/WP/117]).

20. Juichi Katoh, *Guide to Design and Construction of Coastal Aquaculture Ponds*, new ed. (Tokyo: Japan International Cooperation Agency, 1980).

21. C. R. de la Cruz, *Fishpond Engineering: A Technical Manual for Small- and Medium-Scale Coastal Fish Farms in Southeast Asia* (South China Sea Fisheries Development and Coordinating Programme, 1983 [SCA Manual 5]).

22. J. F. Muir and R. J. Roberts, *Recent Advances in Aquaculture*, Vol. I (London: Croom Helm Ltd., 1982).

Cage Culture System

Cage culture is a method of farming aquatic organisms in a particular type of rearing facility. There are a number of water-based rearing facilities. Pens and enclosures have sides constructed of wooden poles, mesh or netting, with sea bed bottoms, and tend to be bigger, ranging in size from around 0.1 hectares to over 1,000 hectares. Cages are enclosed on the bottom as well as on the sides, are generally smaller (typically having a surface area of less than 1,000 m^2) and are generally more intensively managed than pens.

Cage Aquaculture gives good coverage of cage culture as practiced in Northern Europe, especially for cage systems in protected sites.[23]

Aquaculture Engineering: Technologies for the Future a proceedings compiled and published by the Institution of Chemical Engineers gives an up-to-date review on recent work relating to engineering design of cages in exposed areas.[24]

Integrated Systems

In integrated aquatic farming a number of production components are linked in series to share the production resources. For example, in a simple integrated system, microbiologically clean brackish water form deep wells can be used first to depurate a crop of market sized oyster. The effluent from oyster depuration goes into a shrimp pond and the shrimp effluent is then used to feed the growing oysters. In a somewhat more complicated form, a farm producing land crops can use their crop residues for livestock feed. The livestock waste can then be used for fish culture and the effluent from the fish culture ponds can be used for crop production.

Integrated farming in aquaculture is a finely developed art in China, which has been culturing fish for longer than most nations; it has evolved a form of integration which is also suitable for the specific social, economic and cultural conditions of that country. *A Guide to Integrated Warm Water Aquaculture* gives a good review of integrated warmwater aquaculture and contains useful information for anyone who intends to design and integrate aquaculture production systems in the tropics.[25]

---

23. Malcolm C. M. Beveridge, *Cage Aquaculture* (Farnham, Eng.: Fishing News Books, 1987).
24. Institution of Chemical Engineers [Great Britain], Scottish Branch and Aquaculture Engineering Subject Group and Institute of Aquaculture, *Aquacultural Engineering: Technologies for the Future* (New York: Hemisphere Publishing, 1988).
25. David Little and James Muir, *Guide to Integrated Warm Water Aquaculture* (Sterling, Scotland: University of Sterling, Institute of Aquaculture Publications, 1987).

As the intensity of aquatic production increases, so will the pressure for integrated production systems. An integrated system has the potential for offering the best solutions to efficient use of production resources and effluent management.

E. Effluent Management

One of the major constraints to the growth of brackish water aquaculture has been the cost, time and uncertainty involved in obtaining permits for effluent discharge. Regulators, whose experience has almost exclusively been with discharge that could potentially cause significant environmental impact and whose training has been almost devoid of subjects relating to aquaculture, find little justification in treating aquaculture effluent differently from other wastewater. However, with the increasing demand for brackishwater aquaculture of animals such as shrimp and oysters, effluent and water quality management is becoming increasingly important.

Aquaculture Effluent Discharge Program, Year One Final Report is one of the best publications on aquaculture effluent.[26] It is a report of a project funded through the Center for Tropical and Subtropical Aquaculture by the U.S. Department of Agriculture. The report is largely a literature review and contains an excellent reference list.

Part One of the report contains a characterization of shrimp, freshwater prawn, marine fish, and freshwater fish pond effluent in Hawaii. Part Two gives an excellent review of the existing wastewater treatment processes and their applicability to aquaculture effluent. Part Three cites previous experiences and studies in treating commercial shrimp pond effluent and concludes that there is no existing cost-effective method for treating aquaculture effluent.

The report points out that suspended particulates is the major component in aquaculture effluent management. The report proposes an innovative treatment process for shrimp pond effluent which uses bivalves to concentrate the suspended particulates so as to improve mechanical sedimentation efficiency.

Another excellent brief piece on pond water management is "Rethinking Shrimp Pond Management."[27] The author, however, while covering all the

---

26. David Ziemann, Gary Pruder, and Jaw-Kai Wang, *Aquaculture Effluent Discharge Program, Year One Final Report* (Honolulu, Hawaii: U.S. Dept. of Agric., Center for Tropical and Subtropical Aquaculture, 1990).

27. George W. Chamberlain, "Rethinking Shrimp Pond Management," *Coastal Aquaculture* 5 (2) (1988).

chemical variables important to pond water management, fails to give equal attention to the physical and microbiological characteristics of the effluent. Chamberlain lists dissolved oxygen, pH, ammonia and nitrite, hydrogen sulfide, redox potential, sediment core samples, phytoplankton and type, and bacterial counts as shrimp pond parameters to be monitored. Conspicuous by its absence from the list is suspended solids. The control and management of dissolved oxygen, suspended solids and algal density in the water column are vital to the proper management of shrimp pond water. The total concentration, the size distribution, and the microbiological characterization of the suspended solids are far more important to shrimp pond water quality, shrimp pond productivity, and effluent management than sediment core samples.

F. Feasibility Studies

The American Cyanamid Company conducted a feasibility study entitled *New Engineering Approaches for the Production of Connecticut Oysters*.[28] The research was funded by the Connecticut Research Commission and the stated goal was to revitalize the Connecticut oyster industry through an infusion of new technology, particularly by the development of a factory process in which oyster culture conditions could be controlled at optimum. The report produced is excellent and is still referred to by aquacultural engineers today. It shows how existing information can be weaved together to produce a proposal for an aquaculture factory and represents a major step toward using available knowledge to develop pilot or commercial-scale operating aquaculture systems.

The report can also be used to illustrate the difficulty in applying existing biological data to engineering design. For example, oysters have been the subject of scientific biological studies for over 100 years and have been harvested from long before the days of the Roman Empire, both in Europe and in China. Yet, there is no useful information on an optimum diet—or even what constitutes a balanced diet—and feeding pattern for the oyster from spat to market size. Today, as in 1968, the growth potential of the oyster under different feeding patterns and diets is still unknown, but pioneering aquacultural engineering at the University of Hawaii has shown that with a continuous diet of rich food and high growth temperatures, market sized *C. virginica* can be produced in about six months. The biologists

28. D. R. Goodrich, L. J. Calbo, R. B. Wainright, and A. Perlmutter, *New Engineering Approaches for the Production of Connecticut Oysters* (Hartford, Conn.: Central Research Division, American Cyanamid Co., 1968).

perhaps still think of oyster aquaculture as the seeding and harvesting of oyster beds. In 1968, the engineers were already thinking of an oyster factory.

Planning an Aquaculture Facility—Guidelines for Bioprogramming and Design is directed at managers, not engineers, of aquaculture projects.[29] It offers an uncluttered view of project planning and outlines the procedure which should be followed in the planning of an aquaculture facility. The appendix on biological design criteria can be used by engineers as a reminder of the information required in project design.

"A Mathematical Model of Some Aspects of Fish Growth, Respiration and Mortality" is an interesting long journal article that some of us may still find useful today.[30] The author describes the growth of aquatic animals in terms of physical phenomena which may be modelled. There are numerous examples to lead the reader through the actual development required to produce a useful model which can be used by engineers to design a commercial facility.

G. Conference Proceedings

Aquacultural engineering is an emergent discipline and much of its important literature can be found in conference proceedings. The utilization of heated effluent in intensive aquaculture systems is well explored in a two volume proceedings entitled *Aquaculture in Heated Effluents and Recirculation Systems*, of a conference which was held May 28–30, 1980 in Stavanger, Norway.[31] Many recirculation systems in existence at the time the conference was held are reviewed and the many disease problems in intensive recirculating systems are presented. A number of possible solutions are discussed but no consensus on design guidelines emerges. The use of simple recirculating systems has been on the rise, however, much work is needed to demonstrate both system reliability and economics.

Proceedings of the Bio-Engineering Symposium for Fish Culture is the product of a conference, sponsored by the Fish Culture Section of the American Fisheries Society and the Northeast Society for Conservation En-

---

29. Carol M. Brown and Colin E. Nash, *Planning an Aquaculture Facility—Guidelines for Bioprogramming and Design* (Rome: Aquaculture Development and Coordination Programme, United Nations Development Programme and Food and Agriculture Organization, 1988 [ADCP/REP/87/24).

30. E. Ursin, "A Mathematical Model of Some Aspects of Fish Growth, Respiration and Mortality," *Journal of the Fisheries Board of Canada* 24 (1967): 2355–2453.

31. Klaus Tiews, ed., *Aquaculture in Heated Effluents and Recirculations Systems*, 1, 2. (Berlin: H. Heenemann, 1981).

gineers, that was held in Traverse City, Michigan, in October 1979.[32] The proceedings covers a number of important subjects, such as water quality, instrumentation, and effluent management. These are the areas in which joint effort by biologists and engineers is urgently required.

Report of the National Consultative Meeting on Aquaculture Engineering is the proceedings resulting from a meeting sponsored by the regional office of FAO.[33] Twenty papers were presented and all of the authors were from the Philippines. The papers cover a wide range of items of interest to aquacultural engineers, such as layout and design of pond systems, cage and pen systems, and aeration systems.

Aquaculture Engineering: Technologies for the Future is the proceedings of the symposium organized by the Scottish Branch of The Institution of Chemical Engineers (IChemE) and the Aquaculture Engineering Subject Group of the IChemD, in association with the Institute of Aquaculture.[34] The symposium was held at the University of Sterling in Scotland, June 20–23, 1988. The two sections on cage culture and on tank technologies are especially worth noting.

The *Aquacultural Engineering* journal published a special edition in 1986 containing papers presented at a workshop entitled "Aquacultural Engineering and Simulation."[35] The workshop was sponsored by the United States National Research Council, the Taiwan Committee for Scientific and Scholarly Communication with the United States, and Academia Sinica, and held at the University of Hawaii in January of 1985. The proceedings contains two review papers covering material science in aquacultural engineering and biotechnology in marine aquaculture, which are of general interest to aquacultural engineers. Its section on nutrient inputs and their control is also useful reading.

H. Aquacultural Production

It is well known that the culture of aquatic organisms originated several thousand years ago in Asia. Csavas estimated the total aquaculture produc-

---

32. Loche Jo Allen and Edward C. Kinney, eds., *Proceedings of the Bio-engineering Symposium for Fish Culture* (Bethesda, Md.: Fish Culture Section of the American Fisheries Society and the Northeast Society of Conservation Engineers, 1981).

33. Association of Southeast Asian Nations, *Report of the National Consultative Meeting on Aquaculture Engineering: Regional Small-Scale Coastal Fisheries Development Project* Manila, Philippines: United Nations Development Programme and Food and Agriculture Organization, 1986 [ASEAN/SF/86/GEN/1]).

34. Institution of Chemical Engineers, Scottish Branch and Aquaculture Engineering Subject Group & Institute of Aquaculture, *Aquaculture Engineering: Technologies for the Future* (New York: Hemisphere Pub. Co., 1988).

35. Keith R. Murray, Jaw-Kai Wang, and Gary D. Pruder, eds., "Aquacultural Engineering and Simulation," *Aquacultural Engineering* 5 (2–4) (1986).

tion of the world was 4,738, 713 metric tons in 1975 and 8,004, 766 metric tons in 1983.[36] Of these, more than 77% was produced in Asia. Aquaculture provides 8.9 million tons of fish and shellfish. China is the highest producer of aquaculture products, followed by Japan, Republic of Korea, the Philippines, and the United States. Europe produces about one million tons per year. Carp is the most important species worldwide. Aquaculture provides about 13% to the world supply of fish. Aquaculture could supply 25% of the total world fisheries production of 85 million tons by the year 2000. With a production of 22 million tons evenly divided between developing and developed countries.

The catfish industry in the United States offers an example of the rapidly expanding aquaculture. The first commercial catfish production in the United States was in Arkansas in 1960. By 1989, farmers delivered 155,000 tons of catfish and increased production is expected.

Expansion in the world salmon production is equally phenomenal. Production in Norway, the world leader, was estimated at 265 million pounds in 1989, up from only 15 million pounds in 1980.

Cultured shrimp is another example of the growing importance of aquaculture in the world. Starting in Japan, shrimp farming was adopted and improved by the Taiwanese. It then spread to Ecuador and Brazil and other South and Central American countries and to such countries as China, Thailand, the Philippines, and Indonesia. Cultured shrimp production in the world has been growing at an astonishing annual rate of 36% over the 1984-to-1988 period, supplying almost 39% of the world's shrimp consumption in 1988.

The share of seaweed in the total aquaculture production is an impressive 20.1%. It is seldom realized outside the circle of specialists that aquacultural seaweed accounts for almost half the world seaweed production.

Basic economics dominate the development of aquaculture and the development of technology will set the pace for aquaculture development. Aquacultural engineering will play a major role in increasing production of aquaculture products. The rapid development of the cultured shrimp industry is driven mainly be the desire of many countries to earn foreign currency. For the United States, in 1988 imports of edible seafood grew to 3.0 billion pounds worth about $5.5 billion. The United States Department of Agriculture projects that the trade deficit in fishery products probably will continue

---

36. Imre Csavas, "Problems of Inland Fisheries and Aquaculture," paper presented at the Asian Productivity Organization Symposium on Fishing Industry, Tokyo, Nov. 12–18, 1985, in *Fishing Industry in Asian and the Pacific* (Tokyo: Asian Productivity Organization, 1988).

to increase. The fishery trade deficit was $1.8 billion in 1980 and $3.5 billion in 1986.[37]

I. Summary

There is a vast amount of literature from many countries of the world related to the practice of the aquacultural engineering profession, but only a few can be classified as aquacultural engineering publications. Much of the more recent literature is in the form of journal articles waiting to be incorporated into books or other forms of publication. This is particularly true in the areas of aeration, pond management, harvesting, and post-harvest processing and transportation. Among the aquacultural engineering publications reviewed there is an obvious lack of literature on the functional design of aquacultural systems, the optimal pond size and shape, the water exchange rate, the choice of production process, etc. The choice of production process is an emergent question asked by an emergent profession. How should an engineer decide whether or not the system should be an integrated multicomponent system, a polyculture system, or a single species intensive system?

Effluent management is another question that will frequently be faced by engineers. The management of effluent is a key to efficient production, and it is clear that effluent management is intimately linked to the choice of feed components and the choice of production process. Aquacultural engineers and the entire aquaculture industry will have to learn to look at the whole system instead of at one component at time, and keep in mind that a combination of pond, aeration equipment, and pumps is not necessarily a production system.

In the years ahead, we will see an increasing development of integrated aquatic farming systems for landed aquaculture, including integrated aquaculture-agriculture systems as well. Integrated production offers the best hope in improving efficiency in production resource utilization and in effluent management.

Aquaculture would do well to learn from the experience of agriculture, that in the design and operation of production systems biology cannot be separated from engineering. Agriculture has taught us that in order to develop a good mechanical tomato harvester, the breeder must breed a variety for the system, the agronomist must develop a cultural practice, and the

---

37. United States Department of Agriculture, Commodity Economics, Economic Research Service, *Aquaculture—Situation and Outlook Report* (Washington, D.C.: USDA, 1990 [AQUA-4]).

agricultural engineer—not an engineer who happens to like tomatoes—must design a harvester. The aquaculture industry must also recognize that aquaculture engineering is not simply a mixture of the traditional branches of engineering.

I would like to thank those who participated in the search for aquacultural engineering literature. Special thanks to D. Booth, R. Burley, R. M. Carson, J. E. Colt, R. B. Fridley, R. E. Garrett, S. R. Ghate, G. E. Kaiser, T. B. Lawson III, J. F. Muir, K. R. Murray, G. Oltedal, R. Piedrahita, G. D. Pruder, and F. W. Wheaton.

SECTION III

Publishing and the Classical Literature

10. Textbook Publishing in Agricultural Engineering

GEORGE E. MERVA
Department of Agricultural Engineering
Michigan State University

A. Brief History of Textbooks

Textbooks used in classroom instruction, along with reference texts, form an important part of the literature of agricultural engineering. The necessity of textbooks for agricultural engineering courses was recognized at the time the American Society of Agricultural Engineering (ASAE) was organized in 1907. In a presentation to the textbook committee of ASAE in 1985, Segerlind[1] pointed out that the first article in the 1907 *Transactions of the American Society of Agricultural Engineers* was "The Courses in Agricultural Engineering That Should Be Offered" by H. W. Riley.

The first several texts published were descriptive rather than technical. The first major publisher to offer an engineering oriented textbook series devoted to agricultural engineering was John Wiley and Sons in 1924, followed by McGraw-Hill in 1925. Since that time, many companies have played a major role in publishing and distributing agricultural engineering textbooks. (Table 10.3)

In the 1950s, Wiley made a major contribution to unify the agricultural engineering textbook field by publishing the Ferguson Foundation Series of Agricultural Engineering Texts. This series included texts in farm machinery, tractors and power units, soil and water, farm structures, food process engineering and electricity.

This series was the mainstay of agricultural engineering textbooks until the early 1970s when publishing costs and textbook prices began to rise. At the time, several books in the series suffered from low sales and thus did

---

1. L. J. Segerlind, "A Global View of Textbook Needs," Oral Presentation to winter meeting of ASAE, Chicago, Ill., 1985.

not warrant development costs associated with a new edition. Wiley's commitment to the Ferguson Foundation Series was reduced to texts in soil and water conservation and food process engineering. At about this time also, books in power and machinery and general agriculture formed the nucleus of McGraw-Hill's offering.

In the 1970s other publishers began offering texts in agricultural engineering. One of the most successful of these was AVI Publishing Company, initially known for their publications in food science, that introduced a number of agricultural engineering textbooks, some of which had previously appeared in the Ferguson Foundation Series. Texts in food engineering, farm machinery, tractors and power units, drying of agricultural crops, storage of agricultural crops, plant environment principles, electricity, and animal environment were among the more popular textbooks published by AVI. Van Nostrand Reinhold assumed the AVI label in December 1986.

Wiley, McGraw-Hill and AVI served the agricultural engineering profession well until the late seventies. At that time, the low volume of textbook sales resulted in a decision by AVI not to reissue some texts and to abandon agricultural engineering texts except in subject areas where volume demand was retained. Today, most publishers accept only text or reference books in agricultural engineering that have a wide appeal and therefore can produce volume sales. Such books are usually introductory in nature and are not suitable for upperclass undergraduate and graduate agricultural engineering education.

B. The Textbook Problem

Several surveys of textbook use in agricultural engineering have been performed in the past two decades and concerns have been voiced over the inadequacy and lack of availability of texts for instruction (see e.g., McDow[2] and Merva[3]). Texts in current use are not universally considered to be satisfactory by a majority of users. Even with the most popular three texts, Merva[4] found that only 50% of instructors using the texts rated them as adequate, while 48% found them inadequate. This situation has resulted in many schools using different texts and low sales volumes for each title.

2. J. J. McDow, *Present Availability of Textbooks for Mechanized Agriculture* (1973 [*ASAE Paper* 73-5553]).
3. G. E. Merva, "The Agricultural Engineering Textbook Dilemma," *The Agriculture Engineer* 62 (1981): 10–12.
4. G. E. Merva, *Proposal for an Agricultural Engineering Textbook Series*, mimeographed report to the ASAE A514 Textbook Committee, 1978.

In general, higher volume textbooks were being used in mechanization/management or service type instruction rather than for engineering courses.

The low volume of sales remains a serious problem for the agricultural engineering profession. The higher volume texts that are descriptive in nature appeal more to vocational-type training and do not serve the needs of the professional agricultural engineer. Conversely, books which contain the fundamental technical information needed by the professional agricultural engineer do not appeal to the mechanization/management-oriented curriculum offered in many agricultural engineering departments. The problem is compounded by the changing needs of the profession over the last two decades. This is best understood by considering the educational needs of agricultural engineers.

The present practice in agricultural engineering education is to utilize the same engineering science courses that are taught to mechanical and civil engineers, i.e., engineering mechanics, strength of materials, thermodynamics, and fluid mechanics. These courses are in addition to the basic science subjects of mathematics, physics, chemistry and computer science. Depending on the institution, soil mechanics and/or soil physics may also be required.

In addition to the engineering sciences common to mechanical and civil engineering students, agricultural engineers have needs that are unique to agriculture. The most obvious need results from the emphasis on the biological and physiological aspects of the life science system with which the agricultural engineer deals. While the engineering sciences provide a core of study that is essential for the educated agricultural engineer, the biological and physiological aspects simply are not covered in these courses.

The biological and physiological element that gives uniqueness to agricultural engineering, is also the need that is most difficult to satisfy. This in a large part is the reason for the lack of agricultural engineering textbooks. The relatively small number of agricultural engineering students in colleges throughout the United States and Canada at any one time (as compared to mechanical or civil engineering students) coupled with the fact that no other engineering discipline requires or utilizes extensive biological/physiological training results in a low demand for texts suited to agricultural engineering use.

Agricultural engineering faculty on whom the publishers must largely rely also contribute to the textbook shortage and quality in agricultural engineering. For many young agricultural engineering Ph.D.'s, education in the biological sciences is minimal. This is due in part to university or engineering requirements which determine the coursework required for an accredited undergraduate engineering degree. Virtually all accredited engineering colleges require 120 or more semester credit hours for a bachelor's degree. Once those required courses have been taken, few credit hours remain to educate a student in the engineering aspects of biology and physiology.

Most agricultural engineering teaching faculty received their education under these circumstances. This often results in a deficiency in the biological/physiological area outside their engineering specialty. This imperfection is sometimes remedied through graduate coursework or by self study in a chosen research area. A faculty member whose area is crop drying may be quite unaware of the physiological needs of roots as affected by tillage operations. Thus, to produce texts which are biologically and mathematically based and which could be used to teach basic general biological/physiological principles, becomes a difficult, time-consuming task. This is a significant problem because other activities take precedence and the texts remain unwritten.

It is also unfortunate that most available courses in the biological sciences are lacking a mathematical treatment of the subject so that available courses cannot be used to fulfill the engineering science requirement for accreditation and to meet the needs of engineers. These courses are, therefore, not incorporated into the curriculum because this would lengthen the time required to receive a degree.

This does not mean that the agricultural engineering student receives no training in the biological aspects. They do, because of the overall philosophy and the unique teaching approach which has developed in agricultural engineering departments. Rather than teaching basic mathematically-based biological and physiological principles which are then applied in the design sequence, concepts related to particular aspect are reserved for the design sequence of agricultural engineering courses.

This results in a treatment which is largely specific to crops and regions. Principles relating to the tillage, planting, harvesting and storage of potatoes are most likely to be studied at a university located in a region where potatoes are an important crop, while similar concepts for cotton will be taught in a school located in a cotton growing region. Some principles such as tillage are similar between regions while others such as storage and harvesting are not. Thus, a book written by a person whose primary area is potatoes will not find wide demand in the areas where cotton predominates.

A factor contributing to the textbook dilemma is the lack of incentive to produce a text with wide appeal. A textbook requires significant time and recompense is likely to be in terms of author prestige rather than direct financial return. Established faculty do not write while younger faculty find it more beneficial to perform research resulting in several publications rather than a book. Much research today is industry-funded with emphasis on immediate problems and solutions. This discourages preparation of textbooks or monographs.

C. The Situation in the 1980s

Most agricultural engineering departments offer more than one path leading to a degree. In addition to the professional engineering degree, a management-oriented program with the title of agricultural mechanization, agricultural engineering technology, or agricultural technology management is offered. Most university departments also offer service courses that are available to students enrolled in other departments. A two-year technology program aimed at training people for the electric power and machinery service fields may also be offered by some schools. In the section following, these programs will be distinguished as agricultural engineering (AE) and agricultural mechanization programs. While this section will discuss primarily the AE programs, the textbook situation also pertains to the other offerings. Information in this section was gathered from forty land-grant institutions that offer agricultural engineering and is based on a textbook survey performed by the Textbook Committee of the American Society of Agricultural Engineers and reported by Henry.[5]

Typical of most engineering programs, regardless of area, is the large number of courses and subjects that must be offered. In agricultural engineering these areas include freshman survey, computing, electrical power and processing, food engineering, power and machinery, structures and environment, soil and water engineering, and waste management as well as other miscellaneous courses. The latter group include courses which may be offered at all levels including graduate courses such as finite element analysis. Not included are shop courses which generally are not taken by the professional agricultural engineering students, and service courses or courses taught to two year students training for the machinery service area. Table 10.1 presents a synopsis of the overall text situation as it existed in 1985.

Table 10.1 is indicative of the present status of textbook use and publishing in agricultural engineering. Areas which have more recently developed, such as Design, Food Engineering, and Waste Management, have relatively few texts. Texts used must be purchased by students and are uniformly used throughout the United States based on regional variations and the expertise of the instructor.

5. Zachary Henry, *Strategies for Financing Textbooks: The Challenge of Providing Quality Educational Material* (1985 [*ASAE Paper* 85-5543]).

Table 10.1. Areas of study and text material required in forty Agricultural Engineering departments' professional agricultural engineering curricula

	Total students	Texts assigned	% of students[a]	Text required	% of students[b]
Power and Machinery[c]	900	18	90	16	73
Soil and Water	1000	27	94	22	83
Electric Power and Processing	1000	24	82	17	64
Structures and Environment	728	14	76	11	61
Miscellaneous	300	12	52	7	37
Computer Programming and Use	200	7	52	5	39
Waste Management	100	5	100	5	100
Design	100	9	100	5	82
Survey of Agricultural Engineering	300	7	69	7	69
Food Engineering	600	6	100	5	95

[a]Percent of all students taking a course in which the textbook was either required or was optional, based on all students including those in courses in which no text is used, or in which instructor's notes are used.

[b]Percent of students taking a course in which the textbook was required based on all students taking courses including those for which no text was used or instructor's notes were used.

[c]Numbers are rounded to nearest 100.

The power and machinery area is different with eighteen texts written by separate authors being used to teach 90% of the students enrolled in 1985. This averages to a potential of forty-five students per book or author. As column 4 shows, sixteen of these texts were actually required. For the authors of required books, the average was about forty-two book sales per year. The actual figure is probably lower.

The situation is somewhat better in the areas of soil and water, electric power and processing, and structures and environment, where the average is about thirty-eight book sales per author. There is a marked drop in computer programming and use. However, the figure is not indicative of sales of texts because many schools use courses taught by computer science departments rather than in the agricultural engineering departments.

Because the mechanization/management type curriculum is viable in many schools, and represents a substantial portion of enrollment, Table 10.2 presents data similar to that in Table 10.1.

Table 10.2 shows that the potential financial return for a text author is better for the mechanization/management area than for professional engineering. If the number of students is divided by the number of authors used, sales of as many as ninety copies per year (for electric power and processing) might be realized. This is still not adequate to entice authors to write for any reason other than professional continuation and recognition.

Table 10.2. Areas of study and text material required in forty Agricultural Engineering departments' mechanization/management curricula

	Total students	Authors used	% of students	Authors required	% of students
Power and Machinery	1800	34	95	30	87
Soil and Water	500	10	99	9	91
Electric Power and Processing	1000	12	100	11	98
Structures and Environment	600	10	76	9	61
Miscellaneous		3	100	3	100
Computer Programming and Use	300	6	68	3	44
Waste Management	100	3	100	2	86
Survey of Mech.	300	4	83	4	73
Food Engineering	100	4	100	4	100
Design	0	—	—	—	—

There were seventy-four individual publishers for the authors listed in the 1985 survey. Because of space limitations, only major publishers of texts are listed in Table 10.3, i.e., those who published six or more texts used in various courses taught by agricultural engineering faculty.

McGraw-Hill had twenty-three texts by different authors. However, only nine of these were required for specific professional agricultural engineering courses, and only five by mechanization/management curriculum. Wiley fared somewhat better, with ten texts required for engineering and six for mechanization/management courses. Nine of the John Deere series on machinery management and/or repair were required for various mechanization/management courses. While AVI had only eleven texts listed, eight were required in engineering courses and seven were required for mechanization/management type courses.

Table 10.3. Major publishers of agricultural engineering texts

	Total texts	Required AE texts	Required AM texts
McGraw-Hill	23	9	5
John Wiley & Sons	16	10	6
John Deere	12	1	9
AVI	11	8	7
Prentice-Hall	8	3	2
American Society of Agricultural Engineers	8	6	3
Iowa State University Press	8	5	3
Midwest Plan Service	6	2	2

Table 10.4 presents the major course areas in agricultural engineering, along with the number of author's texts, that were used in each area during the 1985 academic year. Column 3 gives the number of different publishers represented among the texts used for each course area. The power and machinery area used texts from eighteen different authors printed by eight different publishers indicating several publishers contributed more than one text to a course area. A similar situation existed for electrical power and processing, soil and water engineering, and structures and environment, while for food engineering, design, and waste management, each publisher contributed a single text, on the average.

Table 10.4. Courses offered, authors of texts used and number of publishers for professional agricultural engineering curricula

Course area	Number of authors	Number of publishers
Power and Machinery	18	8
Electrical Power and Processing	24	16
Soil and Water	27	18
Structures and Environment	14	10
Food Engineering	6	6
Freshman Survey	7	6
Computer Programming and Use	7	6
Design	7	7
Agricultural Waste Management	5	5
Miscellaneous	12	9

It appears in the traditional coursework areas of both the professional agricultural engineering curriculum and the mechanization/management type curriculum (see Table 10.5) that there are no clear-cut texts that apply across the United States. This situation is unfortunate because relatively few books are used per author, resulting in a low return to the author and the publisher. Because sales are low, new texts must be few, which raises the fraction of printing cost which must be recovered for each book sold. Finally, the fraction of development costs that must be prorated to each text must be large in order to recover costs. These conditions keep inventory costs high and the cost to each purchaser high, discouraging students from purchasing a non-required text.

D. *The Textbook Situation Today*

Significant changes have occurred in the availability of textbooks in the seven years since the survey reported by Henry was undertaken. In a recent

Table 10.5. Courses offered, authors of texts used and number of publishers for mechanization/management curriculum

Course area	Number of authors	Number of publishers
Power and Machinery	34	16
Electrical Power and Processing	12	8
Soil and Water	12	5
Structures and Environment	10	9
Food Engineering	4	4
Freshman Survey	4	3
Computer Programming and Use	6	4
Agricultural Waste Management	3	2
Miscellaneous	3	2

survey of thirty-eight universities in the United States and Canada conducted in preparation of this chapter, forty-nine publishers were found to be furnishing over 200 texts written by 136 authors. The survey yielded a listing of 445 agricultural engineering course titles along with the text being used for the course. If not given, the publisher for a text was located in *Books in Print* wherever possible. Of the 200 texts, there were fifty-seven for which no publisher could be found. This number included a number of unpublished texts being distributed to classes as photocopied material. In addition, fifty-six courses were being taught without benefit of text.

The number of publishers supplying three or more titles is given in Table 10.6. These publishers appear to be the primary suppliers of agricultural engineering texts.

Table 10.6. Primary publishers of agricultural engineering textbooks in 1990

Publisher	Titles in use	Locations or courses using text
John Wiley and Sons	17	92
ASAE	11	52
McGraw-Hill	11	16
AVI (Van Nostrand Reinhold)	10	63
Prentice-Hall	6	12
Iowa State University	5	16
Harper and Row	4	5
Midwest Plan Service	3	5
Academic Press	3	5
Elsevier	3	5

Table 10.7 presents the textbook titles, along with the authors and publishers, that are being used by three or more agricultural engineering departments in the United States and Canada. The disparity between the number of texts furnished by a publisher in Table 10.6, and the number of times a publisher is listed in Table 10.7 occurs because many publishers supply fewer than three texts, or the texts are used by fewer than three schools.

Table 10.7. Prevalent textbook titles in use in Agricultural Engineering departments in the United States and Canada in 1990

Title	Author, date	Publisher
Light Agricultural and Industrial Structures	Gordon L. Nelson et al., 1988	AVI
Principles of Farm Machinery	Robert A. Kepner et al., 1978	AVI
Tractors and Their Power Units	J. Bruce Liljedahl et al., 1988	AVI
Irrigation System Design: An Engineering Approach	Richard H. Cuenca, 1989	Prentice-Hall
Engineering Models for Agricultural Production	Donn Hunt, 1986	Iowa State University Press
Erosion and Sediment Pollution Control	R. P. Beasley et al., 1984	Iowa State University Press
Farm Power and Machinery Management	Donn Hunt, 1983	Iowa State University Press
Design of Small Canal Structures	U.S. Bureau of Reclamation, 1978	U.S. Government Printing Office
Drainage Engineering	James N. Luthin, 1973	Krieger Publishing
Midwest Plan Service Handbooks	MWPS	Midwest Plan Service
Physical Properties of Plant and Animal Materials	Nuri N. Mohensin, 1986	Gordon and Breach
Transport Processes and Unit Operations	Christie J. Geankoplis, 1983	Allyn and Bacon

E. The Future

Significant changes in the publishing of agricultural engineering texts are doubtful in the near future given the present belief of the profession that the agricultural engineering student can be educated using the same engineering science courses that are common to the mechanical and civil engineer. This concept is flawed for two reasons. Firstly, it presumes that agricultural engineering is purely a combination of mechanical and civil engineering. Secondly, it assumes that today's agricultural engineering undergraduates enter the university with the same agricultural background as their predecessors, a strong farm background.

The first presumption has a stronger basis than the second since certain principles of statics, mechanics, material science, thermodynamics, and fluid mechanics apply to the agricultural engineer as they do to civil and mechanical engineers. However, this premise along with the second assumption, ignores the uniqueness of the agricultural engineer, who works with living biological materials.

In the past, students entering agricultural engineering came to the discipline largely from the farm where they had already gained familiarity with the biological and physiological aspects. This is not so today. Even those coming from a farm may be from a specialty farm. Such an individual may have an in-depth background in one aspect but be totally lacking in another. An example might be a beef farmer's daughter with little or no appreciation of the biological aspects of vegetables.

The above relates directly to textbook publishing in agricultural engineering. The regional/crop specific approach to educating agricultural engineers, along with the need to include both the biological and physiological as well as the mechanical principles has resulted in instructors groping for the text that will fit their own expertise, interest, and situation.

Successful textbook publishing in agricultural engineering requires textbooks that have wide appeal. Such a text must cover basic principles that apply not only regionally, but across the profession regardless of location. This is unlikely to occur. Ignoring the biological/physiological educational needs of the undergraduate agricultural engineer may, at best, lead to maintaining the status quo, while, at worst, it may lead to the demise of the profession at the undergraduate level. Attempting to supply the agricultural engineering student with the unique aspect of the education while attempting to impart principles of design in the coursework. Design education is situation specific and must be built on basic scientific and engineering principles. If the biological and/or physiological aspect must be taught at the same time, they too will be situation specific. Thus, a book which is general enough to cover all aspects of a situation will not be specific enough to be used in all regions.

The only solution to the problem appears to be to introduce, at the engineering science level, mathematically-based coursework steeped in biological and physiological concepts to equip the agricultural engineering student with the basis upon which engineering design can be built. Such texts, if basic in nature, will apply across the entire profession rather than being of regional interest. Only then will textbook sales revenue adequately remunerate the author and the publisher and absorb developmental costs, without pricing the text beyond the student.

Textbook publishing in agricultural engineering today using traditional publishing practices is not cost efficient. New concepts must be developed to ease costs. Fortunately, publishing is at a crossroads and new concepts are forthcoming. Microcomputer-based software packages and a high degree of standardization in laser printing allow many text-preparation activities to occur in faculty offices. For example, the Postscript[6] format is a common output format for many software packages and a common input format for many laser printers. The industry's development and maintenance of such standards facilitates user production of camera-ready copy that includes equations, graphs and illustrations. Such equipment makes it possible to reduce textbook development costs.

The problem of printing a low volume text remains. Some publishers will print from camera-ready copy. As authors become more conversant with the available equipment, the market can expect to see more text material published directly from author's copy skipping the typesetting and some developmental costs.

The ASAE recently agreed to publish agricultural engineering textual material which has not found favor with traditional publishers. For such texts, the ASAE recommends camera-ready copy or copy which requires minimal developmental cost. Texts published by ASAE are distributed via the Society's membership and no effort is made to advertise or otherwise promote book sales. Prior to acceptance by the Society of a text, the material is reviewed by both the ASAE Textbook Committee, as well as several other reviewers selected from the Society's membership. Such texts match the quality of any other textbooks on the market.

---

6. Postscript is a public domain printing language. Implementation and Translation is a copyright process of Adobe and is found on many popular laser printers.

11. Publishing of the American Society of Agricultural Engineers

JAMES A. BASSELMAN
Former Manager of Publications
American Society of Agricultural Engineers

WILLIAM CHANCELLOR
Department of Agricultural Engineering
University of California, Davis

The Society as Publisher

JAMES A. BASSELMAN

A. History and Scope

Formation of the American Society of Agricultural Engineers in 1907 provided a focal point for a profession that until that year did not have a formal name. By 1906 the newly named "tractor" as a rival to steam power had appeared. The first tractor school had been held in 1906 at the University of Minnesota. The age of animal power was beginning to wane. Engineers and blacksmiths were replacing animal breeders. In *Seven Decades That Changed America*, Robert E. Stewart described the time as when "agriculture and engineering were on a collision course."[1]

J. Brownlee Davidson, a young professor at Iowa State College, felt the need for an exchange of views and techniques among those interested in teaching "farm mechanics." If you were an "agricultural" engineer at the time, there were as many as six engineering "Founder" societies available for joining—yet not one that could cover the diversity of agriculture's engi-

---

1. Robert E. Stewart, *Seven Decades That Changed America* (St. Joseph, Mich.: American Society of Agricultural Engineers, 1979).

neering requirements. Davidson stirred up interest among his counterparts at other colleges for a new society and an organizational meeting was held at the University of Wisconsin (Madison) in December 1907. Eighteen charter members formed an association they called the American Society of Agricultural Engineers. Its constitution read: "The object of this Society shall be to promote the art and science of engineering as applied to agriculture."[2] Prior to the formation of ASAE, publication in the science of "farm engineering" was either included in the agricultural literature or among that of the civil engineers or the mechanical engineers. Among the pre-ASAE agricultural publications that promoted the new power age were *Farm Implement News* (1882), *Implement and Tractor* (1876), and *Threshermen's Review* (1892), later incorporating *Power Farming*. The new society through its meetings became an important source of technical information and simultaneously established an official status for agricultural engineering literature.

In its first year of existence ASAE began publishing the *Transactions of the American Society of Agricultural Engineers*. These became a record of the Society's activities and carried printed copies of technical papers presented during meeting sessions. The president's annual address became the barometer of the fledgling organization's progress.

In 1914 a news sheet entitled the *Bulletin* was published several times a year. By 1916 it was expanded to a *News Letter* and published at more frequent intervals. In 1920 the official "journal" of the organization was launched: *Agricultural Engineering*. The magazine was complete with technical papers, detailed accounts of the society/affairs, and advertising. The journal replaced all other society publications except the *Transactions of the American Society of Agricultural Engineers*. These same technical papers appeared in both publications for a few years but soon *Agricultural Engineering* became established as the official mouthpiece. The *Transactions of the American Society of Agricultural Engineers* was discontinued in 1935.

Early promotional material released by ASAE made the following claim:

Publication of the *Transactions of the American Society of Agricultural Engineers* was begun consequent to the founding of the Society in 1907, as a means of disseminating material of permanent value for reference presented in the form of papers and reports at Society meetings. The *Transactions*, together with the Society's monthly journal, *Agricultural Engineering*, constitute the most complete library on agricultural engineering in existence.

2. ASAE, *Constitution*, 1907.

It was not until 1958 when the need for an exclusive technical publication resulted in expanding the society's publications by adding a new periodical entitled *Transactions of the ASAE*. Although the name is similar to the society's first publication in 1907, the new publication was devoted to technical topics only.

The extent to which the ASAE grew and attracted participation by agricultural engineers throughout the world, no doubt, helped the society's publication program attain worldwide recognition and acceptance. Since 1950 an increasing number of agricultural engineers from countries other than the United States became members of the society and participated in the organization's national meetings. Since its beginning, ASAE has scheduled two national meetings—one in the summer and one in the winter. Attendance records from these meetings document the fact that agricultural engineers from all over the world attend in great numbers. The American Society of Agricultural Engineers is held in high regard because it provides its members several benefits, a major one being publishing opportunities. There has been a close tie between the ASAE and the Canadian Society of Agricultural Engineering (CSAE) since its founding in 1959. In 1986, the CSAE affiliated with ASAE, increasing the effectiveness of both organizations, particularly in publications.

The publication program of ASAE began with one editor, who selected material for publication mostly from the society's national meetings. In 1955 when more papers were presented than could be published, a publication committee was appointed to assist the editor in making decisions for approval by the ASAE Board of Directors. Additional committees were formed as new publications were introduced. The peer-review committee members were appointed to function as support editors for the ASAE Technical Editor. One support editor was assigned to each technical division of ASAE. When the peer-review system was introduced in 1960, Division Editors represented (a) Power and Machinery, (b) Soil and Water, (c) Farm Structures and (d) Rural Electrification. As an indication of change in the profession and areas of new interest the current Division Editors are (a) Power and Machinery (including Forest Engineering), (b) Soil and Water, (c) Structures and Environment, (d) Electrical and Electronic Systems, (e) Emerging Technologies, and (f) Food and Process Engineering.

The current ASAE staff consists of three editors: (a) creative publications, (b) technical publications, and (c) membership publications. The Technical Editor is responsible for *Transactions of the ASAE, Applied Engineering in Agriculture*, monographs and textbooks. Each editor has a support committee and the Technical Editor has six Division Editors for review and material selection. Each Division Editor is recognized as editor of a subset publication. One subset is published for each Division of ASAE by collecting all technical articles from that division and published in *Transactions of the ASAE* and *Applied Engineering in Agriculture* during each year and bound as a separate publication.

The fact that ASAE set up a publishing program from its inception is perhaps its strongest determinant in establishing itself as a bellwether force for those to follow. *Agricultural Engineering* became the premier source for the latest developments in coverage of news, new products, technical papers and announcements of meetings and events that chronicled the development of the profession. The ability to publish was a great benefit of membership.

After World War II, society membership grew and attendance increased. Concurrently the number of technical papers increased causing a demand for more publishing capability. With the 1958 introduction of the *Transactions of the ASAE*, review committees were set up to select and publish the most significant developments presented at meetings. Universities (academics) insisted on peer reviews to help gain stature in the profession. Soon a peer-review system was established.

To maximize its publication program by obtaining additional support, ASAE instituted a surcharge or page charge for articles that had been accepted for publication through the peer-review system. ASAE's governing body felt that publication was a mutual responsibility of (a) subscribers, (b) members of the profession, and (c) the organization or project producing information of significance or of lasting value. Subscribers pay for value they receive through subscriptions. Members of the profession have an obligation to publish worthwhile material, so it was felt that part of the cost should be borne by ASAE. Thirdly, sponsors of the research or activity reported also have the obligation to see that the results are published and made accessible to those most likely to use the information. This is a responsibility of public-service agencies. Supporters of the page-charge policy claim major benefits. More information can be published, mutually benefitting sponsors, ASAE members, subscribers, and the profession. Shared author per unit costs are less because many costly duplications are avoided, such as mailing lists, requests for bids, mailing facilities, editorial staffs, printing facilities, and similar manuscripts from various parts of the country.

Other benefits include the effectiveness of information storage and recall because information is recorded in a central location, indexed for retrieval, and made much more accessible than if reported through several sources in various styles and scattered geographically. Distribution of results of research or development is optimized through ASAE publications which go to those reference libraries and individuals especially interested in technical details. Especially important for those who publish in the peer-review program is the prestige that accrues to the authors and affiliated institutions by having a manuscript screened and accepted by a professional society.

Conference Proceedings

Establishment of a program for publishing conference proceedings happened in an evolutionary manner. The society held a special conference titled Materials Handling of Agricultural Products on the Iowa State campus in 1958. The conference had a two-fold purpose: (a) it served to bring together agricultural engineers to compare experiences, and (b) to pool sources of information on the topic. The publication committee of ASAE, which was instrumental in planning the conference, selected the presentations from the conference as material for the monthly publication, *Agricultural Engineering*. As a result the September issue that year became ASAE's first conference proceedings. Reprints of that issue are still in demand. It was soon apparent that conferences grew in importance in society activities, and conference proceedings became an important part of the publication program.

Over seventy ASAE conferences on nearly as many subjects have been held—resulting in publication of conference proceedings. Topics range from "Advances in Drainage" to "Robotics and Intelligent Machines in Agriculture." Papers presented at these conferences, although not peer-reviewed by the society's reviewing system, are prescreened by each conference planning committee and are valuable contributions to the literature. Conference proceedings have special value because they are timely and include current information from many sources in one package. Each paper is indexed in the society's annual index.

Technical Papers

Each year nearly 1,500 technical papers are presented at ASAE national and regional meetings. All papers presented and copies supplied to ASAE by the author are indexed and entered in the society's database. The society maintains a file on all papers presented since 1954 and can supply a hard or microfiche copy upon request.

Indexing

Throughout the years the annual index of ASAE publications has served as the chronicle or official record of the society. ASAE maintains a copy of each annual index from the one in 1907 for the *Transactions of the Ameri-*

can *Society of Agricultural Engineers* to the current one that includes a variety of ASAE publications, reports from international organizations, and several society journals and publications. In 1961, Carl W. Hall combined ASAE indexes with those from *Canadian Agricultural Engineer, Landtechnische Forschung*, and *Institution of Agricultural Engineers Journal and Proceedings* of England to form a cumulative *Agricultural Engineering Index* covering the period from 1907 to 1960. This index has been continued on a ten-year interval. By 1986, indexed material mounted to the point that it was decided to publish the cumulative index on a five-year interval. By then the number of publications included CIGR (Commission Internationale du Génie Rural), *Agricultural Mechanization in Asia, Africa and Latin America, Agricultural Engineering Australia*, and *The IAMFE Journal by the International Association on Mechanization of Field Experiments* in Norway.

ASAE publications are indexed regularly by the National Agricultural Library (*AGRICOLA*), *Engineering Index, Biological and Agricultural Index*, Commonwealth Agricultural Bureau (CAB) and *Chemical Abstracts*.

Applied Engineering in Agriculture

In 1985 ASAE introduced a new peer-reviewed publication, *Applied Engineering in Agriculture*, as a spin-off from *Transactions of the ASAE*. Its purpose was to provide a means of disseminating peer-reviewed, applications-oriented articles as contrasted with research articles. Material published represented original, important contributions to applied agricultural engineering literature that have current, lasting, or apparent future value. The publication was designed to serve as a communications link among practicing agricultural engineering consultants, mechanization people, agribusiness people, and cooperative extension service staff. This publication added a new audience for publishing and using information on applied science and engineering.

Other Publications

Each year ASAE Standards, Engineering Practices and Data are published in the *ASAE Standards Book*. The first ASAE standards were published in the journal or transactions as regular business transactions of the society. In 1954 the *Agricultural Engineers Yearbook* was introduced as the annual record of all current ASAE standards. Included in the publication were membership and committee rosters, constitution and bylaws, award winners, product directory and pertinent society information. Throughout the years the number of standards increased and a separate *ASAE Standards Book* is being published. The society's membership roster is published also as a separate publication. The committee roster is included in both. The product directory has become a special issue of *Agricultural Engineering*. Several bibliographies covering specific subject areas were published after 1970.

The monograph series of the society began with *Compaction of Agricultural Soils* in 1971. By definition a monograph is an in-depth treatise limited to a single subject averaging from 200 to 400 pages. The publication format lends itself for use as a reference or textbook. To date ASAE has published eleven such monographs—the latest *Knowledge Engineering in Agriculture*. Most popular is *Design and Operation of Farm Irrigation Systems*, published in 1981, which has had worldwide requests.

A textbook entitled *Design in Agricultural Engineering* by Christianson and Rohrbach was published by ASAE in 1986. This was the first of a series. The third textbook entitled *Engine and Tractor Power* by Carroll Goering was published in 1989. The ASAE Textbook Committee is encouraging authors to submit prospective titles for consideration.

A *Distinguished Lecture* series was established in 1975 through a grant from Deere and Co. The fund provides for traveling expenses and a stipend for a recognized authority to present a lecture annually during the Winter Meeting of ASAE on a design problem dealing with farm tractors and self-propelled agricultural equipment. The lecture is published and distributed at no cost to agricultural engineering students who attend. The lecture then becomes part of the society's literature and is marketed and indexed by ASAE. Each lecture contains material from a forty-five-minute presentation to provide students with industry experiences.

B. Roadmap for the Twenty-First Century

A Strategic Plan for the Society was launched with a four-day planning conference in 1984. A broad, cross-section of members attended with representation from all elements of the society's structure. Over sixty members participated. The overall challenge of the planning conference was to identify the type of society the members want it to be by the society's 100th anniversary in 2007. The Strategic Plan, called *Project 100—Roadmap for the 21st Century*, was approved by the society in 1986. After key objectives were selected and priority objectives were assigned, six missions were established. First among them was "*Information and Technology*—Be a world resource for engineering information, technologies and ideas." The mission's vision depicts "the Society as a complete source for quality engineering information for agriculture worldwide. It leads the world in its expertise of engineering in agriculture." The priority objective is to make available keyword, worldwide indices of applicable quality information within five years. (The plan is to include 90% of quality, worldwide information within ten years.) Since 1986 several agricultural engineering publications have been added to the society's database and a decision shall be made soon to determine if the mission is on target.

The ASAE database is indexed by keyword, author, title, citation and company or institution location. An annual printout is mailed to all members of the society and to subscribers of the society's three periodicals— *Transactions of the ASAE*, *Agricultural Engineering*, and *Applied Engineering in Agriculture*. The database includes ASAE technical papers, conference proceedings, monographs, distinguished lectures, textbooks, historical books, and issues reports, as well as the three periodicals.

In pursuit of the Project 100 mission objective to establish ASAE as the source for the worldwide database, other publications have been included in the annual indexes. Other periodicals indexed are: *Canadian Agricultural Engineer* by the Canadian Society of Agricultural Engineering, *Journal of Agricultural Engineering Research*, by the British Society for Research in Agricultural Engineering, *Agricultural Engineering Australia* by the Agricultural Engineering Society (Australia), *The Agricultural Engineer* by the Institute of Agricultural Engineers in England and *Agricultural Mechanization in Asia, Africa and Latin America* by the Farm Machinery Industrial Research Corp. in Tokyo. Technical reports and conference proceedings include those from Commission Internationale du Génie Rural (CIGR, formerly in Paris but now in Brussels), Institute of AE-IMAG, Royal Netherlands Meteorological Institute De Bilt, Netherlands, and the U.S. Committee on Irrigation and Drainage.

The five other missions selected for Project 100 are:

Membership: ASAE to be a focal point for people worldwide with interests in diverse engineering technologies applicable to agriculture. To meet this charge the society must be recognized worldwide as the home for experts in engineering topics relating to agriculture, natural resources, and associated industries. Priority objectives in reaching this goal are (a) incorporate emerging technologies or interest areas into the society, (b) expand international connections, and (c) provide membership services that meet the needs of members.

Standards: ASAE to be a world center for engineering standards and practices for agriculture. To meet this charge the society must be known to be a timely developer and publisher of complete and up-to-date Standards and Engineering Practices, and a leader in developing and gaining acceptance of international standards. Priority objectives in reaching this goal are (a) identify and meet areas of need and opportunity for new and revised standards, and (b) expand world awareness and provide increased leadership in international development and acceptance.

Continuing Education: ASAE to offer and support educational and development opportunities worldwide. To meet this challenge the society must offer and support continuing education and development programs for its members and others with engineering interests in agriculture.

Communications: ASAE to support the missions of the society with efficient, worldwide communications systems. To meet this goal the society must have modern communications systems that provide to its members and customers worldwide information on a timely, cost-effective basis and that the society be a spokesman for the profession beyond its membership.

Vitality: ASAE to sustain continuity and vitality of the society. To meet this goal the society must be a dynamic, growing organization of members worldwide, have strong financial resources and be led by an effective management staff.

Publishing Influence

Perhaps partly by choice and partly by fortuitous opportunity ASAE was at the right spot at the right time to establish itself as prime mover in a newly named profession. The publication program of the society was patterned after those of the already established engineering societies, and is being followed by many other agricultural engineering organizations worldwide. An impressive test of ASAE's publishing influence is to observe the number of times that ASAE publications are referenced in other agricultural engineering literature worldwide. It is rare indeed to find an article without one ASAE citation.

Subject-Matter Areas of Technical Articles Appearing in ASAE Periodicals

WILLIAM CHANCELLOR

The material published in ASAE periodicals constitutes a major portion of the body of knowledge which defines the field of agricultural engineering as practiced in the United States. A bibliographic file has been created to assist individuals in locating pertinent information. It covers all technical articles in the periodicals of the American Society of Agricultural Engineers. The database *Computerized Agricultural Engineering Index*[3] lists 9,889 technical articles which appeared in the years 1954 through 1989 in the ASAE periodicals: *Agricultural Engineering, Transactions of the ASAE*, and *Applied Engineering in Agriculture* (Tables 11.1 and 11.2). The *Paper Series* of presentations at ASAE annual meetings are not included in the version of this Index analyzed here. For each of these articles, up to four keywords have been assigned on a free-form basis. The keywords were selected to represent the contents of the article when the title would not be present, thus many are the same as those in the title. The *Computerized Agricultural Engineering Index* has over 6,000 different keywords. Terms were not pre-coordinated although synonyms were not used as keywords. All entries went through a review process to assure unanimity of subject and bibliographic coding.

The existence of the *Index* presented the possibility of analyzing the data to develop information on the publishing history of various subject-matter categories. The objectives for performing such analyses are to gain some insight about which subject matter areas: (1) tend to be covered by a large number of published articles; (2) have been of concern for long or short periods of time; and (3) are "new" or "old" topics.

The analysis was done by: (1) selecting certain keyword roots to represent subject matter areas; (2) counting the number of articles in the *Index* containing the keyword roots; (3) finding the mean publication year of the articles associated with each keyword root; and (4) computing the standard deviation of the publication year data associated with each keyword root. The mean of the publication year provides an indicator as to whether a topic may be an "old" or "new" area, whereas the standard deviation of the publication year data could serve as a measure of degree to which publishing activity in a given topic area has been sustained over time. If data are normally distributed, the mean plus or minus the standard deviation should define a range within which approximately 70% of the input data could be expected to fall.

Of the over 6,000 keywords, 650 were selected that might have meaning to people in the agricultural engineering field when the keyword was presented individually out of context. For each of the 650 keywords, the entries in the *Index* were examined to isolate those that contained the keyword root either in the title or as a keyword or keyword part. From the information obtained the list of keyword roots was reduced from 650 to 400 so that only those associated with ten or more published articles were considered.

3. William J. Chancellor, *Simple Search Systems for Published Articles on Agricultural Engineering* (St. Joseph, Mich.: American Society of Agricultural Engineers, 1989 [*ASAE Paper* no. 89-3514]).

American Society of Agricultural Engineers   185

Table 11.1. Number of technical articles in ASAE periodicals by year

Publication year	Number of ASAE technical articles
1954	109
1955	84
1956	73
1957	70
1958	105
1959	119
1960	155
1961	149
1962	150
1963	174
1964	220
1965	250
1966	351
1967	290
1968	307
1969	303
1970	296
1971	311
1972	297
1973	328
1974	309
1975	277
1976	265
1977	280
1978	264
1979	314
1980	340
1981	354
1982	376
1983	406
1984	418
1985	445
1986	396
1987	413
1988	423
1989	468

Because the database covered a period of only thirty-six years, during which the number of technical articles published annually ranged from 70 to 468, the very simple method of analysis chosen produced results which were subject to the characteristics of this particular set of data. To illustrate the nature of the interaction between the analysis method and the data set characteristics, Tables 11.3 through 11.5 were prepared. Each of these tables shows the mean values of the three analysis parameters for ten (forty

Table 11.2. Technical articles in selected agricultural engineering periodicals (1980–89)

Publication	80	81	82	83	84	85	86	87	88	89
Trans. ASAE	308	322	340	370	361	378	301	308	303	327
Applied Engineering in Agriculture	—	—	—	—	—	22	56	62	62	102
Agricultural Engineering	32	32	36	36	57	45	39	43	58	39
ASAE Total	340	354	376	406	418	445	396	413	423	468
Agr. Mech. in Asia, Africa, and Latin America	59	54	55	59	57	54	57	60	57	60
Can. Agr. Eng.	34	21	26	32	38	24	29	34	52	43
J. Agr. Eng. Res.	39	49	49	47	54	54	78	74	83	73

count) deciles of the 400 keyword root listings—ordered according to number of articles per keyword root (Table 11.3), mean date of publication (Table 11.4), and standard deviation of publication year (Table 11.5).

When the 400 records were ordered according to the number of articles per keyword root (Table 11.4), there was found to be little effect of this parameter on mean publication year, with the exception that for keyword roots having over 100 publications, the mean publication year tended to decrease with increases in the number of publications per keyword root because larger numbers tended to occur only when the subject had been one of high publication activity from the earlier portions of the time span exam-

Table 11.3. Decile means for keyword root listings, sorted by number of articles per keyword

Decile[a]	No. of articles/keyword		Publication year	Std. dev. yr. of pub.
	Mean	Range		
1	10.8	9–12	1977.1	7.90
2	14.1	13–15	1976.6	7.73
3	17.9	16–20	1977.7	7.32
4	23.7	21–26	1976.5	7.56
5	29.8	27–32	1976.4	7.61
6	38.3	33–43	1976.9	7.83
7	53.3	44–64	1976.6	8.05
8	80.5	65–99	1976.6	8.36
9	128.5	100–159	1975.9	8.68
10	325.0	160–940	1974.4	8.63

[a]Forty listings per decile.

Table 11.4. Decile means for keyword root listings, sorted by date of publication

Decile[a]	Publication year		Number of articles/keyword	Std. dev. year of pub.
	Mean	Range		
1	1970.3	1967.0–1972.1	39.5	8.95
2	1973.1	1972.1–1973.8	63.2	9.00
3	1974.4	1973.9–1974.7	99.0	9.12
4	1975.3	1974.7–1975.9	90.0	8.93
5	1976.3	1975.9–1976.8	117.3	8.32
6	1977.2	1976.8–1977.7	96.6	8.10
7	1978.0	1977.7–1978.3	91.6	7.71
8	1978.8	1978.3–1979.4	50.3	7.78
9	1980.4	1979.5–1981.2	51.6	7.07
10	1981.1	1981.3–1987.2	25.5	4.78

[a]Forty listings per decile.

ined. For this same reason, it was found that the standard deviation of the year of publication tended to increase slightly with the number of publications per keyword root when this parameter was above twenty (Table 11.3).

When the 400 listings were ordered according to the mean publication date (Table 11.4), the number of articles per keyword root tended to be low for subjects with very early popularity and very late popularity because the truncation of the database at 1954 and 1989 excluded articles which might otherwise have been counted in a continuum of data. This same truncation

Table 11.5. Decile means for keyword root listings, sorted by standard deviation of publication year

Decile[b]	Std. dev. of pub. year[a]		Number of articles/keyword	Pub. year
	Mean	Range		
1	4.27	1.57–5.45	27.9	1980.9
2	6.15	5.46–6.62	33.5	1979.3
3	6.95	6.62–7.29	71.8	1977.9
4	7.54	7.31–7.77	73.0	1976.8
5	7.97	7.77–8.19	83.8	1976.4
6	8.41	8.19–8.62	66.3	1975.8
7	8.85	8.64–9.04	70.4	1975.6
8	9.29	9.06–9.50	152.6	1975.3
9	9.76	9.51–10.06	81.1	1973.9
10	10.44	10.07–12.57	66.7	1972.8

[a]Standard deviation of publication year indicates the time span of publication activity in the subject-matter are represented by the keyword root.
[b]Forty listings per decile.

effect was responsible for reduced standard deviation of publication year values particularly for the topics associated only with recent articles.

When the 400 listings were ordered according to the standard deviation of the publication date (span of publishing activity for each subject matter area) (Table 11.5) it was found that the keyword roots with low numbers of associated publications (mainly the most recently popular topics) had the lowest standard deviation values because their span of publishing activity had been limited to the last few years. Keyword roots representing topics that had strong publishing activity over the full period of the database tended to have high numbers of articles per keyword root associated with high values for the standard deviation of publication year. The highest standard deviation values were associated with keyword roots which had a bimodal popularity (numerous articles published in the 1950s and 1980s, for example). The mean publication year decreased steadily with increases in the standard deviation of the publication year. Part of this latter effect was due to the truncation of the database and the increasing number of publications per year as time progressed.

The mean parameters for the articles associated with the 400 keyword roots having 10 or more articles per keyword root are:

Total number of article keyword links	9,834
Mean year of publication	75.27 (1975)
Standard deviation of publication year	9.39 Yrs.

The above standard deviation was comparatively high because the full database time span was represented.

The findings relative to the individual keyword roots are presented in Table 11.6. Also of interest in Table 11.6, in which the database is presented ordered according to the number of articles per keyword root, are the listings for very high frequency topics. The topics with over 200 associated articles tend to be basic elements of the agriculture of North America. A unique commentary on the role of agricultural engineers is the appearance of "system (not expert)" as the second most frequently used keyword root, preceded by "soil" and followed by "water." The presence of Soil, Water, Irrigation, Erosion, Drain, Land, Runoff and Tillage (in order of decreasing frequency) in this "above 200" group indicates the significant component of soil and water articles in the published agricultural engineering literature.

When the database records are considered according to the mean year of publication, both the very early topics and the very late topics are of particular interest. Keyword roots such as "wafer (hay)," "pellet," and "radial (ply tires)" were signs of the times. It is not surprising to see keyword roots

Table 11.6. Database listings ordered by number of articles per keyword

Keyword root	Number of articles	Mean pub. year	Pub. span	Keyword root	Number of articles	Mean pub. year	Pub. span
BALE (BIG, ROUND)	9	1982.88	3.69	TRI-AXIAL	11	1979.90	8.42
CARRIER	10	1981.80	5.47	BONE	12	1977.75	7.42
CHISEL	10	1975.30	6.04	BROODER	12	1979.41	10.08
ELEVATOR	10	1979.40	10.66	CABBAGE	12	1970.75	8.30
LUBRICATION	10	1974.10	10.82	COMPOST	12	1978.75	5.58
MACADAMIA	10	1978.40	3.83	CRUSH	12	1971.58	11.66
MOUNTAIN	10	1976.30	8.56	CURTAIN	12	1976.66	7.84
ORGANIC (MATTER)	10	1984.30	2.87	DAM	12	1974.66	12.57
ORGANIZATION	10	1977.10	9.72	DEHYDRAT....	12	1974.75	7.44
PAPAYA	10	1975.30	7.35	DESICCANT	12	1982.00	5.10
REPAIR	10	1979.20	7.98	GLUE	12	1969.33	9.52
SEPTIC (TANK)	10	1980.70	7.12	SHIPPING	12	1982.08	5.03
SOCIAL	10	1978.50	9.62	WEEVILS	12	1977.50	4.76
STILL	10	1980.20	10.37	ACCIDENTS	13	1973.53	7.85
TRELLIS	10	1976.90	10.13	BIBLIOGRAPHY	13	1970.46	9.24
WEAR	10	1970.20	10.02	COOK	13	1980.76	6.81
X-RAY	10	1980.80	8.66	CROSS-FLOW	13	1973.23	5.48
AEROSOL	11	1979.09	8.85	LOESS	13	1976.30	8.82
CALF	11	1978.45	8.90	MACERATION	13	1981.84	7.65
DIAPHRAGM	11	1986.18	3.84	MOLD (NOT MOLDBOARD)	13	1973.76	9.77
DURABILITY	11	1968.00	9.81	NUCLEAR	13	1975.15	10.36
ISOTOPE	11	1969.90	7.63	PALM	13	1969.38	8.70
ONION	11	1978.63	6.47	RAPE-SEED	13	1983.61	6.26
RAISIN	11	1975.63	9.28	STRIP-MINE	13	1980.61	3.73
REFRIGERATION	11	1974.90	10.25	THINNING	13	1978.00	7.80
REMOTE SENSING	11	1977.54	6.79	BIOGAS	14	1983.57	3.53
SHADE	11	1968.18	8.43	EXHAUST	14	1976.71	8.26
SIGNAL	11	1977.00	7.31	GULLY	14	1972.85	12.41

Table 11.6. (Continued)

Keyword root	Number of articles	Mean pub. year	Pub. span	Keyword root	Number of articles	Mean pub. year	Pub. span
LASER	14	1981.64	7.55	FELLER	16	1983.75	3.53
OPERATOR	14	1972.85	8.53	FERMENT . . .	16	1983.00	3.53
OYSTER	14	1978.85	5.81	LABOR	16	1973.18	8.52
ROBOT	14	1985.92	2.20	NAIL	16	1971.12	12.26
SALINE, SALT (SOIL)	14	1973.78	7.89	PEPPER	16	1979.93	8.17
SAW	14	1979.35	8.43	SWEET-SORGHUM	16	1985.00	2.80
SIEVE	14	1981.21	7.14	TELEMETERING	16	1976.62	6.48
TEACHING	14	1978.35	8.65	ASPARAGUS	17	1974.52	8.19
ALGORITHM	15	1983.20	5.62	CLOD	17	1977.00	8.36
CIRCUIT	15	1976.13	7.97	CONVECT	17	1976.29	9.26
CLOUD	15	1976.93	7.24	HYDROSTATIC	17	1974.70	7.67
COFFEE	15	1967.40	4.37	OPTICAL	17	1977.70	6.37
CULTIVAT . . .	15	1973.31	8.56	RISK	17	1983.82	5.82
EXTRUSION	15	1978.46	5.11	AQUA (CULTURE), (TIC)	18	1978.72	6.22
FWD (FWA)	15	1975.60	8.19	CHARGE	18	1976.55	8.31
MOWER (NOT LAWN)	15	1970.86	9.36	CHERRY	18	1971.22	8.06
NEMATODE	15	1978.33	6.82	DEWATER	18	1977.50	6.58
PTO	15	1974.73	9.34	HOME	18	1974.22	7.89
RADIAL (PLY TIRES)	15	1967.46	9.77	IMAGE	18	1985.22	5.02
RUBBER	15	1979.13	8.45	ISSUES	18	1980.44	6.66
SPILLWAY	15	1975.13	12.21	ORIFICE	18	1977.94	11.12
THREE-POINT HITCH	15	1978.80	9.17	TRACERS	18	1973.38	9.77
TROPICS	15	1979.53	8.17	BOLL	19	1976.15	5.79
TURKEYS	15	1976.00	9.88	BRUISE	19	1978.63	8.69
CHOPP	16	1973.50	9.97	EMITTER	19	1981.31	4.52
CUBE	16	1972.93	5.06	SNOW	19	1974.94	9.51
DECISION	16	1983.12	6.62	SONIC (INCL. ULTRA)	19	1972.94	7.84
EMISSION	16	1980.43	4.38	SUBSOILER	19	1978.37	11.04

Table 11.6. (Continued)

Keyword root	Number of articles	Mean pub. year	Pub. span	Keyword root	Number of articles	Mean pub. year	Pub. span
THREE-DIMENSIONAL	19	1980.36	7.97	COLUMN	24	1975.95	8.31
ARID	20	1977.20	9.01	DEMAND	24	1978.33	8.61
BEAM	20	1970.15	8.29	FOAM	24	1975.29	6.85
BLUEBERRY	20	1978.15	5.95	JUICE	24	1980.00	5.47
CHIP	20	1979.90	6.62	MATURITY	24	1977.00	6.93
VISION	20	1986.05	4.13	METAL	24	1972.87	10.22
AXIAL	21	1976.04	8.62	ROAD	24	1975.12	8.51
CENTRIFUGE	21	1972.33	7.87	ORCHARD	25	1977.88	6.70
COMBUST	21	1982.09	8.02	RHEOLOGY	25	1978.36	6.54
EXPERT (SYSTEMS)	21	1987.19	1.57	RIVER	25	1970.76	8.47
FARMING	21	1974.23	11.37	SUGARBEET	25	1977.72	7.45
LAWN (TURF)	21	1975.90	8.28	WEED	25	1974.68	9.41
OIL (VEGETABLE)	21	1983.42	5.54	COLD	26	1971.80	9.51
SILT	21	1980.80	7.78	DITCH	26	1970.80	8.92
SLUDGE	21	1982.19	6.53	MEAT	26	1978.03	6.48
TIMBER	21	1973.95	8.59	NURSERY	26	1980.65	6.30
ATOMIZ	22	1979.22	8.94	URBAN	26	1974.61	7.76
COMMINUTION (GRIND)	22	1969.63	8.82	BARK	27	1978.86	7.72
CUCUMBER	22	1975.36	6.62	BEET	27	1978.14	7.32
PINE (NOT PINEAPPLE)	22	1980.54	7.10	CHICK....	27	1972.96	9.04
DISPOS . . .	23	1973.21	6.95	GRAPE	27	1972.33	7.50
HONEY	23	1971.56	8.96	ORANGE	27	1977.59	5.45
METHANE	23	1982.08	3.87	ROCK	27	1978.88	8.08
SLURRY	23	1976.08	6.95	TRACKS (CRAWLER)	27	1975.59	10.25
WORLD	23	1974.69	8.18	WELLS	27	1970.00	9.73
ANALOG	24	1970.04	5.09	ATMOSPHERE	28	1969.53	8.92
BORDER	24	1973.70	9.41	CABLE	28	1982.03	5.01
C-O-TWO	24	1979.75	8.21	ODOR	28	1975.92	5.10

Table 11.6. (Continued)

Keyword root	Number of articles	Mean pub. year	Pub. span	Keyword root	Number of articles	Mean pub. year	Pub. span
POLE	28	1974.85	10.60	PROTEIN	32	1978.03	4.84
RILL	28	1982.64	4.85	BIOLOG	33	1973.27	6.54
STONE	28	1977.07	10.93	BURN	33	1979.00	8.93
CONFINEMENT	29	1971.93	7.36	HUMAN (FACTORS)	33	1971.87	8.60
HAND (NOT HANDLING)	29	1979.24	6.33	LINT (COTTON)	33	1982.06	4.08
OSCILLAT . . .	29	1972.10	7.49	PHOTO (NOT.SYN OR.VOLT)			
PADDING	29	1981.37	7.17	WAFER (HAY)	33	1972.87	7.41
PEACH	29	1977.00	7.22	STRAW	33	1967.00	6.72
PROGRAMMING	29	1974.75	8.72	COAT	34	1978.35	6.66
SWEET-POTATO	29	1976.75	6.26	HERBICIDE	35	1979.74	8.95
ALTERNATE (IVE)	30	1980.93	5.73	SKIDDER	35	1978.14	9.23
CAB	30	1974.70	7.78	TRANSPLANTING	35	1978.31	7.01
CEMENT	30	1975.86	9.95	AG-ENG	35	1982.14	6.45
ELECTROSTAT	30	1977.10	7.47	MARKET	36	1979.02	7.45
NO-TILL	30	1981.86	4.99	NUT	36	1977.27	7.42
STATISTICS	30	1980.86	9.85	ROOF	36	1976.61	7.87
STORM	30	1979.70	9.07	TRANSDUCERS	36	1973.25	9.68
VINE	30	1972.13	10.47	AERIAL	36	1976.91	9.87
CANNING	31	1975.00	9.04	DRIP	37	1978.00	7.77
COLOR	31	1977.33	7.55	EQUILIBRIUM	37	1983.48	4.89
CONCRETE	31	1969.09	7.55	FAILURE	37	1976.67	9.96
MICROWAVE	31	1980.58	6.48	NOZZLE	37	1977.67	9.07
MULCH	31	1970.87	9.06	CENTER-PIVOT	37	1979.86	7.88
ROOT	31	1979.51	4.42	SENSOR	38	1979.15	6.58
TRICKLE	31	1980.58	5.24	SPECTR . . .	38	1983.00	5.87
CANOPY	32	1981.09	7.49	STEEL	38	1975.63	8.17
COW	32	1974.65	10.46	AIRCRAFT	38	1974.26	9.77
FLOWER	32	1980.59	6.50	FISH	39	1975.64	9.57
PELLET	32	1967.68	7.85		39	1976.87	5.46

Table 11.6. (Continued)

Keyword root	Number of articles	Mean pub. year	Pub. span	Keyword root	Number of articles	Mean pub. year	Pub. span
NOISE	39	1974.92	4.74	PLANS (PLANNING)	50	1973.96	9.18
DIELECTRIC	40	1977.57	8.08	ALCOHOL	51	1982.68	4.86
DIGIT	40	1977.45	9.12	DIESEL	51	1983.49	4.08
EDUCATION	40	1974.65	9.31	FAN	51	1977.88	9.00
RADIO	40	1970.25	8.97	BARN	52	1970.21	10.03
AMMONIA	41	1976.73	9.08	BASIN	52	1976.86	8.16
RIDGE	41	1980.41	6.30	NUTRI	53	1979.20	6.74
ASAE	42	1974.64	9.34	POND	53	1973.54	9.85
SORTING	42	1976.88	6.66	ANAEROBIC	54	1981.33	5.98
DIFFUS	43	1977.83	8.18	BROILER	54	1978.68	7.54
FINITE (ELEMENT)	43	1981.13	6.60	MIXING	54	1974.33	8.19
VEHICLE	43	1975.93	7.43	CLIMATE	55	1973.96	9.00
CONSTRUCT	44	1973.50	10.44	STREAM	55	1975.70	8.98
DIMENSIONAL	44	1977.68	8.64	CLEAN	57	1978.78	7.18
INDUSTR	44	1974.88	9.67	FLOOD	57	1972.26	9.19
LAGOON	44	1977.04	6.85	FREEZING (FROZEN)	57	1975.17	9.32
STANDARD	44	1973.36	10.47	POLLUTION	58	1977.63	6.96
TERRACES	44	1971.06	10.18	HEAT TRANSFER	61	1978.37	6.96
WAVES	44	1980.22	7.11	NITR	61	1981.42	6.78
BIOMASS	45	1983.60	3.02	PEST	61	1980.22	9.10
LETTUCE	45	1975.31	6.90	DIGEST	63	1983.00	4.77
PLOW	45	1972.24	9.80	HYDROLOG	63	1976.19	8.66
BALE (NOT BIG, ROUND)	46	1975.91	9.00	BEEF	64	1974.01	7.21
EGG	46	1973.67	8.84	CHANGE	64	1978.01	8.47
FARM (LAND, SIZE)	46	1970.45	10.06	MILK	64	1972.32	9.58
FILTER	46	1973.78	9.75	PUMP	65	1971.90	10.34
SUGAR (NOT BEET)	48	1979.04	8.73	SORGHUM	65	1978.03	7.29
COB	49	1978.14	9.47	WEATHER	65	1976.06	8.87
LEAF	50	1978.54	7.23	GRADING	68	1975.35	8.07

Table 11.6. (Continued)

Keyword root	Number of articles	Mean pub. year	Pub. span	Keyword root	Number of articles	Mean pub. year	Pub. span
PIPE	68	1971.29	10.22	TILE (DRAIN TUBES)	92	1972.17	10.91
MICRO (COMPUT) (PROCESS)	70	1983.28	2.90	ANIMAL	93	1976.50	7.89
PACK	70	1976.14	8.16	INSECT	93	1974.72	7.73
SHAKER	70	1978.21	7.17	ENSILING, SILO, SILAGE	94	1974.18	11.40
VEGETABLE (NOT OIL)	70	1974.37	8.19	APPLE	96	1976.50	7.89
LOG	71	1977.95	7.04	COMPACTION	97	1975.39	10.95
WALLS	72	1980.80	8.60	PLASTIC	98	1974.67	8.96
GIN (COTTON)	73	1980.58	5.63	FUEL	99	1980.77	6.15
CONSERV	74	1976.67	10.16	ALFALFA	102	1977.29	8.20
ELECTRON . . .	74	1977.67	9.00	EFFICIENCY	102	1974.84	10.42
PEANUT	74	1976.16	8.49	FOREST	102	1976.77	7.72
CLAY	75	1975.45	8.52	SEPARATOR	104	1978.55	7.76
ENGINE	75	1978.98	8.92	TREE (NOT FOREST)	105	1977.53	7.70
TOMATO	75	1977.24	6.30	TRACTION	109	1976.89	8.77
POTATO (NOT SWEET)	76	1977.97	8.66	RADIATION	110	1971.49	9.00
WHEELS	76	1974.89	8.54	TOBACCO	112	1975.27	8.02
EVAPORAT . . .	77	1975.98	9.77	WHEAT	113	1977.38	9.89
BULK	78	1976.15	9.80	IMPACT	115	1979.46	7.71
CELL	78	1974.21	8.27	COMBINE	116	1974.12	9.66
SAND	79	1976.24	8.60	COST	118	1974.73	8.76
DUST	80	1979.52	8.50	CATTLE	119	1973.66	7.70
BERRY	81	1976.50	6.99	CHEM . . .	119	1978.24	8.28
TRANSPORT	81	1978.29	7.49	FORAGE	119	1973.47	9.46
TIRES	82	1976.01	10.48	INFILTRATION	120	1978.50	9.46
GRASS	84	1974.94	9.26	LIVESTOCK	122	1973.80	9.73
EVAPOTRANSPIRA . . .	85	1974.62	10.99	DAIRY	123	1974.31	9.30
LIGHT	85	1973.07	8.50	WIND	123	1976.98	8.06
CITRUS	92	1976.22	7.31	WATERSHED	124	1972.86	9.49
GREENHOUSE	92	1978.26	6.59	WOOD (NOT FOREST)	127	1974.61	9.92
				SOYBEAN	129	1980.75	5.44

Table 11.6. (Continued)

Keyword root	Number of articles	Mean pub. year	Pub. span	Keyword root	Number of articles	Mean pub. year	Pub. span
ECONOM...	131	1977.06	9.53	HOG, PIG, SWINE	212	1976.78	9.32
HAY	132	1970.00	9.55	PHYSICAL PROPERTIES	215	1976.72	7.17
RICE	132	1977.84	7.98	COMPUT...	240	1979.09	7.38
FOOD	134	1976.54	7.45	COTTON	245	1977.78	7.77
MATERIAL	137	1971.43	9.73	RUNOFF	252	1976.62	9.10
PLANTER	137	1977.67	7.66	STORAGE	254	1975.51	9.84
QUALITY	138	1979.33	7.47	MACHINERY	266	1975.19	9.74
SEDIMENT	139	1979.21	7.86	MANURE	268	1978.02	6.82
HYDRAULIC	143	1974.65	10.15	ENGINEER (ING)	272	1971.68	10.07
SOLAR	143	1978.30	7.58	LAND	276	1975.79	9.54
RESEARCH	144	1973.19	10.52	SEED	278	1977.19	8.31
COOL	145	1974.54	10.18	TRACTOR	296	1972.32	10.07
POULTRY	147	1975.25	9.11	DRAIN	313	1973.86	10.13
OPTIM...	149	1978.06	7.07	ENERGY	328	1978.14	7.64
SPRINKLER	149	1974.58	9.49	EROSION	345	1977.91	8.68
HANDLING	155	1973.33	9.24	GRAIN	353	1974.56	9.50
RAIN	159	1977.16	10.17	MOISTURE	360	1975.90	9.36
WASTE	159	1978.59	6.27	SIMULATION	364	1977.76	7.26
ENVIRONMENT	163	1974.76	8.93	AGRICULTURE	379	1974.22	9.44
FEED	165	1972.87	8.83	CORN	387	1977.64	8.61
PHOSPHOROUS	166	1980.68	7.46	DRYER (ING)	424	1974.97	8.70
MANAG...	173	1978.95	8.33	MEASUREMENT	428	1975.77	9.90
BEAN	174	1978.29	8.18	HARVEST	545	1974.55	8.12
AUTOMAT	179	1974.47	8.14	MODEL	553	1980.63	6.65
LOSS	197	1976.80	9.30	IRRIGATION	694	1976.48	9.43
FRUIT	200	1974.19	8.23	WATER	759	1976.00	9.34
SPRAY	200	1977.96	8.79	SYSTEM(NOT EXPERT)	769	1977.41	7.92
PROCESSING	203	1977.42	8.58	SOIL	940	1976.24	9.40
TILLAGE	207	1977.27	9.36	ZZ WHOLE DATABASE	9834	1975.27	9.39

such as "expert (systems)," "vision," "robot," "image," "sweet-sorghum," "rape seed," "biomass," "biogas," "algorithm," and "sensor" among the most recent topics to receive emphasis in publication. When these same recent topics are considered according to the size of the standard deviation of publication dates these recent topics are prominent because they have been receiving active publication attention for only a very limited number of years.

Keyword roots representing subject-matter areas which have consistent attention in publication for many years—particularly at the beginning of the time period examined when there were fewer articles published annually—are prominent among those with high standard deviation values. Some of these technical topics are "dam," "gully," "silo," "compaction," "tile," "tires," " and "nuclear." Along with these, some more general categories of interest and effort are noticeable in this long-span group—"farming," "research," "standard," "efficiency," "conservation," "engineering," "transducers," and "statistics."

Keywords observed individually can sometimes be misleading, as can averages; however, as used here, the keywords can give some general impression of the technical matters that agricultural engineers have shared with their peers through the periodical publications of the ASAE. The overall result is the impression of a strong concentration on the basic elements of agriculture coupled with an amazing diversity of technologies and a strong emphasis on the understanding and designing of systems to link agriculture and technology.

12. The Place of Standards in Agricultural Engineering

JIMMY L. BUTT
Former Executive Vice-President
American Society of Agricultural Engineers

A. Standards

The standards developing process captures the consensus judgment of carefully chosen specialists as to the state of the art for a specific segment of technology at a given time. Current standards therefore are extremely valuable to those who need to bring current expertise to bear on a subject covered by a standard. Superseded or outdated standards are useful in documenting consensus requirements and practices that were considered state of the art at various times in the past and are valuable in litigation and for historical documentation.

Standards have been variously defined as follows:

A standard is a common language that promotes the flow of goods between buyer and seller and protects the general welfare.[1]

Standardization is the process of formulating and applying rules for an orderly approach to a specific activity for the benefit, and with the cooperation, of all concerned.[2]

Standards are specifications and definitions developed and adopted because of a need for action on a common problem.[3]

---

1. Page L. Bellinger, *Improving Product Safety and Productivity through Standards* (St. Joseph, Mich.: American Society of Agricultural Engineers, 1987 [*ASAE Paper* no. 87-5530]).
2. Ibid.
3. Russell H. Hahn, from personal conversation and oral presentations, St. Joseph, Mich., American Society of Agricultural Engineers, 1989.

Standards are basically engineering specifications prepared to define materials, products, processes, tests, testing procedures, and performance criteria in an effort to achieve certain specified purposes. Therefore, standards must accurately and specifically define the properties required without unnecessary, restrictive specifications which thwart originality or progress.[4]

In fact, standards are all of the above and more as perhaps best expressed by a review of some of the reasons given for standards:

(1) To provide *interchangeability* between products manufactured by two or more companies, thus improving safety and performance for users.[5]
(2) To provide a *sound technical base* for codes, regulations, and education relating to agri-industry.[6]
(3) To establish *performance criteria* for equipment, materials, systems.[7]
(4) To improve degree of *personal safety* in operation and servicing of agri-products, materials, equipment.[8]
(5) To *reduce inventory* in variety of components required to serve an industry, thus improving availability and economy.[9]
(6) To define *common basis for testing*, describing or informing about performance and characteristics of products, methods, materials, or systems.[10]
(7) To provide *design data* in readily available form.[11]
(8) To *increase efficiency* of engineering effort and facilitate communications.[12]
(9) To *identify and number products* and major components to help customers and law enforcement officers deal with theft of machinery.[13]
(10) To *enhance international trade* and understanding through an established set of guidelines which help break down language barriers and promote constructive discussion.[14]

B. Early Developments

The need for standards has been recognized in various ways through the ages. Some early examples:

---

4. Bellinger, *Improving Product Safety*.
5. *Why Your Support of CSP Is Good Business*, undated brochure (St. Joseph, Mich.: American Society of Agricultural Engineers).
6. Ibid.
7. Ibid.
8. Ibid.
9. Ibid.
10. Ibid.
11. Ibid.
12. Ibid.
13. Page L. Bellinger, "The Deere and Company Engineering Standards Program," paper presented at ANSI seminar on company standards policies at Itasca, Ill. (Moline, Ill.: Deere and Company, 1985).
14. Bellinger, *Improving Product Safety*.

(1) A cylindrical Egyptian cubit stone used as a unit of measure about 7000 B.C.[15]
(2) The order of Charles I in 1631, following confusion with assorted arms and armors, that "hereafter there shall be but one uniform fashion of armors . . . throughout our said Kingdome of England and Dominion of Wales."[16]
(3) The Boston Standard on building bricks, specifying standard dimensions (9" x 4" x 4") so they could rebuild the city more quickly with compatible bricks following the fire of 1689.[17]
(4) Thomas Jefferson's 1800 contract with Eli Whitney, considered this country's earliest proponent of standardization, to produce 10,000 muskets. Whitney divided the production process into simple operations and demonstrated that interchangeability of parts was feasible and economically advantageous. Whitney later achieved fame as inventor of the cotton gin.[18]
(5) ASAE, in 1927, adopted a standard for tractor power takeoffs specifying 540 rpm, spline dimensions, and direction of rotation thus helping eliminate a potentially confusing, dangerous, and costly variety of power sources.

In more recent times and in step with the rapid acceleration in the development of technology, the need for standards has grown proportionately. Standard gear shift patterns, highway signs, window and door dimensions, oil viscosity ratings, electric power voltages and cycles, land measures and descriptions, standard units of measure, are common everyday examples. A recent ISO standard, ISO 8601, addresses the confusing numeric date designation problem. It specifies that the system proceed from larger to smaller elements: first the year, then month, then day. Thus 1991-08-06 would clearly signify August 6, 1991. Present usage 8/6/91 could mean August 6, 1991, or June 8, 1991 depending on where one lives. To be more specific, time can be added, using twenty-four hour clock as follows: 1991-08-06T20:15:10 meaning August 6, 1991 at 8:15 and 10 seconds in the evening.[19]

There are more than 400 organizations in the United States that develop and maintain some 28,000 voluntary standards.[20] These figures do not include the additional thousands issued by companies and governmental agencies. One farm machinery manufacturer in 1989 reported company standards based on some 1,800 standards of other organizations, and a total of some 3,500 company standards.[21]

15. Ibid.
16. Extract from a Commission by Charles I on the Subject of Arms and Armor, June 1631, *Standards Engineering* 41 (1) (1989): 10.
17. Bellinger, *Improving Product Safety*.
18. *Standards Make the Pieces Fit*, ASTM undated brochure about standards (Philadelphia, Pa.)
19. "Universal System of Representing Dates and Times," *ISO Bulletin* 20 (1) (1989): 4.
20. John L. Donaldson, "Laboratory Accreditation in a Pluralistic Society," *Proceedings, 37th Annual Conference*, Standards Engineering Society, Dayton, Ohio, 1988.
21. Michael D. Morrell, "A Manufacturer's Need for Electronic Delivery of Standards," paper presented at ANSI Public Conference, Washington, D.C. (Moline, Ill.: Deere and Company, 1989).

C. The Need for Standards in Agriculture

In agriculture, the power take-off (PTO) standard ASAE S203.10, originally adopted in 1927 and used worldwide today provides a good example of how standards serve the user. Without this standard various tractor manufacturers might produce tractors with different PTO speeds, sizes, spline dimensions or directions of rotation. Farm machinery to be powered by the tractors, and manufactured by different companies, might carry entirely different fittings. Thus users would be unable to connect their equipment to their power source. The result would be that users would need to buy several power sources, or buy equipment manufactured by only one company, or adapt their farming practices to the equipment they could afford. The result: higher costs, more inconveniences, higher food and fiber costs to the public.

Other examples of agricultural engineering standards that have made significant contributions to safety and to the efficiency, productivity and quality of agricultural products:

(1) Lighting and Marking of Agricultural Field Equipment on Highways (ASAE S279.9) provides specifications for lighting and marking agricultural field equipment whenever such equipment is operated or traveling on a highway. The objective is to help reduce highway accidents involving farm equipment.
(2) Volumetric Capacity of Forage Wagons, Wagon Boxes, and Forage Handling Adaptations of Manure Spreaders (ASAE S238.1) provides a uniform method for calculating and expressing volumetric capacity. This assures that competing manufacturers rate their products in a uniform manner with obvious advantages to the consumer.
(3) Interchangeability of Disk Halves for Agricultural Equipment Press and Gage Wheels (ASAE S221) establishes dimensions and relationships to insure interchangeability of disk halves supplied by different manufacturers. This gives users much greater flexibility in finding and replacing parts.
(4) Agricultural Tractor Test Code (ASAE S209.5) defines test conditions, tests to be made, data to be obtained, formulas and calculations, so that performance data on various makes and models of tractors will be comparable. This enables users to evaluate different makes and models to make wiser buying decisions.
(5) Safety for Electrically Heated Livestock Waterers (ASAE EP 342.1) promotes safety as applied to the construction and installation of electrically heated waterers by specifying wiring and grounding details, service connections, and protective steps. This reduces the likelihood of electrical shock to humans and animals.
(6) Uniform Terminology for Bulk Milk Handling (ASAE S254.3) establishes uniformity in terms used in bulk milk handling. Without such uniformity, various companies, organizations, individuals would use different names in identifying similar devices or processes resulting in confusion and misunderstanding.
(7) Procedure for Sprinkler Testing and Performance Reporting (ASAE S398.1) defines test procedure for data such as flow rate and radius of throw and provides methods for interpreting sprinkler performance to aid in evaluation. This helps assure performance satisfaction.

(8) Dimensions for Compatible Operation of Forage Harvesters, Forage Wagons, and Forage Blowers (ASAE S328.1) provides dimensions to assist manufacturers of this equipment make the operation of any one compatible with the others. This helps users by assuring compatibility of equipment even among competing manufacturers.

These examples are representative of the 189 standards published by the American Society of Agricultural Engineers in *ASAE Standards 1989*.[22]

D. Organizations Producing Standards

Standards used in agriculture may be subdivided into five categories on the basis of scope of coverage: (1) international, (2) regional, (3) national, (4) industry, and (5) company. Increasingly, priority of usage is in that order, for good reasons.

International standards are preferred, when applicable, because they provide maximum cost-effectiveness in design, manufacturing, and serviceability through utilization of parts and materials commonly used worldwide.[23] The major international standards bodies are the International Organization for Standardization (ISO) and the International Electrotechnical Commission (IEC), the latter serving the electrical and electronics field. ISO, with some ninety member and observer countries, has over 150 technical committees of which TC 23-Tractors and Machinery for Agriculture and Forestry addresses most of the issues pertinent to agricultural engineering. Many other committees have some relationship to agricultural engineering: TC 31-Tires, Rims and Valves; TC 34-Agricultural Food Products; TC 127-Earthmoving Machinery; TC 131-Fluid Power Systems; TC 147-Water Quality; and TC 159-Ergonomics are examples.

The ISO lists the following five basic objectives to indicate how their standards are an essential source of information for a company:[24]

(1) To unify the technical language used by economic partners and to clarify contractual relations.
(2) To reduce costs, in particular by solving recurrent problems.
(3) To update and maintain the common fund of manufacturing experience which serves as the starting point for innovation.
(4) To provide the essential technical data for developing industrial and trade strategies.
(5) To propose the necessary tools for achieving targeted quality.

---

22. *ASAE Standards 1989* (St. Joseph, Mich.: American Society of Agricultural Engineers, 1989).
23. Morrell, "A Manufacturer's Need for Electronic Delivery of Standards."
24. "Standards: An Essential Source of Information for Your Company," *Access to Standards Information* (Geneva: International Standards Organization, 1986).

Regional standards organizations are formed for the purpose of harmonizing standards among those nations creating the regional group. Those standards relating strictly to the uniqueness of the geographic area or to trade agreements among countries may never be advanced as international standards although they could be. Some examples of regional standards organizations: ARSO, CAC, CEN, and COPANT.[25] In any thorough literature search, the standards of regional organizations encompassing an area included in the search, or involving products used in that area, should be reviewed. The 1992 activation of the European Community (EC) will create major standardization pressures among member countries and with non-EC trading interests.

National standards are those adopted by organizations representing various nations. The American National Standards Institute (ANSI) is the organization that represents the United States. ANSI is a federation of voluntary organizations and companies interested in standards and representing virtually every technical discipline as well as all facets of trade and commerce, organized labor, consumer interests and representatives of government.[26] Numbered among its members are the major standards developing organizations of the nation. Other national standards producers are: SCC (Canada), BSI (British), DIN (German), AFNOR (French), JISC (Japanese), UNI (Italian), IRATRA (Spain), GOST (Russian), SAA (Australia), to name a few.

Industry standards are produced by organizations whose technical or trade interests encompass an area of technology needing standards. In agricultural engineering, for example, the technical scope of the American Society of Agricultural Engineers (ASAE) encompasses farm tractors and machinery, irrigation and drainage, agricultural electrification and electronics, food engineering, farm structures and buildings, and allied technologies. It follows that ASAE would be interested in standards related to these areas of technology. But there also are other technical societies whose scope overlaps some of these same areas: SAE has interest in tractors, ASCE in irrigation, AIChE and ASME in food engineering. There also are trade associations that have strong interest in standards for commercial reasons, and they want inputs into the standards developing process: Equipment Manufacturers Institute (EMI) in farm tractors and machinery, Irrigation Association (IA) in irrigation equipment, Outdoor Power Equipment Institute (OPEI) in lawn and garden equipment, are examples. Most trade associations are not

---

25. *Directory of International and Regional Organizations Conducting Standards-Related Activities* (Washington, D.C.: NBS, U.S. Department of Commerce, 1983 [*USDC Special Publication* no. 649]).

26. *American National Standards Institute Procedures for Management and Coordination of American National Standards* (New York: ANSI, 1987).

producers of standards but they usually have knowledge of those that impact their industry. And they often have direct involvement in the proposal for or development of standards as will be seen in the paragraph on *Proposal and Development of Standards.*

Company standards are specific to the operations of a given company and may be considered privileged information. They are developed to establish internal uniformity and consistency in design,parts and materials specifications and instructions. Company standards will draw heavily from international, national and industry standards when such are available. In fact, those exterior standards are frequently quoted, indexed or incorporated as part of a company standard. It would be wasteful of resources for a company to develop its own internal standard if one already existed elsewhere. Furthermore, economies are realized by following design practices and using parts and materials that are commonly used by other industries, worldwide.[27] Larger companies may have upwards of 400 internal standards, depending upon size and number of product lines, and may reference 1,500 or more external standards and have several thousand plant or component standards.[28]

E. Proposal and Development of Standards

Proposals for new standards, or for revisions of standards, may originate from (a) committees of the standards developing organization itself; (b) other interested organizations, companies, or associations; or (c) individuals. Once presented to a standards developing organization, a proposal must be processed through well defined procedures to allow various points of view to be expressed and resolved, and to assure that all interested parties are given an opportunity to participate.

For example, a proposal to ASAE is first referred to a technical division standards committee which determines if the proposal is pertinent to the society and that division. It then assigns responsibility for developing the standard to a committee (existing or appointed for the purpose). Such committees are made up of individuals known to be highly competent and experienced in the area of technology encompassing the proposed standard. This includes persons with industry, government, manufacturer, educator, and user interests. Each brings a unique perspective and area of expertise to the

27. Bellinger, The Deere and Company Engineering Standards Program."
28. Michael D. Morrell, "Productivity in Standards Development," paper presented at meeting of Standards Engineering Society Meeting, Portland, Oreg., Aug. 1985 (Moline, Ill.: Deere and Company, 1985).

discussion table. Educators, for example, may conduct research to test the feasibility of proposals before adoption; users may provide feedback to help identify needed enhancements. Draft documents or comments supplied by the proposer of the standard are studied and may or may not form the basis of the final document.

ASAE rules specify that a standards developing committee "shall strive to obtain broad representation and shall provide an opportunity for qualified individuals from substantially interested producer, consumer and general interest groups to participate and vote."[29] Committees are responsible for technical accuracy and correctness of information bearing on the particular area of standardization. And all ASAE standards use SI units of measurement to encourage United States transition to the SI system and to conform to practices followed in essentially all nations of the world. Most standards also carry inch-pound units (dual dimensioning) to assist in the transition. Use of SI units helps United States manufacturers market their products around the globe.

To assure that there is substantial agreement on a standard, it must be advanced through several developmental and review stages. First, a 3/4 favorable vote of the originating committee is required for it to move forward. Next, the technical division standards committee verifies technical accuracy and, by 3/4 vote, sends it to the Society Committee on Standards. Third, this committee reviews the standard from the viewpoint of policy and interdivisional coordination. The Committee on Standards reports to the ASAE Board of Directors through its Technical Council, which is the fourth and final approval level before the standard is accepted and published.

Once accepted, a standard is automatically reviewed at least every five years to assure currency. ASAE standards are published annually in a hard-bound volume, *ASAE Standards* (Year), indexed (a) by subject and (b) numerically, and presented in seven technology groupings. It is critically important that standards by published (to assure availability), dated (including each revision), and made available to the interested public and to various indexes and databases so anyone having use for the standard will be able to find it. (See *Searching for Pertinent Standards*.) Each edition of *ASAE Standards* carries additional information such as a listing of related international standards, new or revised standards currently under development, a full explanation of ASAE standardization procedures, and a listing of all ASAE technical committees providing names of committee members. Most

---

29. *ASAE Standards 1985*, published by ASAE, p. 609

standards developing organizations follow procedures similar to those outlined above.

F. Voluntary Consensus Standards

In the United States most standards are developed as "voluntary standards," meaning that compliance is voluntary rather than mandatory. In some countries they are developed by governmental agencies, often with enforcement powers. Why are voluntary standards the preferred option in the United States? Terrence Scanlon, Chairman, U.S. Consumer Product Safety Commission states, "The history of mandatory standards at the CPSC is that they take approximately four years to develop, during which time the consumer is not protected. Moreover, they are frequently the subject of expensive and extensive litigation and, once in place, they are difficult to change. Indeed, one of the real drawbacks of mandatory standards is that they lack the flexibility to keep up with changing technology."[30]

John Pendergrass, Assistant Secretary of Labor for Occupational Safety and Health, on why OSHA relies heavily on voluntary ANSI standards said, "we like to get information about a product or process from the people who develop it . . . . In other words, the most knowledgeable people about potential hazards in a workplace are those who conceived it, designed it, installed it, and work in it every day."[31] Donald C. J. Gray, Commissioner, Federal Supply Service, General Services Administration, noted that "consensus is a judgement arrived at by most of those concerned. In the case of standards, users, technical groups, industry, government, and even affected individuals are part of the consensus process." He went on to say "when the system works well, standards define levels of quality, safety, performance, and engineering which support a healthy competitive marketplace. When the system breaks down—the marketplace becomes stagnant—the competition is reduced—there is a repression of technological advancement."[32]

Most voluntary standards are developed by organizations with little or no financial support from public funds. Some agencies provide a degree of support through memberships, time donated toward standards research or development, purchase and distribution of standards. Rather, most financial support of standards organizations comes primarily from (1) member dues,

---

30. Terrence M. Scanlon, March 25, 1987 address at ANSI conference on self regulation.
31. John A. Pendergrass, Asst. Secretary of Labor for Occupational Safety and Health, in an address at an ANSI Public Conference, March 25, 1987.
32. Donald C. J. Gray, address at an ANSI public conference, March 25, 1987.

(2) voluntary contributions from interested companies, agencies and organizations, and (3) sale of the standards and related publications.

G. Benefits of Standards

Essentially all segments of society feel the impact of standards although many beneficiaries may not realize it. Economic gains from the application of standards reduce production costs which results in lower costs to consumers. Standards directed toward quality enhancement such as testing, grading, sizing, packaging, and defining quality levels result in higher quality products to consumers. Standards establishing uniform terminology help avoid confusion and misunderstanding in the marketplace. Performance standards enable users to make informed judgements regarding the products they buy. Thus the ultimate beneficiary of standards is the consuming public even though only some will realize how standards contributed to their well-being.

But there are other beneficiaries who are closer to standards and more apt to realize the benefits, the consequences of non-use, and who will become directly involved in the application. These are the affected industries, companies, governmental agencies, educational faculty, associations, and those who apply the standards.

Industries and individual companies have much to gain in being able to buy standard components produced in large quantities at lower costs. They gain market opportunities by manufacturing equipment that can be interchanged with other makes or models, thanks to standard dimensions. They are assured of equal opportunity in the marketplace as a result of standard ratings, test procedures, and terminology. They save time by utilizing data that have been assembled by standards committees. They can be reasonably sure that their products are state-of-the-art when current standards are met. They know that failure to conform weakens their case in litigation and in the marketplace.

Governmental agencies use external standards to take full advantage of outside expertise and experience. Enforcement agencies use them in training compliance officers, quality control inspectors, safety review teams, and to make sure knowledge and skills are up-to-date. They are used or referenced in specifications to assure quality and to minimize costs. Agencies benefit by utilizing the consensus knowledge of specialists with varied experiences, rather than relying upon the comparatively limited knowledge resources of a single organization.

Among educational institutions standards, engineering practices and data are useful to research workers, teachers and students. For example, the ASAE standard ASAE EP 433, "Loads Exerted by Free-Flowing Grains on Bins," provides state-of-the-art design information, terminology, equations, and references to related standards that are useful to student and teacher in design courses. Research workers in food engineering, processing or structures and environment find the information in ASAE D 271.2, "Psychrometric Data," a time-saving resource. Learning to make full use of the wealth of accumulated data in standards is a vital part of the educational process for students.

There are other groups that can benefit from a thorough knowledge of standards as they pertain to their work: farmers who can rely on standard test procedures to provide reliable comparative information about the equipment they buy; consultants who draw upon the information in designing their projects, comfortable in the knowledge that it represents consensus state-of-the-art; the insurance industry, knowing that safety standards will help reduce the incidence of accidents; dealerships that are required to stock less inventory thanks to standards that specify standard parts that can be used on a variety of models and products.

These are but examples of how standards may be used. The uses are as varied as the potential applications of the data. Perhaps these examples will suggest still other uses. Standards that relate to a searcher's problem should be included in any comprehensive search of literature.

H. Publication of Standards

Standards may be published as hard copy separates or in book form, or may be available on microfiche, CD-ROM, or in electronic data banks.

Separates are useful to those who need many copies of a single standard. They may be distributed to sales persons to aid in on-site designs, such as ANSI/ASAE S376.1, "Design, Installation and Performance of Underground, Thermoplastic Irrigation Pipelines." Or they may be distributed among equipment designers whose products need to match standard tractor hitches such as ASAE S217.10, "Three-Point Free-Link Attachment for Hitching Implements to Agricultural Wheel Tractors."

Books of standards are useful to those who want to be sure their products conform to all the latest consensus standards. Only by having a current standards book can one be certain of using the most current standard. Others may use the books as a means of having access to a major storehouse of information on a variety of subjects and to maintain a file of what was state-of-the-art at a given time.

Individual standards also are available on microfiche and in major databases. During the 1980s, there was a developing trend toward relying on current searches of databases for the most up-to-date information. For example, Information Handling Services of Englewood, Colorado, maintains a file of standards and specifications that includes ISO and IEC standards, most national standards, those of over 350 United States standards developing organizations, and most military and U.S. Government standards documents.

I. Searching for Pertinent Standards

To locate a specific standard, one contacts the originating organization, provides necessary identification, and receives the copy by mail, fax, microfiche, or other agreed upon process. These organizations also can provide advice as to standards applicable to a search. Most libraries will have copies of standards books that pertain to the area of technology encompassed by the library.

When one wishes to "search the field" for applicable standards, the procedure is more complex and may require an extensive search through the standards books of many organizations, or consultation of comprehensive indexes or electronic databases.

Most of the organizations listed under "Names and Addresses of Major Standards Organizations and Sources" publish standards books containing their own standards, including indexes. The American Society of Agricultural Engineers, for example, publishes standards devoted exclusively to agricultural engineering subjects in a single, annual hardbound volume of approximately 600 pages entitled *ASAE STANDARDS* (YEAR).

Indexes covering the standards of many organizations also are available to aid in comprehensive searches. For example, *The Index and Directory of Industry Standards*, published by Global Engineering Documents, a division of Information Handling Services, comprised (in 1989) three volumes of some 20,000 international standards, over 35,000 domestic standards, covering some 360 major organizations.

Similar coverage is available online through Information Handling Services. These files may be accessed through Dialog Information Services (DIALOG file 92), telephone: 800/3-DIALOG. The following section lists the names, addresses, telephone numbers and major areas of coverage of these and other information sources carrying standards relating to agricultural engineering.

Names and Addresses of Major Standards Organizations and Sources

Agricultural Engineering Related Standards Organizations in the United States

American Society of Agricultural Engineers (ASAE)
2950 Niles Road
St. Joseph, Mich. 49085
Telephone: 616/429-0300
> ASAE is the only United States organization whose standards are devoted exclusively to agricultural engineering. ASAE has some 180 current standards (many used as examples in this chapter) which are published annually in *ASAE Standards* (Year), a hardbound book of about 700 pages covering these areas: soil and water, farm power and machinery, electrical and electronic systems, food and process engineering, agricultural structures and environment, and some emerging technologies (agriculture, forest engineering, bioengineering, knowledge systems).

American National Standards Institute (ANSI)
1430 Broadway
New York, N.Y. 10018
Telephone: 212/642-4900
> ANSI is the United States national organization of standards, and it promotes the voluntary use of standards for industry, engineering and safety design. It has publications, indexes and catalogues pertaining to United States national standards, and it represents the United States in the international standards arena.

Society of Automotive Engineers (SAE)
400 Commonwealth Drive
Warrendale, Pa. 15096-0001
Telephone: 412/776-4841
> SAE publishes standards covering the field of mobility (land, air, water). This includes standards for the basic agricultural tractor. ASAE and SAE jointly publish certain standards pertaining to tractor safety and the tractor-implement interface. SAE's many standards are published annually in *SAE HANDBOOK*, a softbound book in four volumes.

ASTM (Formerly American Society for Testing and Materials)
1916 Race Street
Philadelphia, Pa. 19103
Telephone: 215/299-5400
> ASTM develops and publishes voluntary consensus standards for materials, products, systems and services. They produce several journals, a magazine and a *Book of Standards*. Their coverage of the indicated areas is extensive and their standards are also recognized worldwide.

American Society of Heating, Refrigerating and Air Conditioning Engineers (ASHRAE)
1791 Tullie Circle, N.E.
Atlanta, Ga. 30329
Telephone: 404/636-8400
> ASHRAE has several publications including *ASHRAE Handbook* containing stand-

ards relating to heating, refrigerating and air conditioning, many pertaining to agricultural engineering applications.

National Fire Protection Association (NFPA)
Batterymarch Park
Quincy, Mass. 02269
Telephone: 617/770-3000

NFPA promulgates standards in fire protection, prevention and safety. It has several publications and handbooks, many having applications to agricultural engineering.

American Society of Civil Engineers (ASCE)
345 East 47th Street
New York, N.Y. 10017
Telephone: 212/705-7496

ASCE publishes several journals and produces standards, some of which relate to agricultural engineering (dams, groundwater, irrigation).

American Institute of Chemical Engineers (AIChE)
345 East 47th Street
New York, N.Y. 10017
Telephone: 212/705-7338

AIChE publishes many publications and develops standards relating to food and chemical processing.

American Society of Mechanical Engineers (ASME)
345 East 47th Street
New York, N.Y. 10017
Telephone: 212/705-7722

ASME has extensive programs in the development of safety codes, product and equipment standards and publishes many journals. Some of their standards affect agricultural engineering applications or processes.

International Standards Organizations

International Organization for Standardization (ISO)
1, rue de Varembe
Case Postale 56
CH-1211 Genève 20
Switzerland
Telephone: +4122341240

ISO coordinates the development of international standards and has published indexes, catalogues, bulletins, memento, names and addresses of member organizations. For example, its *Access to Standards Information*, a 100 plus page softbound book, discusses standards networking, lists addresses of specialized information centers, tells how to access on-line information systems, and provides lists of standards books published in various countries.

International Electrotechnical Commission (IEC)
3, rue de Varembe
CH-1211 Genève 20
Switzerland
Telephone: +4122340150

IEC works closely with ISO and handles electrical and electronic standards. Most agricultural engineering standards are coordinated through ISO committees rather than through IEC. This could change as agriculture employs more electrotechnical applications.

Index and Database Sources

Global Engineering Documents (Global)
2805 McGaw Avenue
Irvine, Calif. 92714
Telephone: 800/854-7179
   Global maintains an extensive collection of technical documents from all parts of the world. Their *Index and Directory of Industry Standards* comprises three volumes of some 20,000 international standards, 35,000 domestic standards, covering over 360 major organizations. They also index government standards and specifications and other information sources. (See also IHS below)

National Standards Association (NSA)
1200 Quince Orchard Boulevard
Gaithersburg, Md. 20878
Telephone: 800/638-8094
   NSA lists agricultural engineering related standards of some organizations in its indexes and catalogues. At this writing these data are not available electronically.

National Center for Standards and Certification Information (NCSCI)
National Institute of Standards and Technology (NIST), USDC
Administration Building 101
Mail Stop A-625
Gaithersburg, Md. 20899
Telephone: 301/975-4038
   NCSCI provides information on United States, foreign and international voluntary standards, government regulations, and serves as a referral service and focal point for information about standards. It does not provide copies of documents; rather, it helps locate and identify sources. It distributes several publications related to standards and trade including *Standards Activities of Organizations in the United States*.

Information Handling Services (IHS)
15 Inverness Way East
P.O. Box 1154
Englewood, Colo. 80150
Telephone: 800/525-7052
   IHS, in cooperation with Global Engineering Documents and Dialog Information Services, maintains an extensive database of standards and specifications. These include both United States and international standards as well as those of many foreign countries. They offer hardcopy, microfilm, CD-ROM, tape or on-line services. Standards also are offered through DIALOG File 92. Instructions (search guide) on how to access File 92 are available from IHS.

Acronyms of Some Other Standards Developing Organizations That Relate to Agricultural Engineering

Domestic (from *Index and Directory of U.S. Industry Standards—IHS*)

AASHTO - American Association of State Highway and Transportation Officials
ACI - American Concrete Institute
AIA - American Institute of Architects
AOAC - Association of Official Analytical Chemists
AWPB - American Wood Preservers Bureau
DFISA - Dairy and Food Industries Supply Association
IEEE - Institute of Electrical and Electronics Engineers
IES - Illuminating Engineering Society
ISA - Instrument Society of America
NEMA - National Electrical Manufacturers Association
NFPA - National Fluid Power Association
NFSA - National Fertilizer Solutions Association
TAPPI - Technical Association of the Pulp and Paper Industry
UL - Underwriters Laboratories Inc.
WQA - Water Quality Association

Foreign National and Regional Organizations

(1) Regional
   ARSO - African Regional Organization for Standardization
   CEN - European Standards Committee
(2) National
   AFNOR - Association Française de Normalisation
   BSI - British Standards Institution
   DIN - Deutsches Institut fur Normung
   GOST - Russian Standards Organization
   JIS - Japanese Industrial Standards
   SAA - Standards Association of Australia
   SCC - Standards Council of Canada
   UNI - Italian Standards Organization

13. Characteristics of Agricultural Engineering Literature

WALLACE C. OLSEN
Mann Library, Cornell University

Agricultural engineering has always had the difficulty of belonging both to agriculture and to engineering but clearly not fully to either. This dichotomy is evident in the literature, professionalism, and education as delineated throughout this book, but particularly by Hall in Chapter 2, and Edwards in Chapter 3. The body of knowledge comprising agricultural engineering literature involves a thousand connections between points scattered throughout the full range of engineering subjects with points scattered throughout the full range of agricultural subjects. The beginnings of agricultural engineering as an organized field of study came within the established engineering professions, particularly civil and mechanical. Thus the earliest related agricultural engineering literature is in those engineering disciplines.

The American Society of Agricultural Engineers began in 1907 because of dissatisfaction with membership in other engineering societies. It appears that the concept of a literature base, usually considered one hallmark of a profession, was very much on the minds of the founders. The very first *Transactions* of the new society, December 1907, has an article concerned with the literature of the field. It is historically interesting and photographically reproduced here.[1] W.M. Hummel was at Colorado Agricultural College, an earlier name for Colorado State University.

---

1. W. M. Hummel, "The Literature of Agricultural Engineering," in *Transactions of the American Society of Agricultural Engineers* 1 (1907): 81–85.

THE LITERATURE OF AGRICULTURAL ENGINEERING

W. M. HUMMEL

We are sometimes told that there is no literature on rural engineering and farm mechanics, so recent has been development and progress along those lines. And, at first thought, this seems to be almost true. When we consult the catalogs of our Agricultural College libraries under the subject headings, Agricultural Machinery, Agricultural Engineering, and the like, we find little to reward us for our search. But this is not because nothing has been published on these subjects. On the contrary, not a little has been printed, considering the recent development of interest in the matter. But, in general, the material is not in book form. We find it scattered here and there, in experiment station publications, in United States Department of Agriculture publications, and in technical and other periodicals. And the search for it is sometimes a weary task. But we have many means, if we will but use them, of helping ourselves to find not only periodical and government publications on these subjects, but to enable us to find out what books are being published from year to year and from month to month, and to assist us in securing them.

It is the purpose of this paper to treat more particularly of these means of "getting at" literature on agricultural engineering and allied topics, rather than to mention specific works on the topic, either reference or textbook.

Of the literature on rural engineering published by the state experiment stations, little is general in nature. Nearly everything relates to experiments with specified machines and tools, descriptions of buildings and equipment at the various stations, etc., with comments as to their convenience, practicability, etc.

To find out what has been published on agricultural engineering by the stations we have but to consult the card index issued by the Office of the Experiment Stations. This is furnished to each agricultural college and experiment station and to boards or commissioners of agriculture in each state. Or it may be bought at the rate of $2.00 for 1,000 cards, or $1.25 for a set of cards on a given division or subject. The index is arranged by subjects according to a numerical scheme fully explained by a pamphlet issued by the Office of Experiment Stations, and each card gives title of the article, name of author, reference to the experiment station publication containing the article, and to the Experiment Station Record, with a brief abstract showing the nature and scope of the article. It enters agricultural engineering topics under the headings, drainage, irrigation, farm implements, roads and bridges, fences, and farm buildings.

For information as to current publications of the state experiment stations we may also consult the Experiment Station Record, the most important periodical publication of the Department, since 1889. This now gives not only abstracts of the bulletins and reports of state experiment stations, but of publications of the United States Department of Agriculture, and other books, journals, etc., giving reports of agricultural science in different countries of the world. Each number has a full table of contents arranged by departments, as agricultural engineering, rural economics, etc., and each department is in charge of a special editor. Every volume contains detailed author and subject indexes and a general index has also been issued for volumes 1–12, 1889–1901. This is sent out, without cost, to institutions for agricultural education and to many individuals who co-operate with the Department. Or, it may be secured from the Superintendent of Documents at ten cents a number or one dollar a volume.

Most of the work of the United States Department of Agriculture in agricultural engineering is done, as is well known, by the irrigation and drainage division of the Office of Experiment Stations, and includes investigation and experimentation not only in drainage and irrigation, but in application of power to farm work, use of various tools and implements, use of certain materials for farm building construction, suggestions as to outlines of courses in agricultural engineering education, etc. Other valuable material may also be found in the publications of the Bureau of Road Inquiries, annual reports of the Department, the Yearbooks, etc.

One of the most comprehensive indexes to Agricultural Department publications is the "List of Publications of the Agricultural Department from 1862–1902, with analytical index," published in 1904, which can be secured from the Superintendent of Documents at Washington. The Yearbooks are fully indexed from 1894–1900 by Bulletin No. 7 of the Division of Publications. For material in the later volumes the indexes for each year must be consulted. For information as to current publications of the Department we may consult both the monthly lists of publications sent free to all applicants, and the Experiment Station Record. For this purpose the monthly publication called "Government Publications," issued monthly at Washington by Mary Greathouse, is useful. This periodical also gives information as to ways of securing United States publications, and notes on important municipal and state publications. Its cost is fifty cents per year.

The Bureau of Public Roads Inquiries issues at regular intervals special lists of its works on public roads, and the Department issues, from time to time, subject indexes to the Farmers' Bulletins. The Agricultural Department library lists of publications and accessions are also valuable for consultation, and Bulletin No. 41 of the Library is an extensive bibliography of 181 pages on publications relating to irrigation and drainage.

Some of the best and latest information as to various engines, motors, etc., adapted to agricultural uses may be found in engineering periodicals. And though we may not have these at hand to consult, we may find out what is in them and secure the articles on topics of special interest to us by watching the Engineering Index. This is a monthly publication, at three dollars yearly, indexing the contents of nearly two hundred of the leading Engineering journals of the world, in English, French, German, Dutch, Italian, and Spanish. Each entry gives author and title of article, its length in words, and when and where published, together

84

with a descriptive abstract. The general arrangement of entries is by subjects, and the editor will supply any articles entered, if desired. For the year 1906, the Annual Engineering Index, issued by the Engineering Magazine and arranged in the same way, may be consulted. For earlier years the bound and cumulated volumes of the Index may be used. Of these, Volume I covers the years 1884-1891; Volume II, 1892-1895; Volume III, 1896-1900; and Volume IV, 1901-1905. The monthly index, used for the current year, is printed in successive issues of the Engineering Magazine as well as separately.

For less technical articles we may go to the general periodicals. And these, too, we find well indexed. Poole's Index practically covers general periodical literature for 1802-1902, and is arranged by subjects in such a manner as to be most easy to use. The Annual Literary Index indexes one hundred and fifty periodicals yearly and may be used for periodical literature of the years 1903-1906, and later as volumes are issued. For the current year we have the "Reader's Guide," published monthly, with a cumulative yearly number; and the "Library Index," issued monthly with quarterly cumulative numbers. All of these have their entries arranged by subjects, give definite reference to the periodicals in which the various articles appear, and are easy to use.

For information as to books in print we may consult works found in almost any public or college library. The last five volumes of the American Catalog include, in alphabetical arrangement by author, title, and subject, all books copyrighted in the United States from 1876-1905. The Annual American Catalog, arranged in the same way, gives us information as to books of 1906. In addition, we may consult the United States Catalog of books in print, for 1902 and 1902-05, under subject, and the "Monthly Cumulative Book Index," and the "Book Review Digest" for 1907. The last two publications give good notes on the general character of the new books included. Price and publisher, with other information, is also given in all the trade catalogs. If we wish to keep track of foreign publications on the topic we may, of course, consult the equally well arranged and valuable foreign trade catalogs.

Of special reference and text books we have no time to treat. All interested in Farm Mechanics are familiar with the periodicals: "Farm Implement News," published in Chicago; "Implement Age," published in Philadelphia; "Farm Implements," published in Minneapolis; "Farm Machinery," published in St. Louis; "Implement Trade Journal," published in Kansas City; and with the other periodicals dealing especially with this topic. There is much improvement to be desired, both in their make-up and the material included, but they are valuable because they are the best we now have. From catalogs of farm implement manufacturing companies we may also get much that is helpful, and in Bailey's new "Cyclopedia of Agriculture" we have excellent reference articles on agricultural engineering topics, with good, though brief, bibliographical lists.

Much of the material most valuable to us we may find in books classed under engineering and architecture. Some of these books are too technical for our purposes, but there are many handbooks which we may find useful for class as well as reference use along the lines of sanitary engineering, house drainage, motors, water supply, roads and bridges, etc. In architectural books and periodicals we find considerable material as to country homes and buildings, though it is apt to be less practical than we could wish. In consular reports we find information as to farm implements in foreign countries, and in various census bulletins we may get most valuable statistics as to manufacture of implements and machines in our own land.

Material on the subject of our interest seems meagre, and it is to a degree. Yet there is considerable if we can but get at it, and use it. We need to have material collected, revised, and put in convenient form for use, little less than we need additional study of the subject and publications about it. It is, evidently, the work of agricultural engineering students and instructors to do this as well as to make new investigations and to print their results.

Reproduced from *Transactions of the American Society of Agricultural Engineers* 1 (1907), with the permission of the American Society of Agricultural Engineers, St. Joseph, Mich.

A. Engineering in Agriculture

Engineering aspects occur in all subdisciplines of the agricultural sciences. This dependency of one discipline on another or the interaction of disciplines is common in an applications science such as agriculture. Therefore, organizing the literature of agricultural engineering becomes rather complex since the subjects are not machinery and power alone but involve, for example, soil science, food storage and processing, and forestry. This is further exemplified by the emerging and recently influential areas as reviewed in Chapters 5–9. The application of engineering to many aspects of rural activities requires great subject diversity. And this diversity must be represented in the formal organization of the literature in the abstracting and indexing tools of agriculture.

Three large databases concentrate on agriculture: AGRICOLA, AGRIS, and CABI (Commonwealth Agricultural Bureaux International). All are available on digitized tapes or compact disks, as well as in printed abstracting or indexing publications. All aim to cover the world literature of the agricultural sciences in all subject areas. The AGRICOLA and CABI files are selective of citations to be included based on the nature, subject, and value of the individual pieces or articles of literature. CABI and AGRIS add about l00,000 citations a year while AGRICOLA has been providing fewer entries during each of the past five years, and in 1991 entered only 84,685. AGRIS is an international cooperative effort accepting citations on a geographic basis from food and agriculture organizations around the world, and makes few judgments on the value of the literature indexed.

Engineering Index is the abstracting service which approaches agricultural subjects from the total engineering context.[2] It has the basic topics important to agricultural engineering such as grain elevators, machinery, farms, soils, irrigation, farm structures, etc. Many agricultural engineers use this index for searching the literature of their field; the abstracts are very good and the literature coverage is extensive. It is not, however, as complete a database as those dedicated to agriculture nor can it be analyzed as easily for the data in this chapter.

Agricultural Engineering Abstracts (AEA), compiled and published by CABI, is probably the most accurate source for getting numbers of indexed items per year. This CABI abstracting tool pulls citations from all pertinent literature and puts them in one publication, *AEA*. The subject categorization is done for us which is also the case with the CABI online files. Table 13.1

---

2. *Engineering Index* has been published since 1906 by The Engineering Societies, New York. Since 1970 the file has been available in digital form, and online.

Table 13.1. Agricultural engineering as a part of all agriculture, derived from online files, 1984–88

	AGRICOLA	CAB Abstracts
Agricultural engineering	16,197[a]	25,073 citations
Total agriculture database	467,500	640,506 citations
Agricultural engineering as % of total database	3.5%	3.9%

[a]For subjects included, see Table 13.3.

contrasts the agricultural engineering coverage with the total online files of both databases.

This is a decidedly lower production of all agricultural literature than the 7.6 to 7.9% of agricultural economics and rural sociology from an analysis of the same two data sources.[3] It must be noted that these two databases have the same basic policies on the inclusion of agricultural literature.

Agricultural Engineering Abstracts has been published since 1976 with roughly the same coverage of subject and document formats over the years. The printed abstracts mirror the online file, particularly more closely in recent years. Table 13.2 provides data on the number of abstracts published yearly since the inception of the *AEA*. It must be noted that a separate abstracting journal is published by CAB International entitled *Irrigation and Drainage Abstracts* which adds about 30% more citations which are central to agricultural engineering.

Previous studies of the coverage of the major agricultural bibliographic databases indicates that about 20% of each database has distinct material not in either of the others.[4] It is clear that CABI does not enter a quantity of items which appear in the other two databases because of differing policies centered on geography and the anticipated value to readers. Based on past studies the publication of agricultural engineering literature is probably close to 30% above the *AEA* annual numbers. This puts the estimated annual production of articles, monographs, reports, patents, and dissertations at about 5,900 citations per year.

B. Subject Concentrations

In the AGRICOLA bibliographic system prepared by the National Agricultural Library of the United States, ten major subject categories were

---

3. Wallace C. Olsen, "Characteristics of Agricultural Economics Literature," chapter 2 in *Agricultural Economics and Rural Sociology: The Contemporary Core Literature* (Ithaca, N.Y.: Cornell University Press, 1991).

4. Olsen, "Characteristics of Agricultural Economics Literature."

Table 13.2. Abstract entries in *Agricultural Engineering Abstracts*

Year	Number of citations/abstracts	
1989	4,804	
1988	4,804	
1987	4,910	
1986	4,215	
1985	5,621	
1984	5,523	
1983	4,820	
1982	4,489	
1981	4,394	
1980	5,272	
1979	3,336	
1978	3,639	
1977	2,637	
1976 (v. 1)	2,941	
	61,405	Average 1976–89 = 4,386

searched in order to get data on the subject concentrations of agricultural engineering literature. These subject categories were searched in AGRICOLA for Table 13.3 as well as data following. These are the subject categories of agricultural engineering as categorized in AGRICOLA with minor omissions. For example, two possible categories, J200: Soil Chemistry and Physics; and W000: General Pollution, were examined and excluded. The engineering aspects were so minor that these categories were discounted as of little importance to the agricultural engineering counts.

The category codes used for searching (e.g., K130, N000, etc.) are assigned at the time of indexing as an indication of the subject matter of the citation whether an article, chapter, report or book. More than one subject category in agricultural engineering can be assigned depending on the subject coverage of the item. A random sample drawn from the data in Table 13.3 was used to derive an average of 1.1 agricultural engineering category codes per citation. This means that the citations in the table are about 10% duplicated throughout Table 13.3. The continuing decline of indexed items from 1980 in all subject categories represents a policy at the National Agricultural Library to concentrate more and more on United States publications instead of worldwide coverage. No decline is represented in the CABI agricultural engineering records; neither is an increase which is probably also a management decision to stay within a fixed range per year.

Because of the evident decline in the numbers of total entries into AGRICOLA since 1984, comparison of subject concentrations is difficult. Table

Table 13.3. Subject categories and indexed items for agricultural engineering in the AGRICOLA database

Subject & category code	1980	1981	1982	1983	1984	1985	1986	1987	1988	TOTAL
1. Forest Products, Engineering & Harvesting (K130)	479	530	298	303	358	235	194	148	134	2,679
2. General Agricultural Engineering (N000)	144	216	161	147	181	144	94	127	133	1,347
3. Structures & Structural Equipment (N100)	1,016	1,006	816	820	706	385	307	333	299	5,688
4. Farm Equipment (N200)	4,064	3,520	1,908	1,713	1,609	1,021	718	686	387	15,626
5. Biomass Energy Resources (P120)	1,231	1,333	1,081	938	875	804	610	419	296	7,587
6. Alternative Energy Resources (P130)	700	709	682	432	327	154	137	112	79	3,332
7. Drainage & Irrigation (P210)	2,164	2,269	1,481	1,511	1,471	1,187	1,021	1,154	851	13,109
8. Food Storage (Q110–116)	1,002	886	996	835	556	585	485	475	385	6,205
9. Feed Processing & Storage (R100)	764	730	497	469	358	251	239	194	196	3,698

13.4 has the ten subject categories of Table 13.3 in percentages of the totals on an annual or cumulative basis.

The major shift in indexing worldwide literature in AGRICOLA to concentration on United States literature is represented by numbers after 1984. This is true in the database as a whole. Therefore, the subject concentrations for worldwide literature are probably best represent by the 1980–84 period of Table 13.4. On these five years as well as 1980–88, the most prolific literature groups are: first, Farm Equipment; second, Irrigation & Drainage; third, Biomass Energy Resources. It is essential to remember that

Table 13.4. Percent of subject concentrations within agricultural engineering citations: nine years of AGRICOLA

Concentration	1980	1988	1980–88	1980–84
1. Forest Products, Engineering & Harvesting (K130)	4.1%	4.6%	4.9%	4.3%
2. General Agricultural Engineering (N000)	1.2	4.6	2.2	1.9
3. Structures & Structural Equipment (N100)	8.6	10.4	9.3	9.6
4. Farm Equipment (N200)	34.4	13.4	25.6	28.2
5. Biomass Energy Resources (P120)	10.4	10.3	12.4	12.0
6. Alternative Energy Resources (P130)	5.9	2.7	5.6	6.3
7. Drainage & Irrigation (P210)	18.3	29.5	21.5	19.8
8. Food Storage (Q110–116)	8.5	13.6	10.2	9.4
9. Feed Processing & Storage (R100)	6.5	6.8	6.6	6.2
Total Citations	11,564	2,760	59,271	44,284

AGRICOLA is an agricultural sciences database and that the literature covered has proven supportive of the agricultural sciences. The first two subject groups are fairly conclusive with nearly 50% of all indexed citations. Biomass and alternative energy resources when combined total 18.0 to 18.3% or very near the irrigation and drainage percentage.

C. Language Concentrations

The worldwide scope and subject inclusiveness of the AGRICOLA and CABI databases provide the best source for information on the languages most heavily used in agricultural engineering literature. Table 13.5 provides this data and a comparison with the language coverage for all citations in both databases. As shown in Table 13.1, the agricultural engineering percentage of the total CABI database is a small .4% greater than that in AGRICOLA. But the language coverage in CABI is considerably more widely distributed. It seems likely that the CABI data on language distribution probably is more representative of worldwide publishing in agricultural engineering.

A previous study of these two databases concerning agricultural economics and rural sociology had similar variations between the databases with English (87.6%) and French (2.5%) in AGRICOLA, while CABI showed English (64.1%) and German (8.0%) as the two prime languages.[5]

5. Norbert Deselaers, "The Necessity for Closer Cooperation among Secondary Agricultural Information Services: An Analysis of AGRICOLA, AGRIS, AND CAB," *Quarterly Bulletin of the International Association of Agricultural Librarians and Documentalists* 31 (1) (1986): 19–26.

Table 13.5. Languages in primary agricultural databases derived from online files, 1984–88

	AGRICOLA			CABI Online		
	Agricultural engineering		Total file	Agricultural engineering		Total file
English	11,954	77.8	88.4%	15,233	64.4%	70.9%
Russian	765	5.0	3.0	1,123	4.7	5.6
German	495	3.2	2.8	2,115	8.9	5.5
French	232	1.5	1.6	927	3.9	3.7
Spanish	48	0.3	0.6	298	1.3	2.1
Japanese	168	1.1	0.6	569	2.4	1.5
Italian	54	0.4	0.2	734	3.1	1.5
Portuguese	35	0.2	0.3	101	0.4	1.1
All other	1,616	10.5%	2.5	2,559	10.8	8.0
Total	15,367	100.0%	100.0%	23,659	100.0%	100.0%

Examination of the literature demonstrates clearly that the English-language percentage is continuing to increase when dealing with research and advanced educational material. There is no doubt, that local languages are gaining in agricultural engineering at the beginning college level and in materials prepared for the local farmer or engineer.

D. Types of Publications

Both AGRICOLA and CABI code their bibliographic entries so that one is able to determine, not always precisely, the type of document from which the citation emanated. CABI has four format codes: Numbered Parts (journal articles); Numbered Wholes (mostly reports in a series); Unnumbered Parts; and Unnumbered Wholes. The last three categories match the definitions of monographs and reports or parts thereof as used in the studies in this and the following two chapters. Although the AGRICOLA database uses different descriptions, the divisions may be grouped similarly. Table 13.6 compares these two sources as well as that identified in the citation analysis of 8,184 references in peer-reviewed literature of agricultural engineering published from 1965 through 1989. The latter category is discussed in detail in the two chapters following.

The bibliographic databases include few dissertations, standards or patents. What few are indexed are counted as monographs and reports in both the CABI database and AGRICOLA. Citations in research journals sometimes vary from those in textbooks, literature reviews, and advanced ex-

Table 13.6. Sources of agricultural engineering literature

	Journals	Monographs & reports	Dissert.	Patents & standards
CABI[a]	57.0%	43.0%		
AGRICOLA[a]	49.2%	50.8%		
Analyzed publications	58.1%	38.2%	3.1%	0.6%

[a]Data from online databases, 1984–88.

pository monographs. Given a large enough sample, these potential skews are removed or minimized.

This range of 49.2–58.1% of the literature coming from journals is decidedly different than that of 74–85% journal literature reported for all of agriculture,[6] or the 72.1–80% journal literature in agricultural economics and rural sociology.[7] This is a direct result of the heavy publishing in the report format in engineering. The use of graphics and computation in agricultural engineering often for field applications lead to small, easily produced reports.

A Master's thesis study of citation practices in agricultural engineering was done by Bryan A. Morgan in 1977 at Loughborough University of Technology. The study included analysis of twenty theses (1972–1976), 38 unpublished research reports and subsequently published papers. There appeared to be "no statistically significant differences between citations in the unpublished and published material."[8] The analysis provided data on form of publication, country of origin, language and date.

   6. Olsen, "Characteristics of Agricultural Economics Literature," Table 4.
   7. Ibid., Table 5.
   8. B. A. Morgan. *A Comparison of Citation Practices in Published and Unpublished Materials in Agricultural Engineering.* Master's Thesis. Loughborough University of Technology, Oct. 1977. 79 p. Quote, p.iv.

14. Citation Analysis and the Core Monographs

WALLACE C. OLSEN
Mann Library, Cornell University

Citation counting and analyses have been used extensively in libraries and scholarship to provide answers to some questions, or to observe patterns and trends in literature use. Beginning in the 1950s, citation analysis grew dramatically as a result of two major thrusts: 1) the implementation of computing storage and speeds in compilation and analysis, and 2) publication of immense citation databases, *Science Citation Index* and the *Social Science Citation Index,* which index the citations in articles in approximately 5,500 journals.[1] Bibliometric techniques have been used in a variety of applications within the publishing, library and scholarly community. They have proven useful in measuring research and education productivity, determining a scholar's output and impact, and as indicators of social and economic growth. The reader is referred to a discussion of citation analysis and its applications in the agricultural sciences in a recent publication.[2]

A structured citation database is necessary from which correlations, data, and conclusions can be obtained. This study began by examining the Institute for Scientific Information's print and online tools.[3] It is clear upon examination, however, that these publications do not adequately index the literature of agricultural engineering and, therefore, have limited use. Some useful data from ISI sources concerning the journal literature will be given in the chapter following. One common problem with the ISI databases is that they deal with the research literature. Citations at the ends of largely research articles are reflective of the point of view and approach of a re-

---
1. *Science Citation Index* and *Social Science Citation Index* are products of the Institute for Scientific Information, Philadelphia, Pa.
2. Wallace C. Olsen, *Agricultural Economics and Rural Sociology: The Contemporary Core Literature* (Ithaca, N.Y. and London: Cornell University Press, 1991).
3. *Science Citation Index.*

searcher. This may or may not reflect the literature used by an educator or applications practitioner, particularly in the agricultural sciences. It was necessary to establish another path for identification of titles, and for quantifying what we wanted to know. Findings then had to be related to other studies or databases wherever valid.

A. Purpose and Methods

The primary aim of this study is to determine the core literature of agricultural engineering of the past thirty years which still has impact in academic teaching and research today. A further aim is to determine the relative rank and merits of the titles for both the worldwide academic community, but also for the developing or Third World countries if this proved different. This focus required a careful analysis of the tools and citation analysis methods to determine the core. As mentioned earlier, the research literature analyses are relatively numerous and useful, but they do not measure that literature of greatest value to advanced students and beginning researchers. The aim is to identify the literature in the middle of a continuum from undergraduate college education, to advanced and graduate students, to post-doctoral research literature. There is, of course, an overlap from the center of the continuum into undergraduate and advanced research literature.

A variety of literature tools exist such as specialty abstracting tools, *Agricultural Engineering Abstracts* being a good illustration.[4] The tools present comparison problems caused by differing definitions, time periods, and formats of coverage. Literature studies already published offer some help, but they are few and tend to concern themselves with journals only, while the aim here was to examine all formats. The one aspect missing in compilations and studies was the qualitative evaluation by professionals in agricultural engineering.

This led logically to literature reviews, selective reading brought together in book form, reserve readings used in academic departments as adjuncts to the classroom, and landmark agricultural engineering monographs. This overview literature proved most nearly to meet the aims of the project as a valid method to establish our base and to identify literature use patterns. Overview literature incorporates two desired quality factors: the works are classics in their own right, and they are peer-reviewed.

In surveying the citations to literature in many monographs and landmark

4. *Agricultural Engineering Abstracts* published from 1975 by Commonwealth Agricultural Bureaux International.

works it became clear that agricultural engineers are not as securely tied to the literature and citation justification as are the agricultural economists. In fact the reverse is usually true. Agricultural engineering literature cites few predecessor sources related to the writings inhand. In several cases of textbooks and histories no citation aids were provided. This may speak to the state of scholarship among agricultural engineers, who are better known for calculations, drawings, and data analysis. It is fairly clear that the American Society of Agricultural Engineers has been influential in changing this pattern in both its journals and books. This point was discussed with a Steering Committee (see the Preface) established to guide the Core Agricultural Literature Project and decisions made concerning those titles to be used for establishment of the initial database of citations.

A recent literature review from China published as a separate document of 190 pages points up the earlier standard method of reviewing in agricultural engineering in the developed countries. The thirty-two articles averaging near six pages each are a mixture of original articles by Chinese authors, and fifteen translated articles originally appearing in journals published in English, Japanese and Russian. The text is entirely in Chinese although the few literature references are in mixed languages and alphabets. Only three of the articles have more than ten references; most have none. The title of this work is *Chung-ching shih, Ko hsueh ci shu wen* (Literature on the Trends of Farm Mechanization) and was issued by the Chongqing Branch of Chung-kuo ko hsueh chi shu ching pae yen chiu so (Science and Technology Document Publishing House) in 1981. The volume aims to bring Chinese agricultural engineering up to date on latest happenings in agricultural mechanization in advanced countries with most illustrations drawn from the United States, Japan, and the USSR. Monographs such as this devoted entirely to a broad status report or literature review are uncommon in agricultural engineering.

Citation analysis has pitfalls in methods and applications and must be carefully executed within different disciplines, by databases used, and by time periods. These are summarized and discussed in a previous work.[5] Potential problems can be eliminated or minimized through a variety of techniques, some statistical, some empirical. An extensive knowledge of the literature of the field can obviate false starts and misdirection.

An effort was made to locate those monographs which would fit established criteria, and eliminate possible skews. This included adequate coverage of different dates of publication, subject scope to match the major areas of agricultural engineering in the past thirty years, and geographic distribu-

---

5. Olsen, *Agricultural Economics and Rural Sociology*, chapter 4.

tion of publications or the authors. Attention was also paid to publishing organizations with heavy influence and credibility in the discipline. The Core Agricultural Literature Project had a particular interest in determining if the developed country literature was differently oriented than that for the Third World or developing nations. Therefore, this bias was included for the purposes of comparison of these two groups.

Sources of Citations in Agricultural Engineering

Monographs

Advances in Evapotransporation: Proceedings of the National Conference on Advances in Evapotransporation, Dec. 1985, Chicago. St. Joseph, Mich.; American Society of Agricultural Engineers, 1985. 453p. (*ASAE Publication 14-85*)

Ahmed, Iftikhar and Bill H. Kinsey, eds. *Farm Equipment Innovations in Eastern and Central Southern Africa*. United Kingdom; Gower Publishing Company, 1984. 368p.

American Society of Agricultural Engineers. *Advances in Drainage: Proceedings of the ASAE Fourth National Drainage Symposium, December, 1982, Chicago*. American Society of Agricultural Engineers, 1985. 750p.

American Society of Agricultural Engineers. *Agricultural Waste Utilization and Management*. American Society of Agricultural Engineers, 1985. 750p.

Asian Productivity Organization. *Farm Mechanization in Asia*. Tokyo; Asian Productivity Organization, 1983. 510p.

Booher, L. J. *Surface Irrigation*. Rome; Food and Agriculture Organization of the United Nations, 1974. 160p. *(FAO Agricultural Development Paper no. 95)*

Crossley, C. P. and J. Kilgour. *Small Farm Mechanization for Developing Countries*. New York; John Wiley, 1983. 253p.

Food and Agriculture Organization. *Tillage and Seeding Practices and Machines for Crop Production in Semiarid Areas*. Rome; Food and Agriculture Organization, 1971. 53p.

Gensler, William G., ed. *Advanced Agricultural Instrumentation Design and Use*. Dordrecht and Boston; Martinus Nijhoff Pub., 1986. 480p.

Grimmova, Helga and Milan Stastny. *Uspora Energie ve Sklenicich, Studijni zprava = Energy Savings in Glasshouses, A Review*. Prague; Ustav Vedeckotechnickych Informaci pro Zemedelstvi, 1983. 57p. (*Studijni Informace; Zemedelska Technika*, 83, 4).

Hancarova, Daniela. *Melioracni Opatreni Pro Zvyseni Urednosti Ulehlych Pud*. Prague; Ustav Vedeckotechnickych Informaci pro Zemedelstvi, 1983. (*Studijni Informace; Zemedelska Technika*).

Hillel, David, ed. *Advances in Irrigation*, Vol. 4. Orlando; Fla; Academic Press, 1987. 405p.

Hunt, Donnell R. *Engineering Models for Agricultural Production*. Westport, Conn.; AVI, 1986. 260p.

Kennedy, W. K. and T. A. Rogers. *Human and Animal-Powered Water-Lifting Devices*. London; Intermediate Technology Publications, 1985. 111p.

Molenaar, Aldert. *Water Lifting Devices for Irrigation*. Rome; Food and Agriculture Organization, 1956. 76p. (*FAO Agricultural Development Paper* no. 60)

Quick, Graeme R., and George F. Montgomery, eds. *Bibliography on Combines and Grain Harvesting; Citations from the International Literature on the Engineering,*

Biological and Economic Aspects of the Harvesting of Crops for Grain and Seed. St. Joseph, MI; American Society of Agricultural Engineers, 1974. 71p. *(ASAE Special Publication SP-0274).*

Stout, B. A. *Equipment for Rice Production.* Rome; Food and Agriculture Organization, 1966. 169p. *(FAO Agricultural Development Paper* no. 84)

Stout, B. A. *Handbook of Energy for World Agriculture.* Barking, Essex, England and New York; Elsevier, 1989. 506p.

Smith, Harris P. and Lambert Henry Wilkes. *Farm Machinery and Equipment.* 6th ed. New York; McGraw Hill, 1976. 488p.

United Nations Industrial Development Organization. *Monographs on Appropriate Industrial Technology.* New York, 1979. No. 3: *Appropriate Industrial Technology for Paper Products and Small Pulp Mills.* 149p.

United Nations Industrial Development Organization. *Monographs on Appropriate Industrial Technology.* New York, 1979. No. 4: *Appropriate Industrial Technology for Agricultural Machinery and Implements.* 160p.

United Nations Industrial Development Organization. *Monographs on Appropriate Industrial Technology.* New York, 1979. No. 5: *Appropriate Industrial Technology for Rural Requirements.* 169p.

United Nations Industrial Development Organization. *Monographs on Appropriate Industrial Technology.* New York, 1979. No. 7 *Appropriate Industrial Technology for Food Storage and Processing.* 120p.

United Nations Industrial Development Organization. *Monographs on Appropriate Industrial Technology.* New York, 1979. No. 9: *Appropriate Industrial Technology for Oils and Fats.* 50p.

The Journal of Agricultural Engineering Research has published review articles since 1982; all were analyzed through 1988.

1982 (v. 27).

J. R. Sharp. "A Review of Low Temperature Drying Simulation Models." pp. 169–190.
M. J. O'Dogherty. "A Review of Research on Forage Chopping." pp. 267–289.
D. J. Campbell. "A Review of the Clod Problem in Potato Production." pp. 373–395.
D. V. Rees. "A Discussion of Sources of Dry Matter Loss During the Process of Haymaking." pp. 469–479.

1983 (v. 28).

A. M. Ramsay. "Mechanical Harvesting of Raspberries—A Review with Particular Reference to Engineering Development in Scotland." pp. 183–206.
L. M. Kahn and M. A. Hanna. "Expression of Oil from Oilseeds—A Review." pp. 495–503.

1984 (v. 29).

H. Göhlich. "The Development of Tractors and Other Agricultural Vehicles." pp. 3–16.
N. R. Scott. "Livestock Buildings and Equipment: A Review." pp. 93–114.
H. Kuipers. "The Challenge of Soil Cultivations and Soil Water Problems." pp. 177–190.
T. C. D. Manby, O.B.E. "Selected Aspects of Efficient Crop Production." pp. 275–293.

1984 (v. 30).

F. W. Bakker-Arkema. "Selected Aspects of Crop Processing and Storage: A Review." pp. 1–22.
C. P. Schofield. "A Review of the Handling Characteristics of Agricultural Slurries." pp. 101–109.
M. F. Diprose and F. A. Benson. "Electrical Methods of Killing Plants." pp. 197–209.

1985 (v. 31).
C. W. Plackett. "A Review of Force Prediction Methods for Off-road Wheels." pp. 1–29
P. J. Kettlewell and M. J. B. Turner. "A Review of Broiler Chicken Catching and Transport Systems." pp. 93–114.
D. C. McRae. "A Review of Developments in Potato Handling and Grading." pp. 115–138.
D. H. Rackham and D. P. Blight. "Four-Wheel Drive Tractors—A Review." pp. 185–201.

1985 (v. 32).
J. L. Parry. "Mathematical Modelling and Computer Simulation of Heat and Mass Transfer in Agricultural Grain Drying: A Review." pp. 1–29.
K. L. Hughes and A. R. Frost. "A Review of Agricultural Spray Metering." pp. 197–207.
S. Gunasekaran, M. R. Paulsen and G. C. Shove. "Optical Methods for Nondestructive Quality Evaluation of Agricultural and Biological Materials." pp. 209–241.

1986 (v. 33).
A. J. Hunter. "Thermodynamic Criteria for Optimum Cooling Performance of Aeration Systems for Stored Seed." pp. 83–99.
G. A. Carpenter. "Dust in Livestock Buildings—Review of Some Aspects." pp. 227–241.

1986 (v. 34).
M. A. Foale, R. Davis and D. R. Upchurch. "The Design of Rain Shelters for Field Experimentation: A Review." pp. 1–16.
T. R. Cumby. "Design Requirements of Liquid Feeding Systems for Pigs: A Review." pp. 153–172.

1986 (v. 35).
B. C. Ball. "Cereal Production with Broadcast Seed and Reduced Tillage: A Review of Recent Experimental and Farming Experience." pp. 71–95.

1987 (v. 36)
T. R. Cumby. "A Review of Slurry Aeration: 1. Factors Affecting Oxygen Transfer." pp. 141–156.

1987 (v. 37)
M. P. Foster and M. J. Down. "Ventilation of Livestock Buildings by Natural Convection."

1988 (v. 39).
R. S. Billington. "A Review of the Kinetics of the Methanogenic Fermentation of Lignocellulosic Wastes." pp. 71–84.

1989 (v. 44)
M. J. O'Dogherty. "A Review of the Mechanical Behaviour of Straw when Compressed to High Densities." pp. 241–65.

These sources include two systematic reviewing publications in agricultural engineering. The one is a Czechoslovakian series, *Studijni Informace; Zemedelska Technika* and the other is literature reviews in the highly respected *Journal of Agricultural Engineering Research*. Czechoslovakia, Germany, Scandinavia, Netherlands, and the USSR have long had active agricultural engineering programs particularly in research. Agricultural engineering publications from these countries are represented in a higher percentage than in the agricultural economics and rural sociology disciplines.[6] The Czech state-of-the-art and literature reviews, however, are as parochial as much United States agricultural engineering literature by giving few citations in some instances, and those being concentrated in publications from Central Europe. Two of these, however, were considered adequately broadly-based to be valuable for analysis.

The *Journal of Agricultural Engineering Research* has mostly British or Commonwealth literature reviewers, but is much more inclusive of literature published outside the reviewer's country. These articles also do not cite literature exensively in a few cases due to some of the esoteric subjects.

B. Compilation and Citation Analysis

The 8,184 citations in these overview and literature publications were systematically counted, and the following data gathered: title of publication; date of publication; format of publication (e.g. journal or monograph); and category of publisher (e.g. commercial press, university, government). During this process early in the study, the titles of monographs were noted and entered into a computerized list. Each time the same monograph or a chapter in it was cited, a tally was made for that title. Similarly, each time a journal or report series was cited, a tally was made for it. The end result is a systematic count of which journals, report series, and monographs were cited and the numbers of times. Additional, select data were gathered which provide the basis for some analysis in this and the following chapter.

Before examining the results, some definitions must be understood. Distinctions between journals, serial works, and monographs are necessary. Series issues, short works, and books were treated as monographs when they were cited as distinct works with an author or editor, when a title was distinctive, and the item was complete in itself. Therefore, a work such as an *FAO Irrigation and Drainage Paper* in which the subject was complete in the issue and was distinct from the next issue in the series was counted and identified as a monograph. The same was true of proceedings volumes when given distinctive titles which varied from one year to the next. Those with no distinct titles or special subject focus were counted as journals. Journals follow the pattern of having several articles on different and specific subjects and are usually only a few pages long. Chapters in books were counted as monographic titles. These definitions worked well since they follow the citation patterns of the agricultural engineering authors.

6. Ibid., chapters 4 and 5.

In the compilation process select materials were excluded:

(1) Very short monographs, fifty pages or fewer which were cited only once and evaluated by reviewers as less than critical.
(2) Country, State, or Provisional documents of brief pagination often highly specialized and site specific.
(3) Select esoteric works *when* in a difficult-to-read language, or *when* very limited in geographic scope.
(4) Select geographic materials when not in a national or international context and which got cited only once.
(5) Early background, technique, and statistical working tools, primarily the core of the engineering sciences. Numerous works of this type were kept in the lists based on the number of times cited or their high ratings.
(6) Early editions were combined with the latest edition although information is usually given on both.

These analyses offered some overall patterns of publishing which are briefly noted here.

Format of Literature Cited

The purpose of an article or book dictates the nature of the literature cited. Therefore, a journal article reporting the results of research will tend to cite the research literature, supporting methodologies, and overview earlier works. As mentioned, this study has attempted to identify advanced university instructional literature plus the basic research literature. The choice of the literature in the sources of citations list which was analyzed had to be systematically chosen. The literature and state of the art reviews analyzed from the *Journal of Agricultural Engineering Research* are heavily oriented toward latest developments and overviews of current research. It is logical to expect citation differences between this corpus and that of the monographs.

The diversity is not as great as might be anticipated. Variations are shown in Figure 14.1. The journal literature cited in both groups ranged from 54.8% in the monographs to 67.7% in the journals. Citations from the two sources average to 58.1% to journals, 38.2% to monographs, and 3.7% to dissertations, patents, and standards.

Monographic Publishers

The 38.2% of citations to monographs were analyzed by type of publisher. Figure 14.2 indicates these publisher types and numerical variations by the journal articles and the monographs analyzed. Governments are strong; these include state or province, federal or national, UN, and FAO. The FAO clearly is very influential in agricultural engineering literature.

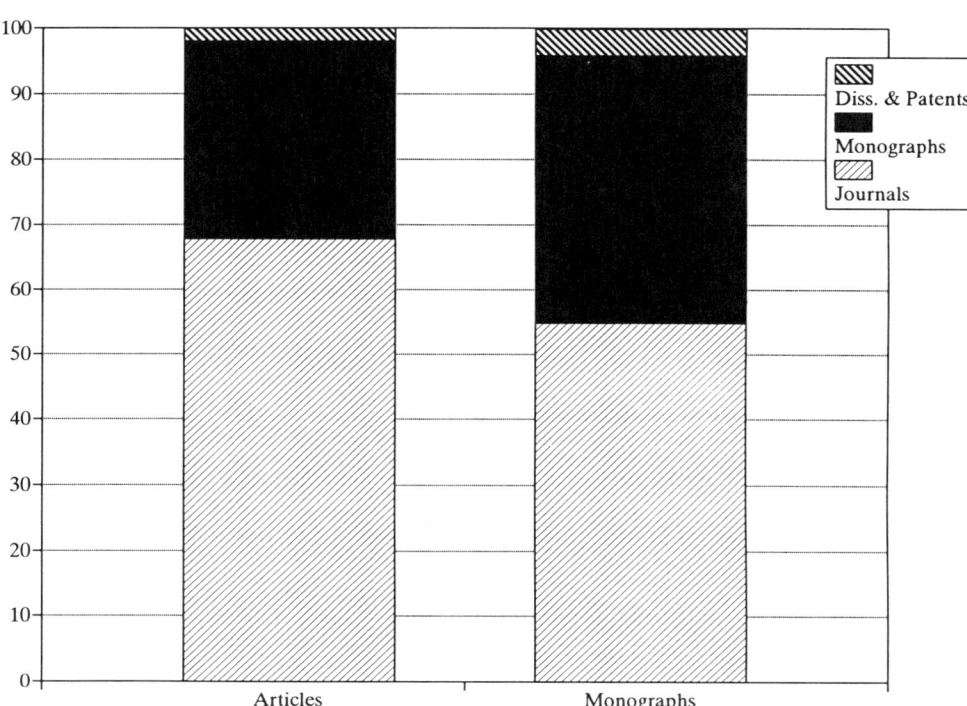

Figure 14.1. Literature formats of citations analyzed in journal articles and monographs.

Independent organizations which are strong in the journal articles analyzed are exemplified by private institutes and international agricultural research centers, but primarily by societies.

C. Monograph Peer Evaluation

The monograph list compiled by citation analyses was perused by the Steering Committee prior to being sent for peer evaluation. Two agricultural engineers working in developing countries were also asked to look at the list and determine which titles had application for the Third World. Additional feedback on Third World appropriateness was sought from two scholars from developed countries with wide experience in agricultural engineering in the Third World. These processes resulted in:

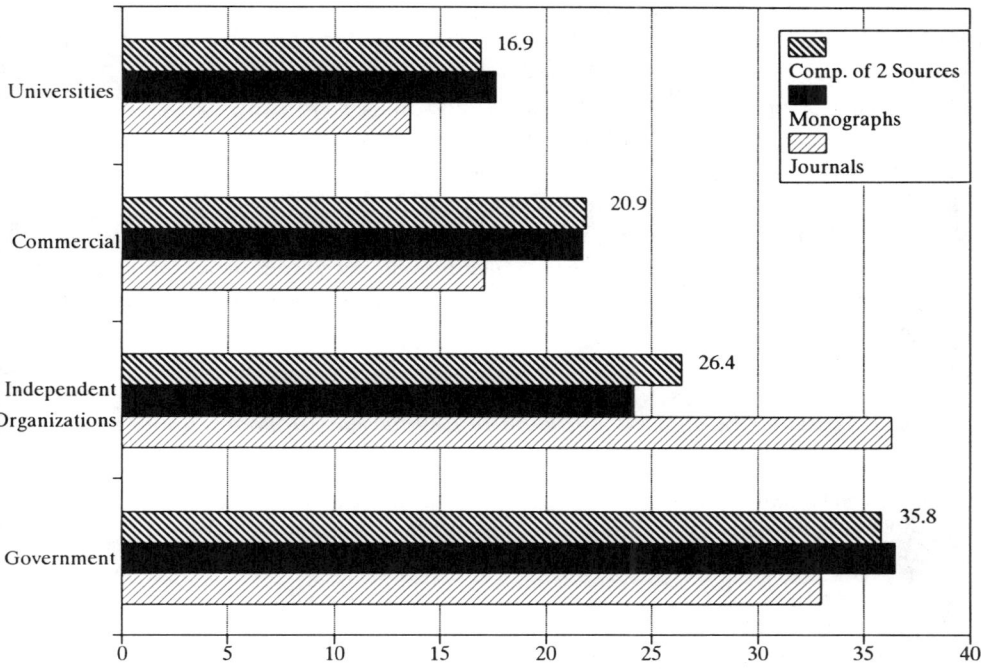

Figure 14.2. Publishers of monographs and reports cited.

(1) Breaking the master list into two portions, one for the advanced countries, and a second for the developing nations. There was extensive overlap between them.
(2) The developed countries list was near 1,500 titles. Since the next step was to have these titles ranked by agricultural engineers, the Steering Committee felt that evaluations would be unrealistic unless the titles were placed into subject groups and distributed to specialists in those subjects. This resulted in eight subject lists using the six categories of the *Transactions of the American Society of Agricultural Engineers* to which were added two general categories at the beginning of the set. These first two subject lists were sent to all reviewers along with a list in their subject specialty. Lists were issued in these categories:
  (1) Basic Agriculture, Engineering and Science;
  (2) General Agricultural Engineering, including International Development;
  (3) Ag Eng.: Electrical and Electronic;
  (4) Ag Eng.: Food and Process Engineering;
  (5) Ag Eng.: Power and Machinery;
  (6) Ag Eng.: Soil and Water, including Conservation and Irrigation;
  (7) Ag Eng.: Structures and Environment, including Animal Waste;
  (8) Ag Eng.: Emerging Technologies: Bioengineering, Forest Engineering, Robotics, and Computer Simulation.

These eight lists for developed countries agricultural engineering went through several iterations. Titles were dropped and titles were added upon the recommendation of reviewers as well as through the citation analysis process. Approximately 2,500 monograph titles were evaluated in the developed countries list of which 795 made the final compilation. In a similar manner a maximum of 1,300 monographs were evaluated in the Third World listing which resulted in a core of 623. Instructions to the reviewers carefully discussed the aim of identifying monographs of academic and instructional value for Third World institutions, or those which should be represented in a university library in the developed world. Following are the names of reviewers for each type of list.

A. Developed Countries Agricultural Engineering Monograph Reviewers

Donald B. Brooker
  Columbia, Missouri
Robert H. Brown
  University of Georgia
Joseph K. Campbell
  Cornell University
William Chancellor
  University of California, Davis
Ashim K. Datta
  Cornell University
M. P. Douglass
  Silsoe College, England
D. P. Froelich
  South Dakota State University
Kifle G. Gebremedhin
  Cornell University
Carroll E. Goering
  University of Illinois
Carl W. Hall
  Arlington, Virginia
C. Gene Haugh
  Virginia Polytechnic Institute and State University
Harry Henderson
  University of Wisconsin, Platteville
Edward A. Hiler
  Texas A & M University
Donnell R. Hunt
  University of Illinois
Gil Levine
  Cornell University
John B. Liljedahl
  Purdue University

Arne Moller
  Hundested, Denmark
Bryan A. Morgan
  Silsoe College, England
Robert M. Peart
  University Of Florida
Sverker Persson
  State College, Pennsylvania
John Perumpral
  Virginia Polytechnic Institute and State University
Allan L. Phillips
  University of Puerto Rico, Mayaguez
H. B. Puckett
  Urbana, Illinois
Lyle G. Reeser
  East Peoria, Illinois
Adrianus G. Rijk
  Asian Development Bank Manila
Errol D. Rodda
  University of Illinois
Glenn O. Schwab
  Powell, Ohio
B. C. Stenning
  Silsoe College, England
Dennis Stombaugh
  Ohio State University
Dwayne A. Suter
  Texas A & M University
Arthur Teixeira
  University of Florida
Gerald E. Thierstein
  Kansas State University

Jaw-Kai Wang
   University of Hawaii

James Young
   North Carolina State University

B. Third World Agricultural Engineering Monograph Reviewers

Pierre F. J. Abeels
   Université Catholique de Louvain,
   Belgium
Luis A. Balastreire
   ESALQ, Universidad de Sao Paulo
   Piracicaba, Brasil
Luis A. Barbosa Cortez, and Paulo A. Martins Leal
   UNICAMP
   Campinas, Brasil
Mohd. Zohadie Bardaie
   Universiti Pertanian Malaysia
   Selangor Darul Ehsan
   Malaysia
Rupert Best
   Centro Internacional de Agricultura Tropical
   Cali, Colombia
William Chancellor
   University of California, Davis
Dante DePadua
   International Development Research Centre,
   Singapore
Ir. Eriyatno
   Bogor Agricultural University
   Bogor, Indonesia
LeVern W. Faidley
   FAO
   Rome, Italy
L. O. Gumbe
   University of Nairobi
   Nairobi, Kenya
Z. M. Kasomekera
   University of Malawi
   Lilongwe, Malawi
R. N. Kaul
   Ahmadu Bello University
   Zaria, Nigeria
Karel J. Lenselink
   University of Nairobi
   Kabete Campus
   Nairobi, Kenya
Gil Levine
   Cornell University
P. A. M. Misiko
   Egerton University
   Njoro, Kenya
Lyle G. Reeser
   East Peoria, Illinois
Adrianus G. Rijk
   Asian Development Bank
   Manila, Philippines
R. C. Sachan
   ICRISAT, Patancheru,
   India
K. L. Sahrawat
   ICRISAT, Patancheru,
   India
Salim Said
   Universiti Pertanian Malaysia
   Selangor Darul Ehsan,
   Malaysia
Gajendra Singh
   Asian Inst. of Technology
   Bangkok, Thailand

Joint evaluation by:
Norbert Fritsch F.
Walter Luzio L.
Jose D. Opazo A.
Edgardo Ossandon P.
Mario Peralta P.
Gabriel Selles v. Sch.
   Universidad de Chile,
   Santiago, Chile

Weighting the Monograph Lists

Several devices were used to rank the titles. Numeric scores were assigned to these elements and each title computed. The data used are:

(1) The counts made each time a monograph or a chapter of a monograph was cited in the sources of citations list, above, were coded into the working lists. Reviewers were unaware of the counts. This element was given the weight of 1 per citing.
(2) Rankings by reviewers were coded for each of the two top rankings. These were graded two and one and multiplied times the number of recommendations in each category. A statistical equalization on the 8 developed countries lists was required because the first two, general lists were sent to each reviewer resulting in 10 times as many evaluations as for the specialty lists.
(3) If a title was reprinted it was given a score of 1.
(4) If a title went through more than one edition, it was given a score of 2.

The equation for the computations is:

$$((\#) \times 1) + (\# \times 2 + \# \times 1) + (Rpnt. \times 1) + (Ed. \times 2)$$

This formula was used for both lists of monographs which were ranked separately (see Figure 14.3). By this equation, the peer evaluations accounted for between 83% and 93% of all the scores in the two lists.

Within each list the scores were broken into three logical ranges based on accrued value. The fewest titles in both lists are in the first ranked or most important titles; the greatest number of titles group around the middle rank or median. Percentages of titles in each list by ranking are:

Developed Countries List		*Third World List*
10.4%	FIRST ranking	21.8%
52.4	SECOND ranking	50.7
37.2	THIRD ranking	27.5

There is a clear case of high ratings being assigned by reviewers of the Third World titles. The same pattern was true in a similar evaluation in agricultural economics and rural sociology.[7] Some reviewers indicated that if there was any prospect of getting the titles in machine-readable form, they wanted as many as possible. This, of course, only increased the ranking level without significantly influencing the numbers of titles in the list.

Readers are reminded that all the titles in this Core are highly valuable monographs for instruction and research today. Those which are less valuable were never entered for consideration in many cases, and the peer-evaluations had a drop rate over 60%. Therefore, all the titles in the two lists should be viewed as valuable. Rankings in three categories within each list are provided to aid in making decisions for purchase, preservation, or collection assessment.

7. Ibid., chapter 6.

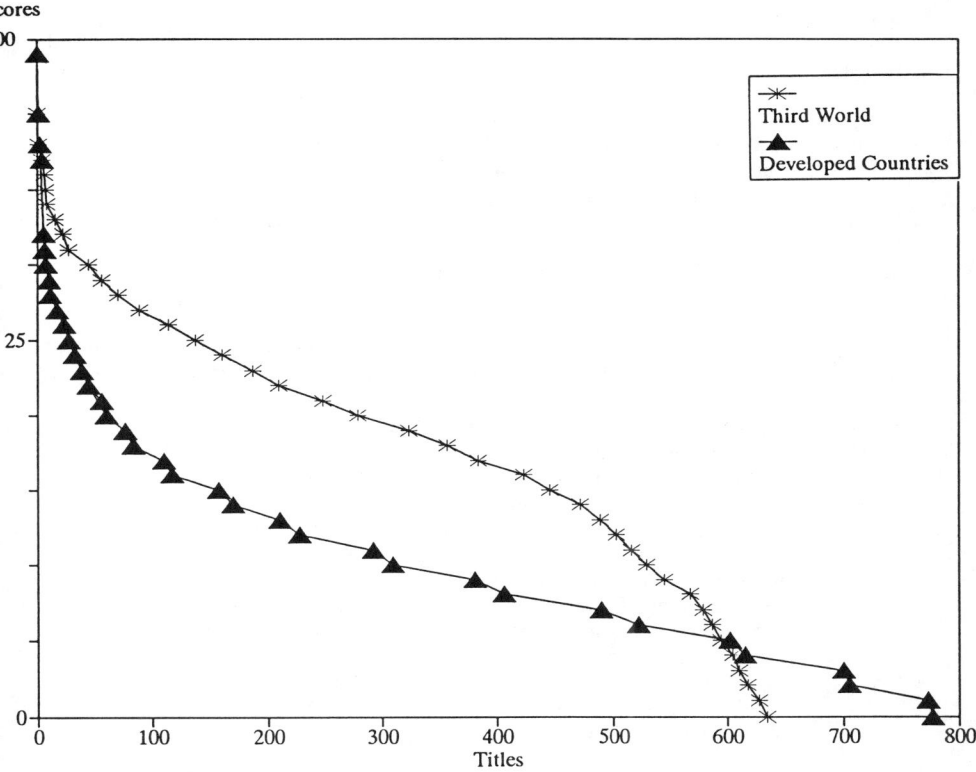

Figure 14.3. Final scoring on monographs.

The two lists have been merged in the compilation following. Rankings for a title are within its own listing only, i.e. developed countries list, or Third World list. Where no ranking is shown, the title is not in that list. There are 1,054 unique titles in the combined list of which 388 are common to both. These 388 constitute the most valuable Core titles.

Core List of Monographs for Agricultural Engineering

Developed Countries Rankings		Third World Rankings

A

Third — Abbenhuis, C. J. E. Administratieve Automatisering in de Zuegenhouderij = The Automation of the Administrative Management on Pig Breeding Farms. Wageningen; IMAG, 1981. 79p. (Instituut Voor Mechanisatie, Arbeid en Gebouwen. Publikatie, no. 161). (Summary in English).

Second — Advances in Cereal Science and Technology, Vol. 1–10. St. Paul, Minn.; American Association of Cereal Chemists, Biennial, 1976–90.

Agarwal, Bina. Mechanization in Indian Agriculture: An Analytical Study Based on the Punjab. New Delhi; Allied, 1983. 290p. — Second

Second — Ageikin, Ia. S. Off-the-Road Wheeled and Combined Traction Devices: Theory and Calculation. (Vezdekhodnye Kolesnye i Kombinirovannye Dvizhiteli: Teoriia i Raschet.) Translated from Russian by A. Jaganmohan. New Delhi: Amerind Publishing Co. Pvt. Ltd, 1987. 202p. (U.S. Dept. of Agric. Translation Ag. TT 85-1-1122). (Also issued in Rotterdam; A. A. Balkema Pub., 1988. 211p.) — Second

Third — Agrawal, K. K. Handbook of Farm Structures for the Semiarid Tropics. Dire Dawa, Ethiopia; College of Agriculture, 1973. 250p.

First — Agri-Mation, 1 & 2; Proceedings of Agri-Mation Conferences and Expositions held in Chicago, 1985 + 1986. St. Joseph, Mich.; American Society of Agricultural Engineers, 1985–86. 2 vols.

First — Agricultural Technology for Developing Nations: Farm Mechanization Alternatives for 1–10 Hectare Farms; Proceedings, Special International Conference, University of Illinois . . . 1978. Urbana-Champaign; Board of Trustees, University of Illinois; Moline, Ill., 1978. 105p.

Second — Ahmed, Iftikhar and Bill H. Kinsey, eds. Farm Equipment Innovations in Eastern and Central Southern Africa. Aldershot, Hamshire England; Brookfield, Vt.; Gower Pub., 1984. 345p. — Second

Third — Aiba, Shuichi, Arthur E. Humphrey and Nancy F. Millis. Biochemical Engineering. 2d ed. New York; Academic Press, 1973. 434p.

First — Akesson, Norman B. and Wesley E. Yates. The Use of Aircraft in Agriculture. Rome; Food and Agriculture Organization of the United Nations, 1974. 217p. (FAO Food and Agriculture Development Paper 94). — Second

Third — Albery, Wyndham John. Electrode Kinetics. Oxford; Clarendon Press, 1975. 184p.

First — Alcock, Ralph. Tractor-Implement Systems. Westport, Conn.; AVI, Publishing Company, Inc., 1986. 161p. — Second

Third — Alekseeva, T. V. et al. Machines for Earthmoving Work: Theory and Calculations. Translation of Mashiny dlia Zem-

	lianykh Rabot: Teoriia i Rashet. 3d, rev. ed. New Delhi; Amerind Pub. Co. for The U.S. Dept. of Agriculture and the National Science Foundation, 1985. 515p. (Originally published: Moscow; Mashinostroenie, 1972).	
	American Society of Agricultural Engineers. Agricultural Technology for Developing Nations: Farm Mechanization Alternatives for 1–10 Hectare Farms; Proceedings, Special International Conference, University of Illinois . . . 1978. Urbana-Champaign; Board of Trustees, University of Illinois, 1978. 105p.	First
Second	American Society of Agricultural Engineers. Computers and Farm Machinery Management; A Symposium of Papers Presented During a Conference Dealing with Computer Applications in Farm Machinery Management, Chicago, 1968. St. Joseph, Mich.; American Society of Agricultural Engineers, 1968. 55p.	Second
Third	American Society of Agricultural Engineers. Conference on Crop Production with Conservation in the 80's, 1980, Chicago. Crop Production with Conservation in the 80's; Proceedings of the . . . St. Joseph, Mich.; American Society of Agricultural Engineers, 1981. 281p.	
Second	American Society of Agricultural Engineers. Conference Papers from Two Conferences on Standardizing Properties and Analytical Methods Related to Animal Waste Research, 1972 + 1974. St. Joseph, Mich.; American Society of Agricultural Engineers, 1975. 355p.	
Second	American Society of Agricultural Engineers. Drainage for Efficient Crop Production; Conference Proceeding . . . Chicago, 1965. St. Joseph, Mich.; American Society of Agricultural Engineers, 1965. 77p.	First
Third	American Society of Agricultural Engineers. Electromagnetic Radiation in Agriculture; Conference Proceedings; A Symposium of Papers Presented During an International Conference Dealing with the Effects of Light and its Associated Spectrum (Ultraviolet, Infrared, and Gamma) on Plants, Livestock, Insects, and Man Held at Roanoke, Virg. St. Joseph, Mich.; American Society of Agricultural Engineers, 1965. 70p.	
Second	American Society of Agricultural Engineers. Energy Management and Membrane Technology in Food and Dairy Processing; From the Special Food Engineering Symposium Held in Conjunction with Food and Dairy Expo. '83, Chicago. St. Joseph, Mich.; American Society of Agricultural Engineers, 1983. 116p.	Second
Second	American Society of Agricultural Engineers. Engineering a Safer Food Machine: A Collection of Agricultural Safety Papers and Speeches. St Joseph, Mich.; American Society of Agricultural Engineers, 1980. 163p.	
	American Society of Agricultural Engineers. Engineering Properties of Biological Materials. St. Joseph, Mich.; American Society of Agricultural Engineers, 1967. 103p.	First
Third	American Society of Agricultural Engineers. Farmstead Engi-	

DC	neering Conference, Proceedings . . . Chicago, 1980. St. Joseph, Mich.; American Society of Agricultural Engineers, 1981. 200p. (ASAE Publications, Vol 8–81).	TW
First	American Society of Agricultural Engineers. Fruit, Nut and Vegetable Harvesting Mechanization. St. Joseph, Mich.; American Society of Agricultural Engineers, 1984. 411p.	Second
Second	American Society of Agricultural Engineers. Guide for Investigation of Subsurface Drainage Problems on Irrigated Lands. St. Joseph, Mich.; American Society of Agricultural Engineers, 1967. 57p.	
	American Society of Agricultural Engineers. Livestock Waste Management and Pollution Abatement; Proceedings of the International Symposium on Livestock Wastes, Ohio State University, 1971. St. Joseph, Mich.; American Society of Agricultural Engineers, 1971. 358p.	Second
Second	American Society of Agricultural Engineers. Modeling Agricultural, Forest, and Rangeland Hydrology; Proceedings of the 1988 International Symposium . . . Chicago. St. Joseph, Mich.; American Society of Agricultural Engineers, 1988. 510p.	
Second	American Society of Agricultural Engineers. Nursery and Greenhouse Mechanization Committee (PM-59). Nursery and Greenhouse Mechanization Equipment and Manufacturers. Rev. ed. St. Joseph, Mich.; American Society of Agricultural Engineers, 1984. 72p.	
Second	American Society of Agricultural Engineers. Sprinkler Irrigation: A Compilation of Papers. St. Joseph, Mich.; American Society of Agricultural Engineers, 1983. 214p.	Third
	American Society of Agricultural Engineers. Status of Harvest Mechanization of Horticultural Crops. Prepared by the Fruit and Vegetable Harvesting Committee (PM-48). St. Joseph, Mich.; American Society of Agricultural Engineers, 1983. 98p. (ASAE Publication no. 3–83).	Second
Second	American Society of Agricultural Engineers. Trickle Irrigation: A Compilation of Published Papers. 11 Published Papers from the 1981 and 1982 Transactions of the ASAE. St. Joseph, Mich.; American Society of Agricultural Engineers, 1983. 72p.	Third
Second	American Society of Civil Engineers. Committee on Operation and Maintenance of Irrigation and Drainage Systems. Operation and Maintenance of Irrigation and Drainage Systems. New York; American Society of Civil Engineers, 1980. 277p. (ASCE Manuals and Reports on Engineering Practice no. 57).	
Second	American Society of Civil Engineers. Engineering and Construction in Tropical and Residual Soils: Proceedings of the ASCE Geotechnical Engineering Div. Specialty Conference, Honolulu, 1982. New York; American Society of Civil Engineers, 1982. 735p.	Second
Second	American Society of Civil Engineers. Ground Water Management, Prepared by the Task Committee of the Committee on	

Citation Analysis and the Core Monographs 243

	Ground Water of the Irrigation and Drainage Division, ASCE. New York; American Society of Civil Engineers, 1972. 216p. (ASCE Manuals and Reports on Engineering Practice no. 40).	
Third	American Society of Civil Engineers. Irrigation and Drainage Division. Water Today and Tomorrow; Proceedings of the Specialty Conference . . . Flagstaff, Ariz., 1984. New York; American Society of Civil Engineers, 1984. 732p.	
Second	American Society of Civil Engineers. Irrigation and Drainage Division. Environmental Aspects of Irrigation and Drainage: Proceedings of Specialty Conference, 1976. New York; American Society of Civil Engineers, 1976.	
Second	American Society of Civil Engineers. Irrigation and Drainage Division. Water Management for Irrigation and Drainage: Proceedings of the Irrigation and . . . Specialty Conference, Reno, Nevada, 1977. New York; American Society of Civil Engineers, 1977. 2 vols.	
Second	American Society of Civil Engineers. Irrigation and Drainage in the 1980's. New York; American Society of Civil Engineers, 1979. 447p.	
Second	American Society of Civil Engineers. Legal, Institutional, and Social Aspects of Irrigation and Drainage and Water Resources Planning and Management. Proceedings . . . Specialty Conference . . . Virginia Polytechnic Univ., 1978. New York; American Society of Civil Engineers, 1979. 895p.	
Third	American Society of Civil Engineers. Methods for Estimating Evapotranspiration: Irrigation and Drainage Specialty Conference, Las Vegas, 1966. New York; American Society of Civil Engineers, 1966. 236p.	
Second	American Society of Civil Engineers. Research Conference on Shear Strength of Cohesive Soils . . . Univ. of Colorado, 1960. New York; American Society of Civil Engineers, 1961. 1164p.	
	American Society of Civil Engineers Staff, ed. Consumptive Use of Water and Irrigation Water Requirements. New York; American Society of Civil Engineers, 1974. 227p.	First
Third	American Society of Civil Engineers Staff. Selected Papers from Agricultural and Urban Considerations in Irrigation and Drainage Specialty Conference, Fort Collins, Colorado, 1973. New York; American Society of Civil Engineers Staff, 1973. 800p.	
	American Society of Civil Engineers Staff. Social and Ecological Aspects of Irrigation and Drainage: Selected Papers. New York; American Society of Civil Engineers, 1970. (Reprinted by Books on Demand).	Second
Second	American Society of Heating, Refrigerating and Air Conditioning Engineers. ASHRAE Handbook; Fundamentals (Inch-Pound Edition). Atlanta; American Society of Heating, Refrigerating and Air Conditioning Engineers, 1989. (Previous edition: ASHRAE Handbook of Fundamentals. New York; American Society of Heating, Refrigerating and Air Conditioning Engineers, 1967–1972. 2 Vols).	Second

Second	Anderson, Edward E. Fundamentals of Solar Energy Conversion. Reading, Mass.; Addison-Wesley Pub. Co., 1983. 636p.
Third	Andert, A. Vyuziti a Premena Energii v Zemedelstvi. (Energy Utilization and Transformation in Agriculture; A Review). Prague; Studijni Informace; Ziemedelska Technika, 1975. 71p. (English Summary)
Third	Andreason, Ingmar and Kenneth Strom. Datorstyrning av Vaxthusklimat : en Forstudie = Computer Control of the Greenhouse Climate : an Introduction. Lund, Sweden; Institutionen for Lantbrukets Byggnadsteknik, Avdelningen for Tradgardsnaringens Byggnadskonstruktioner, 1983. 103p. (Sveriges Lantbruksuniversitet. Institutionen for Lantbrukets Byggnadsteknik. Rapport, 35.)
Third	Animal Welfare Institute. Comfortable Quarters for Laboratory Animals. Rev. ed. Washington, D.C.; Animal Welfare Institute, 1979. 108p.
	Ansari, Nasim. Economics of Irrigation Rates. New York; Asia Publishing House, 1968. 360p.	Third
Third	Antipin, V. G. Voprosy Tekhnologii i Mekhanizatsii Proizvodstva Produktov Zhivotnovodstva: Primenenie Elektroenergii i Sredstv Avtomatki v Sel'skokhoziaistvennom Proizvodstva. (Issues on the Technology and Mechanization of the Production of Animal Products.) Leningrad; NIPTIMESKH Severo-Zapada, 1975. 163p. (Nauchno-issledovatel'skii i Proektho-tekhnologicheskii Institut Mekhanizatsii i Elektrifikatsii Sel'skogo Khoziaistva Servero-Zapada, Nauchnye Trudy, no. 19)
Third	Aravin, V. I. and S. N. Numerov. Theory of Fluid Flow in Undeformable Porous Media. Translation of Teoriia Dvizheniia Ahidkostei i Gazov v Nedeformiruemoi Poristoi Srede from Russian by A. Moscona. Jerusalem; Israel Program for Scientific Translations, 1965. 511p.
Second	Archer, S. G. Soil Conservation. Norman, Okla; University of Oklahoma Press, 1956. 305p.
Second	Arkin, G. F., and H. M. Taylor. Modifying the Root Environment to Reduce Crop Stress. St. Joseph, Mich.; American Society of Agricultural Engineers, 1981. 407p. (ASAE Monograph no. 4).	Third
	Arnal, P. and A. Laguna. Tractores y Motores Agrícolas. 2d ed., revised. Madrid; Ministry of Agriculture and Mundi-Presna, 1989. 429p.	Third
Second	Arnon, Itzhak. Modernization of Agriculture in Developing Countries: Resources, Potentials and Problems. 2d ed. Chichester and New York; J. Wiley and Sons, 1987. 626p. (1st ed., 1981. 565p.)	Third
	Arustamiants, Isaak A. Kratkii Spravochnik Po Elektrifikatsii Sel'skogo Khoziaistva. (A Concise Handbook on Agricultural Electrification). Moscow; Sel'khozgiz, 1959. 250p.	Third
First	ASAE National Energy Symposium. Agricultural Energy: Selected Papers and Abstracts from the 1980 ASAE National Energy Symposium. Vol. 1: Solar Energy and Livestock Produc-

DC	tion, Vol. 2: Biomass Energy, Crop Production, Vol 3: Food Processing. St. Joseph, Mich.; American Society of Agricultural Engineers, 1981. 3 Vols. 671p.	TW
Second	Asian Productivity Organization. Farm-Level Water Management in Selected Asian Countries: Report of a Multi-Country Study Mission. Tokyo; Asian Productivity Organization, 1980. 159p. (APO Project Code OSM/III/79).	First
Second	Asian Productivity Organization. Farm Mechanization in Asia. Tokyo; Asian Productivity Organization, 1983. 510p.	First
Second	Association of Official Analytical Chemists. Official Methods of Analysis of the Association of Official Analytical Chemists. 14th ed. Vol. 1–6: Official and Tentative Methods of Analysis of the Association of Official Agricultural Chemists. Vol. 7–10: Official Methods of Analysis of the Association of Official Agricultural Chemists. Washington, D.C.; Association of Official Analytical Chemists, 1984. 1141p. (13th ed. 1980, 1018p.)	
	At de Saint-Foulc, Jean D'. Irrigation Par Aspension. Paris; Eyrolles, 1967. 231p.	Third
Second	Ayers, R. S. and D. W. Westcot. Water Quality for Agriculture. Revised ed. Rome; Food and Agriculture Organization, 1985. 174p. (1st ed., 1976. 97p.) (FAO Irrigation and Drainage Paper no. 29).	Second
Third	Azovtsev, N. G., P. I. Kozlovskii, and A. M. Shiriaev. Kompleksy Novykh Mashin Dlia Vozdelyvaniia i Uborki Kormovykh Kul'tur. (Groups of New Machines for Cultivating and Harvesting Forage Crops). Moscow; Kolos, 1973. 189p.	Third
B		
Third	Baars, C. Design of Trickle Irrigation Systems. Wageningen, Netherlands; Dept. Irrigation and Civil Engineering, Agricultural University, 1976. 106p.	Second
Third	Babichenko, Liudmila V. Osnovy Tekhnologii Pishchevykh Proizvodstv. (Basic Technology of Food Production.) Moskva; Ekonomika, 1974. 196p.	
Second	Bailey, James E. and David F. Ollis. Biochemical Engineering Fundamentals. 2d ed. New York; McGraw-Hill, 1986. 984p. (1st ed., 1977. 753p.)	
Second	Balls, R. Horticultural Engineering Technology: Field Machinery. Houndmills, Basingstoke, Hampshire; Macmillan, 1985. 216p.	
Third	Balls, R. Horticultural Engineering Technology: Fixed Equipment and Buildings. London; Macmillan, 1986. 246p.	
Third	Balovnev, V. I. New Methods for Calculating Resistance to Cutting of Soil. Translation of Novye Metody Rascheta Soprotivlenii Rezaniyu Gruntov, by P. Datta edited by V. Pandit. New Delhi; Amerind Pub. Co. for The U.S. Dept. of Agriculture and the National Science Foundation, 1983. 103p. (Originally published: Moscow; Rusvuzizdat, 1963.) (Available from Springfield, Vir.; NTIS).	
Third	Banerjee, P. K. and R. Butterfield, eds. Developments in Soil	

DC	Mechanics and Foundation Engineering. London and New York; Applied Science Publisher, 1983–1985. 2 vols.	TW
Second	Bardaie, Z. and W. I. W. Ismail. Agricultural Mechanization in Malaysia. Serdang, Malaysia; Dept. Power and Machinery, Faculty of Agricultural Engineering, University Pertanian, 1981. 143p.	
Second	Barfield, B. J., and J. F. Gerber. Modification of the Aerial Environment of Plants. St. Joseph, Mich.; American Society of Agricultural Engineers, 1979. 538p. (ASAE Monograph no. 2.)	
Third	Barfield, B. J. and R. C. Warner. Applied Hydrology and Sedimentology for Disturbed Areas. Stillwater, Oklahoma; Oklahoma Technology Press, 1981. 603p.	
Third	Barisch, Gottfried, Joachim Glass and Ruth Weigel. Grundlagen der Forsttechnik: Fachwissen des Forstingenieurs (Foundation of Forest Technique). Berlin; VEB Deutscher Landwirtschaftsverlag, 1982. 208p.	
	Barker, John. Dictionary of Soil Mechanics and Foundation Engineering. Longman, Inc., 1981.	Second
	Barker, Randolph and Y. Hayami, eds. Economic Consequences of the New Rice Technology; Proceedings of a Conference . . . Los Baños, Philippines; International Rice Research Institute, 1978. 402p.	Second
Second	Barnes, K. K., et al. Compaction of Agricultural Soils. St. Joseph, Mich.; American Society of Agricultural Engineers, 1971. 471p. (ASAE Monograph no. 1).	Third
Second	Barnes, Maurice and Clive Mander. Farm Building Construction. Alexandria, New York; Diamond Farm Book Publishers, 1986. 224p. (The Farmers Guide Ser.)	Second
Second	Barre, H. J., L. L. Sammet, and G. L. Nelson. Environmental and Functional Engineering of Agricultural Buildings. New York; Van Nostrand Reinhold, 1987. 347p.	Second
Third	Barreto, Geraldo B. Irrigacao: Principios, Metodos e Pratica. 2d ed. Campinas, Brazil; Instituto Agronomico, UNICAMP; 1966. 178p.	Third
Third	Barrett, Eric C. and Leonard F. Curtis, eds. Environmental Remote Sensing, 1 & 2; Bristol University Symposium on Remote Sensing, 1972 & 1974. New York; Crane, Russak, 1976–77. 2 vols.	
Third	Barrett, Eric C. and Leonard F. Curtis. Introduction to Environmental Remote Sensing. 2d ed. London and New York; Chapman and Hall, 1982. 352p. (1st ed., London; Chapman and Hall; New York; Wiley, 1976. 336p.)	
Second	Barrett, J. R. and D. D. Jones. Knowledge Engineering in Agriculture. St. Joseph, Mich.; American Society of Agricultural Engineers, 1989. 214p. (ASAE Monograph no. 8).	
Third	Bartholomai, Alfred, ed. Food Factories: Processes, Equipment, Costs. Weinheim, W. Germany; New York, New York; VCH Publishers, 1987. 284p.	Second
	Bartsch, W. H. Employment and Technology Choice in Asian Agriculture. New York; Praeger, 1977. 125p.	Second

Second	Barwell, I. J., G. A. Edmonds, J. S. G. F. Howe, and J. DeVeen. Rural Transport in Developing Countries. London; Intermediate Technology Publications, Ltd., 1985. 145p.	Second
	Basedow, Ludwig. Bauten der Landtechnik. Berlin; Verlag fur Bauwesen, 1969. 182p.	Third
	Bates, William N. Mechanization of Tropical Crops. London; Temple Press, 1963. 410p. (2d impression, revised). (1st ed., 1957).	Second
Third	Battistelli, Emanuele and Ugo Bobolo. La Meccanizzazione Dell'Azienda Agraria. Torino, Italy; G. B. Paravia, 1972. 2 vols.	
Third	Batty, J. Clair and Steven L. Folkman. Food Engineering Fundamentals. New York; John Wiley and Sons, 1983. 300p.	Second
Third	Bauer, M. E. Remote Sensing of Agricultural Crops and Soils: Final Report. W. Lafayette, Ind.; Purdue University Lab for Applications of Remote Sensing, 1983. 173p.	
Second	Baver, Leonard D., Walter H. Gardner and Wilford R. Gardner. Soil Physics. 4th ed. New York; J. Wiley and Sons, Inc., and London; Chapman and Hall, Ltd., 1972. 498p. (1st ed., 1940. 370p.)	
Second	Bear, J. D., D. Zaslavsky, and S. Irmay, eds. Physical Principles of Water Percolation and Seepage. Paris; UNESCO, 1968. 465p.	
Second	Beasley, R. P., James M. Gregory and Thomas R. McCarty. Erosion and Sediment Pollution Control. 2d ed. Ames, Iowa; Iowa State University Press, 1984. 354p. (1st ed., 1974. 320p.)	
Third	Beecher, Gary R., ed. Research Instrumentation for the 21st Century. Symposium . . . Beltsville Agricultural Research Center, Beltsville, Md.; 1986. Dordrecht and Boston; M. Nijhoff, 1987. 455p.	
Second	Bekker, M. G. Introduction to Terrain-Vehicle Systems. Ann Arbor, Mich.; University of Michigan Press, 1969. 846p.	Second
	Bekker, M. G. Off-the-Road Locomotion. Ann Arbor, Mich.; University of Michigan Press, 1960. 220p.	Third
Second	Bekker, M. G. Theory of Land Locomotion. Ann Arbor, Mich.; University of Michigan Press, 1970. 220p. (Prev. ed., 1956. 520p.)	First
Second	Bell, Brian. Farm Machinery. 3d ed. Ipswich, Eng.; Farming Press, 1989. 265p. (1st ed., 1966; 2d ed., 1979. 249p.).	Second
Second	Bell, Brian. Farm Workshop. Ipswich, Eng.; Farming Press, 1981. 223p.	
Second	Benami, A., and A. Ofen. Irrigation Engineering: Sprinkler, Trickle, Surface Irrigation—Principles, Design and Agricultural Practices. Haifa; Irrigation Engineering Scientific Publications (for Technion-Israel Institute of Technology, Faculty of Agricultural Engineering), 1983. 257p. (Available from Bet Dagan, Israel; International Irrigation Information Center).	First
Third	Bene, J.G., H. W. Beall, and H. B. Marshall. Energy from Biomass for Developing Countries; A State of the Art Report. Ottawa; International Development Research Centre, 1979. 134p.	Third

	Bennett, H. H. Elements of Soil Conservation. 2d ed. New York; McGraw-Hill, 1955. 358p. (1st ed., 1947. 406p.)	First
	Berg, J. C. T. van den. Dairy Technology in the Tropics and Subtropics. Wageningen, Netherlands; Pudoc, 1988. 290p.	Third
Second	Bergmann, Hellmuth and Jean-Marc Boussard. Guide to the Economic Evaluation of Irrigation Projects. Paris; Organisation for Economic Co-operation and Development, 1976. 257p. (1st ed., 1973. 133p.)	Second
Second	Bernacki, H., J. Haman, and Cz. Kanafojski. Agricultural Machines: Theory and Construction. Translation of Teoria i Konstrukcja Maszyn Rolniczych, by Edmund Wiszniewicz. Warsaw; Published for the U.S. Dept. of Agriculture and the National Science Foundation, Washington, D.C., by the Scientific Publications Foreign Cooperation Center of the Central Institute for Scientific, Technical and Economic Information, 1972. 883p. (Available from the United States National Technical Information Service TT 69-50019).	Second
	Bhan, S. Crop Irrigation. New Delhi; Indian Council of Agricultural Research, 1983. 443p.	Second
Second	Binswanger, Hans P. Agricultural Mechanization: A Comparative Historical Perspective. Washington, D. C.; World Bank, 1984. 80p. (World Bank Staff Working Paper, no. 673)	Third
	Binswanger, Hans P. The Economics of Tractors in South Asia: An Analytical Review. New York: Agricultural Development Council; Hyderabad, India; International Crops Research Institute for the Semi-Arid Tropics, 1978. 96p.	Second
Second	Binswanger, Hans P. and Graeme Donovan, with Assistance of Raymund Fabre and Prabhu Pingali. Agricultural Mechanization: Issues and Options. Washington, D. C.; World Bank, 1987. 85p. (A World Bank Policy Study).	Second
	Binswanger, Hans P. and Vernon W. Ruttan. Induced Innovation: Technology, Institutions and Development. Baltimore; Johns Hopkins University Press, 1978. 423p.	Third
Third	Birch, G. G., K. J. Parker and J. T. Worgan, eds. Food from Waste. London; Applied Science Publishers, 1976. 301p.	
Second	Bishop, C. F. H. and W. F. Maunder. Potato Mechanisation and Storage. Ipswich, Suffolk; Farming Press, 1980. 256p.	Second
Second	Biswas, A. K., M. A. H. Samaho, M. H. Amer, and M. A. Abu-Zeid, eds. Water Management for Arid Lands in Developing Countries: Papers from the Training Workshop . . . Organized by the Ministry of Irrigation, Government of Egypt, with the United Nations Environment Programme, Cairo, 1978. Oxford; New York; Pergamon Press, 1980. 262p.	First
Second	Bittner, Richard, et al. Agricultural Safety. 3d ed. Moline, Ill.; Deere and Co. Service Training, 1987. (2d ed., 1974 titled: Agricultural Machinery Safety. 334p.)	Second
Second	Blaxter, Kenneth and Leslie Fowden, eds. Technology in the Nineteen Nineties—Agriculture and Food: Proceedings of a Royal Society Discussion Meeting Held October 1984. London; Scholium Int., 1985. 190p.	
	Bodenstedt, A. A. et al. Agricultural Mechanization and Em-	Third

DC	ployment. Sarrbruecken; Verlag der SSIP—Schrifften Brietenbach, 1977. 151p.	TW
Second	Bodholt, Ole. Construction of Cribs for Drying and Storage of Maize. Rome; Food and Agriculture Organization, 1985. 82p. (FAO Agricultural Services Bulletin no. 66).	First
Second	Bolz, Ray E. and George L. Tuve, eds. CRC Handbook of Tables for Applied Engineering Science. 2d ed. Cleveland, Ohio; CRC Presss, Inc., 1973. 1166p.	
	Boodley, J. W. The Commercial Greenhouse. New York; Van Nostrand Reinhold, 1981. 568p.	Second
Second	Booher, L. J. Surface Irrigation. Rome; Food and Agriculture Organization of the United Nations, 1974. 160p. (FAO Agricultural Development Paper no. 95).	First
Second	Borgman, Donald E., Everette Mainline, and Melvin E. Long. Tractors. 2d ed. Moline, Ill.; J. Deere Service Publications, 1981. 304p. (Fundamentals of Machine Operations, FMO-101B). (1st. ed., 1974. 304p.)	
Third	Borrelli, John; Victor R. Hasfurther and Robert D. Burman, eds. Proceedings of the Specialty Conference on Advances in Irrigation and Drainage: Surviving External Pressures. New York; American Society of Civil Engineers, 1983. 558p.	
Second	Bos, M. G., ed. Discharge Measurement Structures. Presented by the Working Group on Small Hydraulic Structures. 3d ed. Wageningen, Netherlands; International Institute for Land Reclamation and Improvement, 1989. 401p. (ILRI Publication no. 20).	
	Boserup, Ester. Population and Technological Change; A Study of Long-Term Trends. Chicago: University of Chicago, 1981. 255p.	Second
Second	Bosoi, E. S., O. V. Verniaev, I. I. Smirnov, and E. G. Sultan-Shankh. Theory, Construction and Calculations of Agricultural Machines (Teoriia, Konstruktsiia i Raschet Sel'skokhoziaistvennykh Mashin), Vol. 1. (Trans. from Russian.) New Delhi; Oxonian Press, 1987. 314p.	Second
Second	Bottrall, Anthony F. Comparative Study of the Management and Organization of Irrigation Projects. Washington, D.C.; World Bank, Agricultural and Rural Development Dept., 1981. 274p. (World Bank Staff Working Paper no. 458).	Second
Second	Boumans, G. Grain Handling and Storage. Amsterdam and New York; Elsevier Science Publishing Co., Inc., 1985. 436p. (Developments in Agricultural Engineering no. 4).	First
Second	Bouwer, Herman. Groundwater Hydrology. New York; McGraw-Hill, 1978. 480p.	Third
Second	Bowers, Wendell, Benjamin A. Jones, Jr. and Elwood F. Olver. Engineering Applications in Agriculture. 6th ed. Champaign, Ill.; Stipes Publishing Co., 1986. 300p. (1st ed., 1959, 145p. 4th ed., 1973. 273p.)	First
Second	Bowler, D. G. The Drainage of Wet Soils. Auckland; Hodder and Stoughton, 1980. 259p.	

Second	Bowles, Joseph E. Engineering Properties of Soils and Their Measurement. 3d ed. New York; McGraw-Hill Book Company, 1986. 218p. (1st ed., 1970. 2d ed., 1978. 213p.)	Second
Third	Boxall, R. A., et al. The Prevention of Farm-Level Food Grain Storage Losses in India: A Social Cost-Benefit Analysis. Brighton; Institute of Development Studies, University of Sussex, 1978. 239p.	Second
Third	Boyd, James S., ed. Dairy Housing Conference Proceedings. St. Joseph, Mich.; American Society of Agricultural Engineers, 1973. 469p.	
Second	Boyd, James S. Practical Farm Buildings: A Text and Handbook. 2d ed. Danville, Ill.; Interstate Printers and Publishers, Inc., 1979. 277p. (1st ed., 1973. 265p.)	Second
Third	Boyko, Hugo, ed. Saline Irrigation for Agriculture and Forestry. The Hague, W. Junk, 1968. 328p.	
Second	Brace Research Institute, (MacDonald College, McGill University). A Survey of Solar Agricultural Dryers. Ste. Anne de Bellevue, Quebec; Brace Research Institute, 1975. 150p.	
Second	Brater, Ernest F. and Horace W. King. Handbook of Hydraulics: For the Solution of Hydraulic Engineering Problems. 6th ed. New York; McGraw-Hill, 1976. 604p. (1st ed., 1918, by H.W. King; 5th ed., 1963, had slightly variant name.) (Variously Paged).	
Third	Braunstein, Helen M. et al. Biomass Energy Systems and the Environment. New York; Pergamon Press, 1981. 182p.	
Second	British Standards Institution. Glossary of Terms Relating to Agricultural Machinery and Implements. London; The Institution, 1963. 94p.	Second
Third	Bromley, Daniel W. Improving Irrigated Agriculture: Institutional Reform and the Small Farmer. Washington, D.C.; World Bank, Publications Department, 1982. 85p. (World Bank Staff Working Paper no. 531).	Second
First	Brooker, Donald B., Fred W. Bakker-Arkema, and Carl W. Hall. Drying Cereal Grains. Westport, Conn.; AVI Publishing Co., 1974. 265p.	First
First	Brown, Arlen D. and R. Mack Strickland. Tractor and Small Engine Maintenance. 5th ed. Danville, Ill.; Interstate Printers and Publishers, Inc., 1983. 350p. (1st ed. by I.G. Morrison; 2d–3d eds. by A.D. Brown and I.G. Morrison, published as Farm Tractor Maintenance).	Second
	Brown, J. R., ed. Soil Testing: Sampling, Correlation, Calibration, and Interpretation: Proceedings of a Symposium. Madison, Wisc.; Soil Science Society of America, 1987. 144p. (SSSA Special Publication no. 21).	First
Second	Brown, Norman L., ed. Renewable Energy Resources and Rural Applications in the Developing World. Boulder, Colo.; Westview Press, 1978. 169p. (AAAS Selected Symposium no. 6).	Third
Second	Brown, Ray W. and Bruce P. Van Haveren, eds. Psychrometry in Water Relations Research; Proceedings of the Symposium on Thermocouple Psychrometers, Utah State University,	

DC	1971. Logan; Utah Agricultural Experiment Station, 1972. 342p.	TW
First	Brown, Robert H. CRC Handbook of Engineering in Agriculture. Boca Raton, Flor.; CRC Press, Inc., 1988. 3 Vols. 804p.	First
First	Brown, Robert H. Farm Electrification. New York; McGraw-Hill Book Company, 1956. 367p. Also Issued in Bombay; Allied Pacific, 1962.	Second
Second	Brownlee, K. A. Statistical Theory and Methodology in Science and Engineering. 2d ed. New York; Wiley, 1965. 590p. (1st ed., 1960. 570p.)	
Second	Buchele, Wesley F., et al. Symposium on Grain Drying—Damage. Ames, Iowa; Iowa State University, 1968. 161p.	Second
Third	Burget, Stanislav. Automaticke Rizeni Traktoru a Zemedelskych Stroju Studijni Zprava. (Automatic Steering and Control of Tractors and Farm Machinery; A Review.). Prague; Ustav Vedeckotechnickych Informaci, 1971. 84p. (Studijni Informace; Zemedelska Technika, 1971. no. 5)	
Second	Burmistrova, M. F. et al. Physicomechanical Properties of Agricultural Crops = Fizikomekhanicheskie Svostva Selskokhozyaistvennykh Rastenii, translated from Russian. Jerusalem; Published for the National Science Foundation, Washington, D.C. by Israel Program for Scientific Translations, 1963. 250p. (Distr. in UK by Oldbourne Press).	
Second	Burrows, W. C. et al., eds. Proceedings . . . International Conference on Mechanized Dryland Farming, Moline, Ill. and Great Falls, Mont., 1969. Moline, Ill.; Deere and Co., 1970. 344p.	First
Third	Bushuyev, N., N. Alexeyev, and V. Plaksin. Farm Machinery: A Handbook for Agricultural Specialists, Workers and Students. Translated by A. Baikov and V. Kulikov. Moscow; Peace Publishers, 1957 303p.	
First	Butchbaker, Allen F. Electricity and Electronics for Agriculture. Ames; Iowa State University Press, 1977. 391p.	Second
	Butler, Steven. Agricultural Mechanization in China: The Administrative Impact. New York; East Asian Institute, Columbia University, 1978. 58p.	Third
Third	Butorac, Andelko, ed. Proceedings of the 9th Conference of the International Soil Tillage Research Organization, Yugoslavia, 1982. Osijek, Yugoslavia; The ISTRO, 1982. 698p.	Third
	Butterworth, B. Farm Tractors: The Case/David Brown Guide to Tractor Selection, Operation, Economics and Servicing. London; Spon, 1984. 149p.	Second
	Byerlee, Derek, et al. Planning Technologies Appropriate to Farmers: Concepts and Procedures. Mexico, D. F.; Economics Program, Centro Internacional de Mejoramiento de Maiz y Trigo, 1980. 71p.	Third
Second	Byg, George, editor. Grain Damage Symposium. Columbus, Ohio; Ohio State University, 1972. 206p.	Second
Third	Byszewski, Wladyslaw and J. Haman. Soil-Machine-Plant Growth = Gleba Maszyna Roslina, translated from Polish. Karachi, Pakistan; Prepared for the SEA, United States De-	

DC	partment of Agriculture and National Science Foundation by Saad Publications, 1984. 467p.	TW

C

Third	Calvert, N. G. Windpower Principles: Their Applications on the Small Scale. London; C. Griffin, 1979. 122p.	Second
	Campbell, Joseph K. Dibble Sticks, Donkeys, and Diesels: Machines in Crop Production. Manila, Philippines; International Rice Research Institute, 1990. 329p.	Third
	Canada. Agriculture Canada. Farm Machinery—Financial Management. Produced with the British Columbia Ministry of Agriculture. Vancouver; British Columbia; Agriculture Canada, Ministry of Agriculture, 1977. 5 vols.	Third
Third	Candelon, Phillippe. Les Machines Agricoles. 2d ed. Paris; Bailliere, 1973–78. 2 vols.	
Second	Canham, Allan E. Electricity in Horticulture. London; MacDonald, 1964. 199p.	
Second	Cargill, B. F. and G. E. Rossmiller, eds. Fruit and Vegetable Harvest Mechanization; Manpower Implications. East Lansing, Mich.; Rural Manpower Center, Michigan State University and the American Society of Agricultural Engineers, 1969–1970. 360p. 3 vols.	
	Caponera, Dante A. Water Laws in Moslem Countries. 2d ed., revised. Rome; Food and Agriculture Organization, 1973. 229p. (FAO Irrigation and Drainage Papers no. 20/1). (1st ed., 1954; FAO Agricultural Development Paper no. 43).	Third
Third	Carlier, M. Hydraulique Generale et Appliques. Paris; CIGR, 1972. 565p.	Third
	Carruthers, Ian, ed. Aid for the Development of Irrigation. Paris; Organization for Economic Cooperation and Development, 1983. 166p.	Second
Second	Carruthers, Ian, and Colin W. Clark. The Economics of Irrigation. 3d ed. Liverpool; Liverpool University Press, 1981. 300p.	First
Second	Carruthers, Ian., ed. Social and Economic Perspectives on Irrigation. Oxford and New York; Pergamon Press, Inc., 1981. 133p.	Second
Second	Carslaw, Horatio S. and John C. Jaeger. Conduction of Heat in Solids. 2d ed. Oxford; Clarendon Press, 1959. 510p. (1st ed., 1947. 386p.)	
Third	Casey, Hugh E. Salinity Problems in Arid Lands Irrigation; A Literature Review and Selected Bibliography. Tucson, Ariz.; University of Arizona, Office of Arid Lands Studies, 1972. 300p.	
Second	Castle, D. A., J. McCunnall, and I. M. Tring. Field Drainage; Principles and Practices. Batsford, England; David and Charles, Inc., 1984. 224p.	
	Caterpillar Inc. Caterpillar Performance Handbook. 18th ed. Peoria, Ill; Caterpillar Inc., 1979–87. Approx. 760p.	Second
Second	Cedergren, Harry R. Seepage, Drainage and Flow Nets. 3d ed. New York; John Wiley, and Sons, Inc., 1989. 465p. (1st ed., 1967. 489p. 2d ed., 1977. 534p.)	

	Centre D'Etudes et D'Experimentation du Machinisme Agricole Tropical. Manual on the Employment of Draught Animals in Agriculture. (Translation of Manuel de Culture avec Traction Animale.) Rome; by arrangement with the Centre by the Food and Agriculture Organization, 1972. 249p.	Second
Third	Centre National d'Etudes et d'Experimentation du Machinisme Agricole. L'activite Agricole et L'energie. Antony, France; CNEEMA, 1975. (Etudes du CNEEMA no. 408)	
Second	Centre National D'Etudes et D'Experimentation de Machinisme Agricole. Dictionnaire Technique de la Mecanisation Agricole. Dictionnaire Technique de la Mecanization Agricole. Technical Dictionary of Agricultural Mechanization. Technisches Worterbuch der Mechanisierung in del Landwirtschaft. Diccionario tecnico de mecanizacion agricola. Dizionario tecnico di meccanizzazaione Agricola. (1959 ed. Published Under Title: Dictionnaire Technique du Machinisme Agricole). Anthony, C.N.E.E.M.A., 1968–71. 957p.	Third
Second	Chambers, R. and J. Moris, eds. Mwea: An Irrigated Rice Settlement in Kenya. Munich; Weltforum Verlag, 1973. 524p.	Third
	Chambers, Robert. Managing Rural Development: Ideas and Esperience from East Africa. Uppsala, Sweden; Scandanavian Institute for African Studies, 1974. 215p.	Second
Third	Chang, Ch'eng T. and Li C. Shang. T'o Li Chi. (Threshing Machines.) Pei-ching; Chung-kuo Nung Yeh Chi Hsieh Ch'u Pan She, 1985. 300p.	
Third	Chapman, Homer D. and Parker F. Pratt. Methods of Analysis for Soils, Plants, and Waters. Riverside, Calif.; University of California, Citrus Experiment Station, Division of Agricultural Sciences, 1961. 309p. (University of California Agricultural Science Pub. no. 4034, Berkeley; reprinted 1982)	
First	Charm, Stanley E. The Fundamentals of Food Engineering. 3d ed. Westport, Conn.; AVI Publishing Co., 1978. 646p. (1st ed., 1963. 592p.; 2d ed., 1971. 629p.)	Second
Third	Chartier, P. and W. Palz, eds. Energy from Biomass; Proceedings of the EC Contractors' Meeting, Copenhagen, 1981. Dordrecht, Holland and Boston; D. Reidel Pub. Co. for the Commission of the European Communities, 1981. 220p. (Sold and distributed in the U.S. and Canada by Hingham, Mass.; Kluwer Boston)	
	Chatterji, Manes, ed. Energy and Environment in the Developing Countries; Selected Papers Presented at the International Conference at the Indian Institute of Management, Bangalore, India, January, 1979. Chichester and New York; J. Wiley and Sons, 1981. 357p.	Third
Second	Cheremisinoff, Nicholas P. and Paul N. Cheremisinoff, eds. Unit Conversions and Formulas Manual. Ann Arbor, Mich.; Ann Arbor Science Pub., 1980. 171p.	
Second	Cheremisinoff, Nicholas P., ed. Handbook of Heat and Mass Transfer. Houston, Texas; Gulf Pub. Co., 1986. 1300p.	
Third	Cheremisinoff, Nicholas P., Paul N. Cheremisinoff and Fred Ellerbusch. Biomass: Applications, Technology, and Production. New York; M. Dekker, 1980. 221p.	

Third	Cheremisinoff, Nicholas P. Wood for Energy Production. Ann Arbor, Mich.; Ann Arbor Science Publishers, 1980. 152p.	
Third	Chiang-Su Sheng Yang-Chou Shui Li Hsueh Hsiao. Nung Yung Shui Peng. (Agricultural Pumping Machinery.) Peking; Shi Li Tien Li Ch'u Pan She, 1974. 305p.	
Second	Childs, Ernest C. An Introduction to the Physical Basis of Soil Water Phenomena. London, New York; J. Wiley, 1969. 493p.	
	Chowdhury, Bijoy K. Economics of Tubewell Irrigation in West Bengal. Santiniketan; Agro Economic Research Centre, Visva-Bharati, 1971. 75p.	Third
Second	Christensen, Clyde M., ed. Storage of Cereal Grains and Their Products. 3d ed. St. Paul, Minn.; American Association of Cereal Chemists, 1982. 544p. (1st ed. edited by A. Anderson and A. W. Alcock, 1954. 515p.; 2d ed., 1974. 549p.)	Second
Second	Christiansen, J. E. Irrigation by Sprinkling. Logan, Utah; Utah State University, 1972. 124p.	Second
First	Christianson, L. L. and Roger P. Rohrbach. Design in Agricultural Engineering. St. Joseph, Mich.; American Society of Agricultural Engineers, 1986. 310p. (ASAE Textbook no. 1)	First
	Chulalongkorn University. Evolution of Farm Tools and Appropriate Technology in Southeast Asia. Bangkok, Thailand; 1989. 164p.	Third
Second	Chynoweth, David P., and Ron Isaacson, eds. Anaerobic Digestion of Biomass. New York; Elsevier Science Pub. Co. Inc., 1987. 279p.	Second
Second	Clark, J. A., ed. Environmental Aspects of Housing for Animal Production. London, Boston; Butterworth's, 1981. 511p.	Third
Second	Clark, John H. Agricultural Materials Handling Manual. Ottawa; Information Services, Agriculture Canada, 1979. 50p. (Earlier ed.: Canada Committee on Agricultural Engineering of Canadian Agriculture Services Coordinating Committee, Canada Dept. of Agriculture, 1962.)	
Third	Clayton, Christopher R. I., Noel E. Simons, and Marcus C. Matthews. Site Investigation. New York; Halsted Press, 1982. 424p.	
Second	Cloud, Gayla Staples. Agricultural Mechanization in the Third World: A Selective Bibliography, 1975–1985. Monticello, Ill.; Vance Bibliographies, 1986. 41p.	
Second	Collins, A. G., and A. I. Johnson, eds. Ground-Water Contamination: Field Methods. A Symposium sponsored by ASTM Committees D19 on Water, and D18 on Soil and Rock, Florida, 1986. Philadelphia; American Society for Testing and Materials, 1988. 491p. (ASTM Special Technical Publication no. 963).	
Second	Colombia. Ministerio de Agricultura. Biblioteca y Fichero. English-Spanish Technical Vocabulary on Agricultural Mechanization. Bogota, 1964–67. 3 vols.	
	Commonwealth Regional Renewable Energy Resources Information Systems Staff, ed. Wind Powered Electricity Generation Package. Australia; CSIRO, 1985. 139p. (Available from International Specialized Book Services).	Second

Citation Analysis and the Core Monographs 255

Second	Coolman, Fiepko. Who is Who: A Directory of Agricultural Engineers Available for Work in Developing Countries. Paris; Commission Internationale du Genie Rural, 1985. 209p.	
Second	Cooper, Elmer L. and H. Edward Reiley. Agricultural Mechanics: Fundamentals and Applications. Albany, New York; Delmar Pub., 1987. 532p.	Second
Second	Cooper, Turner. Practical Land Drainage. London; Leonard Hill, 1965. 162p.	
Second	Coulson, John Metcalfe and J. F. Richardson. Chemical Engineering. 3d ed. Oxford and New York; Pergamon Press, 1977–79. 3 vols. (2d ed., Macmillan, 1964–75. 3 vols.)	
Second	Coward, E. Walter., ed. Irrigation and Agricultural Development in Asia: Perspectives from the Social Sciences. Ithaca, New York; Cornell University Press, 1980. 369p.	Second
Second	Cox, S. W. R. and D. E. Filby. Instrumentation in Agriculture. New York; Hafner Pub. Co., 1972. 145p.	Second
First	Cox, S. W. R. Farm Electronics. Oxford and Boston; BSP Professional Books, 1988. 310p.	Second
First	Cox, S. W. R. Microelectronics in Agriculture and Horticulture: Electronics and Computers in Farming. Totowa, New Jersey; Allanheld, Osmun, 1982. 230p.	
Second	Craig, R. F. Soil Mechanics. 4th ed. Wokingham, Berkshire, England; Van Nostrand (UK); New York; Van Nostrand Reinhold Co., Inc., 1987. 410p. (1st ed., 1974. 275p.; 3d ed., 1983. 419p.).	
Second	Crook, Charles B. Drainage of Agricultural Lands: An Annotated Bibliography. Washington, D.C.; U.S.D.A., Agricultural Research Service, and National Agricultural Library, 1968. 524p.	Second
Third	Cross, H. R. and A. J. Overby, eds. Meat Science, Milk Science and Technology. New York; Elsevier Science Pub. Co. Inc., 1988. 458p.	
Second	Crossley, C. Peter and John Kilgour. Small Farm Mechanization for Developing Countries. England and New York; John Wiley, 1983. 253p.	First
Second	Crouse, William H. Small Engines: Operation and Maintenance. New York; McGraw-Hill, 1974. 370p.	
Third	Cruz, Ibarra E. Producer-Gas Technology for Rural Applications. Rome; Food and Agriculture Organization of the United Nations, 1985. 97p. (FAO Agricultural Services Bulletin no. 61).	
Second	Csaki, Csaba. Simulation and Systems Analysis in Agriculture, translation of Szimulacio Alkalmazasa a Mezogazdasagban by Ferenc Szirbik and Gyorgy Radovics. Amsterdam and New York; Elsevier Science Publishing Company, Inc., 1985. 262p. (Developments in Agricultural Engineering Ser.: Vol. 2).	
First	Culpin, Claude. Farm Machinery. 12th ed. Oxford; Blackwell Scientific, 1992. 480 p. (1st—9th eds., London, Lockwood, 1938–1976. Variously Paged.; 11th ed., London; Collins, 1986. 450p.)	First

First	Culpin, Claude. Profitable Farm Mechanization. 3d ed. London; Crosby Lockwood Staples, 1975. 307p. (1st ed. under name of Farm Mechanization Management. 1959. 225p.; 2d ed., 1975. 307p.)	Second
Second	Curran, P. J. Principles of Remote Sensing. Harlow, Essex, England; Longman Scientific and Technical; New York; John Wiley and Sons, Inc., 1986. 282p.	
Second	Curtice, David. Handbook of the Operation of Small Wind Turbines on a Utility Distribution System. 2d ed. St. Johnsbury, Vt.; WindBooks, 1984. 192p.49.50.	
Third	Curtis, Stanley E. Environmental Management in Animal Agriculture. Ames, Iowa; Iowa State University Press, 1983. 409p.	Second

D

Third	D'Itri, Frank M., and Lois G. Wolfson, eds. Rural Groundwater Contamination. Chelsea, Mich.; Lewis Pub. Inc., 1987. 416p.	
Third	Dakar, Alberto. Agua na Agricultural; Manual de Hidraulica Agricole. 3d ed. Rio de Janeiro; Livraria Freitas Bastos, 1969. 3 Vol.	
Third	Damme, Egon and Peter Oberlander. Elektrotechnische Anlagen in der Landwirtschaft. (Electrotechnical Installation in Agriculture.) Berlin; VEB Verlag Technik, 1986. 296p.	
Second	Darrow, Ken, Rick Pam (Vol.1 & 2) and K. Keller (Vol.2). Appropriate Technology Sourcebook: For Tools and Techniques that Use Small Local Skills, Local Resources, and Renewable Sources of Energy. Stanford, Calif.; Volunteers in Asia, Vol.1, 1976; Vol.2, 1981. 816p.	Second
Second	Dasberg, S., and E. Bresler. Drip Irrigation Manual. Bet Dagan, Israel; International Irrigation Information Center, 1985. 108p.	
Third	Davies, Cornelius. Considerations and Procedures for the Successful Introduction of Farm Mechanization. Rome; Food and Agriculture Organization, 1954. 36p. (FAO Agricultural Development Paper no. 44).	
Second	Davies, D. B., D. J. Eagle and B. Finney. Soil Management. 4th ed. Ipswich, Suffolk; Farming Press, 1982. 287p. (2d ed., 1972. 254p.; 3d ed., 1977. 268p.)	
Second	Davis, Gene L. Agriculture and Automotive Diesel Mechanics. Englewood Cliffs, N.J.; Prentice-Hall, Inc., 1983. 256p.	
	De Datta, Surajit K. Principles and Practices of Rice Production. New York; John Wiley and Sons, Inc., 1981. 618p.	First
Second	De Ridder, N. A. and A. Erez. Optimum Use of Water Resources. Wageningen; International Institute for Land Reclamation and Improvement, 1977. 250p. (Publication no. 21).	
	De Veen, J. J. The Rural Access Roads Programme: Appropriate Technology in Kenya. Geneva; International Labour Office, 1984. 177p.	Third
Second	Decareau, Robert V. and R. A. Peterson. Microwave Process-	Third

DC	ing and Engineering. Chichester, England; E. Horwood; Weinheim, Federal Republic of Germany; VCH Verlagsgesellschaft, 1986; Deerfield Beach, Fla. 224p.	TW
Third	Deepak, Adarsh and K. R. Rao, eds. Applications of Remote Sensing for Rice Production. Interactive International Symposium on Applications of Remote Sensing for Rice Production, India, 1981. Hampton, Virg.; A. Deepak Publishing, 1984. 449p.	
Second	Dencker, Carl H., et al. Handbuch der Landtechnik. Berlin and Hamburg; P. Parey, 1961. 1046p.	Third
Second	Dencker, Carl H. Landwirtschaftliche Stoff- und Maschinenkunde. Berlin; P. Parey, 1968. 221p.	Third
	Denver Research Institute. Pakistan Council of Scientific and Industrial Research. Appropriate Technology Development Organization (Pakistan). USAID/Pakistan. Village Level Food Processing Project in Pakistan. Karachi, Pakistan; PCSIR, Directorate of Industrial Liaison, 1981. 5 Vols.	Second
Third	Desrosier, Norman W., ed. Elements of Food Technology. Westport, Conn.; AVI Pub. Co., 1977. 772p.	Second
Second	Detraux, Freddy and Otto Oestges. La Mecanisation des Travaux Agricoles. Gembloux; Les Presses Agronomigue de Gembloux, 1979. 428p.	
First	Dieleman, P. J. and D. B. Trafford. Drainage Testing. (Eng., Fr. and Span.). Rome; Food and Agriculture Organization, 1976. 172p. (FAO Irrigation and Drainage Papers no. 28). (Available from Unipub.)	Third
	Dill, H. W., Jr. Worldwide Use of Airphotos in Agriculture. Washington, D.C.; U.S. Economic Research Service, 1967. 23p. (USDA Agriculture Handbook 334).	Second
Second	Dodd, Vincent A. and Patrick M. Grace, eds. Agricultural Engineering; Proceedings of the Eleventh International Congress on Agricultural Engineering, Dublin, Sept.1989. Rotterdam & Brookfield, A.A. Balkema, l989. 4 vols.	Third
Second	Donaldson, Graham F. Farm Machinery Capacity. Ottawa; Royal Commission of Farm Machinery, 1970. 161p.	
Second	Donaldson, Graham F. Farm Machinery Safety. Ottawa; Royal Commission of Farm Machinery, 1968. 137p.	
Second	Donaldson, Graham F. Farm Machinery Testing. Ottawa; Royal Commission of Farm Machinery, 1967. 92p.	
Second	Doneen, L. D., and D. W. Westcot. Irrigation Practice and Water Management. 2d ed. Rome; Food and Agriculture Organization, 1984. 63p. (1st ed., 1971. 84p.) (FAO Irrigation and Drainage Paper no. 1).	First
Second	Doorenbos, J. and A. H. Kanan. Crop Response to Water. Rome; Food and Agriculture Organization, 1979. (FAO Irrigation and Drainage Paper no. 33).	Third
Second	Doorenbos, J. and W. O. Pruitt. Guidelines for Predicting Crop Water Requirements. Rome; Food and Agriculture Organization, 1977. 144p. (FAO Irrigation and Drainage Paper no. 24).	Second
Third	Dorman, Richard George. Dust Control and Air Cleaning. Oxford and New York; Pergamon Press, 1974. 615p.	

Third	Downing, T. Edmond and Gibson McGuire, eds. Irrigation's Impact on Society. Tucson, Ariz.; University of Arizona Press, 1974.	Third
Third	Drablos, Carroll J. W. and Benjamin A. Jones, Jr. Highway and Agricultural Drainage Practice. Urbana, Ill.; University of Illinois, Engineering Publications Office, 1965. 159p. (Illinois Cooperative Highway Research Program Series no. 39). (University of Illinois Bulletin, Vol. 62, no. 107).	
Third	Draper, N. R. and H. Smith. Applied Regression Analysis. 2d ed. New York; Wiley, 1981. 709p. (1st ed., 1966. 407p.)	
Second	Duke, James A. CRC Handbook of Agricultural Energy Potential of Developing Countries. Boca Raton, Flor.; CRC Press, Inc., 1987. 1045p. 4 Vols.	Second
	Dunn, Peter, D. Appropriate Technology: Technology with a Human Face. New York; Schocken Books, 1978, 1979. 220p.	Third
First	Dwyer, M. J., ed. The Performance of Off-Road Vehicles and Machines: Proceedings of the 8th International Society for Terrain-Vehicle Systems Conference, Cambridge University, 1984. Bedford, U.K.; The Society, 1984. 3 vols. (1,306p.) (Selected Papers in the Journal of Terramechanics (1984) 21:97– 235).	Second

E

Second	Earle, R. L. Unit Operations in Food Processing. Oxford and New York; Pergamon Press Ltd., 1983. 207p. (1st ed., 1966. 342p.)	Third
Third	Eastman, Paul. An End to Pounding: A New Mechanical Flour Milling System in Use in Africa. Ottawa, Canada; International Development Research Centre, 1980. 63p. (IDRC-152e)	
Second	Eden, M. J. and J. T. Parry, eds. Remote Sensing and Tropical Land Management. Based on P apers Presented at a Workshop of the Commonwealth Geographical Bureau, McGill University, Montreal, 1983. New York; John Wiley and Sons, Inc., 1986. 365p.	First
	Edwards, Alfred L., Ikewelugo C. A. Oyeka and Thomas W. Wagner, eds. New Dimensions of Appropriate Technology; Selected Proceedings of the 1979 Symposium sponsored by the International Association of the Advancement of Appropriate Technology for Developing Countries. Ann Arbor; University of Michigan, Graduate School of Business Administration, Division of Research, 1980. 251p.	Third
	Edwards, Edgar O., ed. Employment in Developing Nations: Report on a Ford Foundation Study. New York; Columbia University Press, 1974. 428p.	Third
Third	Eggelsmann, Rudolf. Drainage Instruction Book. Hamburg; Verlag Wasser und Boden, 1974. 304p.	
Third	Eggelsmann, Rudolf. Subsurface Drainage Instructions for Agriculture, Civic Engineering and Landscaping. Translation of Dränanleitung für Landbau, Ingenieurbau und Landschaftsbau. Hamburg; Verlag Wasser und Boden Axel Lindow, 1973. 347 leaves. (Translated for National Science	Third

DC	Foundation and U.S. Department of Agriculture, TT75-55041). (2d German ed., 1981)	TW
Second	Eggleston, Jerry, ed. Irrigation and Drainage, Today's Challenges. Proceedings of the 1980 Spe cialty Conference, Idaho, 1980. New York; American Society of Civil Engineers, 1980. 496p.	Second
Third	Egneus, H. and A. Ellegard, eds. Bioenergy '84; Proceedings of an International Conference on Bioenergy, Swedish Trade Fair Centre, Goteborg, Sweden, 1984. London; Elsevier Applied Science, 1985. 5 Vols. (Sole distributor in the USA and Canada: Arlington, Virginia; The Bioenergy Council).	
Third	Elizarov, Vadim P. Avtomatizatsiia Tekhnologicheskikh Protsessov v Polevodstve. (Automation of Technological Processes in Field Cropping.) Moscow; VIM, 1985. 175p. (Vsesoiuznyi Nauchno-issledovatel'skii Institut Mekhanizatisii Sel' skogo Khoziaistva. Sbornik Nauchnykh Trudov no. 104).	
Third	Elliott, L. F., and F. J. Stevenson, eds. Soils for Management of Organic Wastes and Waste Waters. Proceedings of a Symposium at the TVA National Fertilizer Development Center . . . Alabama, 1975. Madison, Wisc.; Soil Science Society of America, American Society of Agronomy, Crop Science Society of America, 1977. 650p.	
	ElMahgary, Yehia and Asit K. Biswas, eds. Integrated Rural Energy Planning. Guildford; Published by Butterworths for the United Nations Environemnt Programme and the International Society for Ecological Modelling, 1985. 200p.	Third
Third	EPA Handbook for Improving POTW Performance. Alexandria, Va.; Water Pollution Control Federation, 1985. 258p.	
Third	Eskin, N. A. M. Biochemistry of Foods. 2d ed. San Diego, Calif.; Academic Press, 1990. 557p. (1st ed. by N. A. M. Eskin, H. M. Henderson and R. J. Townsend. New York; Academic Press, 1971. 240p.)	Third
Second	Esmay, Merle L. and Carl W. Hall. Agricultural Mechanization in Developing Countries. Tokyo, Japan; Shin-Norinsha Co., 1973. 221p.	First
Second	Esmay, Merle L. and John E. Dixon. Environmental Control for Agricultural Buildings. Westport, Conn.; AVI Publishing Company, Inc., 1986. 287p.	Second
	Esmay, Merle L. and Roy E. Harrington, eds. Glimpses of Agricultural Mechanization in the People's Republic of China: A Delegation of 15 ASAE Members Report on their Technical Inspection in China. St. Joseph, Mich.; The American Society of Agricultural Engineers, 1979. 101p.	Second
Second	Esmay, Merle L. Principles of Animal Environment. 2d ed. Westport, Conn.; AVI Publishing Co., 1978. 358p. (1st ed., 1969. 325p.).	First
Third	Euroconsult, editor. Agricultural Compendium for Rural Development in the Tropics and Subtropics. 3d ed. Amsterdam; New York; Elsevier Scientific Pub. Co., 1989. 740p. (1st ed.: Arnhem, the Netherlands; International Land Development Consultants for the Ministry of Agriculture and Fisheries, the	Second

DC	Hague, the Netherlands, 1981. 739p.; 2d ed., International Land Development Consultants, 1985. 738p.)	TW
Third	European Committee of Associations of Manufacturers of Agricultural Machinery (CETIA). Illustrierte Landmaschinen-Terminologie. Illustrated Glossary of Agricultural Machinery. Terminologie Illustreedu Machinisme Agricole. Terminologia Illustrata del Machinisme Agicole. Terminologia Illustrata del Maquinario Agricolo. CETIA; 1966–1971. 278p. (1971 Supplement in Russian)	

F

Second	Falvey, J. Lindsay. An Introduction to Working Animals. Melbourne; MPW Australia, 1987. 196p.	Second
Second	Farley, N. A Handbook of Garden Machinery and Equipment. London; J. M. Dent, 1980. 272p.	Second
Third	Farm Buildings Information Centre. Buildings for Potatoe and Vegetable Storage. Kenilworth, Warwickshire; Farm Buildings Information Centre, 1981. 66p.	
Third	Farm Mechanization in Japan. Tokyo; Shin-Norin-Sha Co., 1958. 136p. (Mechanized Agriculture no. 2476).	Third
Second	Farmer, P. Wind Energy, 1975–1985: A Bibliography. Berlin and New York; Springer-Verlag, 1986. 167p.	
	Farrall, Arthur W. and Carl F. Albrecht, editors. Agricultural Engineering: A Dictionary and Handbook. Danville, Ill.; Interstate Printers and Pub., 1965. 434p.	Second
First	Farrall, Arthur W. and James A. Basselman, eds. Dictionary of Agricultural and Food Engineering. 2d ed. Danville, Ill; Interstate Printers & Publishers, Inc., 1979. 437p.	Second
	Farrall, Arthur W. Engineering for Dairy and Food Products. New York, Wiley, 1963. 674 p.	Second
Second	Farrall, Arthur W. Food Engineering Systems, Volume 1—Operations; Volume 2—Utilities. Westport, Conn.; AVI Pub. Co., 1976. Vol. 1, 1976. 615p.; Vol. 2, 1979. 451p.	Second
	Feachem, Richard, Michael McGarry and Duncan Mara, eds. Water, Wastes and Health in Hot Climates. Chichester, England and New York; Wiley, 1977. 399p.	Third
Third	Feda, Jaroslav. Mechanics of Particulate Materials: The Principles. Translation from Russian of Zakady Mechaniky Partikularnich Latek by Doubravka Hajsmanova. Amsterdam and New York; Elsevier Science Publs., 1982. 446p. (Development in Geotechnical Engineering Ser., Vol. 30).	
Third	Fedosenko, Radii I. and Aleksandr I. Mel'nikov. Ekspluatatsionnaia Nadezhnost' Elektrosetei Sel'Skokhoziaistvennogo Naznacheniia. (Using the Reliability of Power-Supply Systems for Agriculture). Moscow; Energiia, 1977. 320p.	
	Feed and Fuel from Ethanol Production Symposium, Philadelphia, 1981. Proceedings . . . Ithaca, New York; Northeast Regional Agricultural Engineering Service, Cooperative Extension, 1982. 101p. (NRAES 17).	Third
	Feiffer, Peter, and Rosemarie Feiffer. The Combining of Various Crops. Leipzig; Edition Leipzig, 1969. 236p.	Third

Second	Fellows, Peter. Food Processing Technology: Principles and Practice. Chichester, England; E. Horwood; New York; VCH, 1988. 505p.	Second
	Fernandez-Quintanilla, Cesar. Construcciones para el Ganado Vacuno. 2d ed. Madrid; Mundi-Prensa, 1974. 138p.	Third
Second	Ferrero, G. L., G. Grassi and H. E. Williams, eds. Biomass Energy from Harvesting to Storage. London and New York; Elsevier Applied Science, 1987. 327p.	
Third	Ferrero, G. L., M. P. Ferranti and H. Naveau, eds. Anaerobic Digestion and Carbohydrate Hydrolysis of Waste. London and New York; Elsevier Applied Science, 1984. 517p.	
Third	Filippov, Mikhail M. Avtomatizatsiia Elektrosetei v Sel'skoi Mestnosti. (Automation of Power-Supply Systems in Rural Areas). Moscow; Energiia, 1977. 101p. (Biblioteka Elektromontera, no. 454).	
Second	Finkel, Heman J., ed. CRC Handbook of Irrigation Technology. Boca Raton, Flor.; CRC Press, 1982. Vol. 1, 368p., Vol. 2, 223p.	First
Second	Finner, Marshall F. Farm Field Machinery. 2d ed. Madison, Wisc.; American Printing and Publishing, 1973. 226p. (1st ed., 1969)	
First	Finner, Marshall F. Farm Machinery Fundamentals. 2d ed. Madison, Wisc.; American Printing and Publishing, 1985. (1st ed., 1978).	First
Second	Finney, Essex E., Jr. Measurement Techniques for Quality Control of Agricultural Products. St. Joseph, Mich.; American Society of Agricultural Engineering, 1973. 53p.	Second
Second	Fluck, Richard D. and C. Direlle Baird. Agricultural Energetics. Westport, Conn.; AVI Publishing Co., 1980. 192p.	Second
	Foltz, R. D., and T. A. Penn. Protecting Engineering Ideas and Inventions. 3d ed. Cleveland, Ohio; Penn Institute Inc., 1989. 300p.	Third
Third	Food and Agriculture Organization, and United Nations Development Fund. Planning Methodology Seminar: Report of FAO-UNDP Seminar on Methodology of Planning Land and Water Development Projects, Romania, 1971. Rome; Food and Agriculture Organization, 1972. 132p. (FAO Irrigation and Drainage Paper no. 11). (Available from UNIPUB).	Second
	Food and Agriculture Organization/OECD Expert Panel on the Effects of Farm Mechanization on Production and Employment, Rome, 1975. Report on the Meeting . . . Rome; Food and Agriculture Organization, 1975. 406p.	Third
Second	Food and Agriculture Organization. 1983 World List of Seed Processing Equipment. Rome; FAO, 1983. 252p.	Second
Second	Food and Agriculture Organization. A Guide for Instructors in Organizing and Conducting Agricultural Engineering Training Courses. Rome; FAO, 1971. 100p. (3d printing 1985.) 2 Supplements: Elements of Agricultural Machinery, 1977. (Suppl. 1, 241p.; Suppl. 2, 295p.) (FAO Agricultural Services Bulletin no. 12).	Second
	Food and Agriculture Organization. Agricultural Engineering	Third

DC	in Development: Guidelines for Establishment of Village Workshops. Rome; FAO, 1988. 66p. (Agricultural Services Bulletin no. 71).	TW
	Food and Agriculture Organization. Agricultural Engineering; 1945–1971. Rome; FAO. (Available from UNIPUB).	Third
	Food and Agriculture Organization. Agricultural Machinery Workshops: Design, Equipment and Management. Rome; FAO, 1960. 111p. (FAO Agricultural Development Paper no. 66).	First
Third	Food and Agriculture Organization. Agricultural Residues: Bibliography 1975–81 and Quantitative Survey. Residus Agricoles, Bibliographie 1975–81 et enquete Quantitative. Rome; Food and Agricultural Organization, 1982. 160p. (FAO Agricultural Services Bulletin no. 47).	
	Food and Agriculture Organization. Animal Energy in Agriculture in Africa and Asia, Technical Papers Presented at the FAO Expert Consultation held in Rome, Nov. 1982 = Energie Animale en Agriculture en Afrique et en Asie: Documents Techniques Presentes a la Consultation d'experts de la FAO Tenue . . . Rome; Food and Agriculture Organization, 1984. 143p.	Third
	Food and Agriculture Organization. Application of Remote Sensing Techniques for Improving Desert Locust Survey and Control. Rome; Food and Agriculture Organization, 1977. 92p.	Third
Second	Food and Agriculture Organization. Appropriate Technology in Forestry. Rome; FAO, 1982. 137p. (FAO Forestry Paper no. 31).	Second
Second	Food and Agriculture Organization. Automated Irrigation. Rome; FAO, 1971. 93p. (4th printing 1976). (FAO Irrigation and Drainage Paper no. 5).	
Third	Food and Agriculture Organization. Basic Technology in Forest Operations. Rome; FAO, 1982. 132p. (FAO Forestry Paper no. 36).	Second
Second	Food and Agriculture Organization. Compendium of Technologies Used in the Treatment of Residues of Agriculture, Fisheries, Forestry, and Related Industries. Compendium des Technologies Utilisees Dans le Traitement des Residus de L'agriculture, des Peches, des Forets et des Industries ConnexeS. Compendio de las Tecnologias Utilizadas en el Tratamiento de los Residuos Agricolas, Pesqueros, Forestales y de las Industrias Afines. Revised version. Rome; Food and Agriculture Organization of the United Nations, 1982. 624p. (FAO Agricultural Services Bulletin no. 33).	Second
Third	Food and Agriculture Organization. Corrosion and Encrustation in Water Wells. Rome; FAO, 1980. 108p. (FAO Irrigation and Drainage Paper no. 34).	
Second	Food and Agriculture Organization. Dam Design and Operation to Optimize Fish Production in Impounded River Basins. Rome; FAO, 1984. 103p. (CIFA Technical Paper no. 11).	First

Citation Analysis and the Core Monographs 263

Second	Food and Agriculture Organization. Design and Operation of Cold Stores in Developing Countries. Rome; Food and Agriculture Organization, 1984. 80p. (Agricultural Services no. 19/2).	
Second	Food and Agriculture Organization. Drainage Design Factors: Twenty-eight Questions and Answers Based on the Expert Consultation on Drainage Design Factors. Rome; FAO, 1980. 52p. (FAO Irrigation and Drainage Paper no. 38).	Second
First	Food and Agriculture Organization. Drainage Machinery. Rome; Food and Agriculture Organization, 1973. 104p. (FAO Irrigation and Drainage Paper no. 15).	First
First	Food and Agriculture Organization. Drainage Materials. Rome; FAO, 1972. 126p. (4th Printing 1976). (FAO Irrigation and Drainage Paper no. 9).	First
First	Food and Agriculture Organization. Drainage of Heavy Soils. Rome; FAO, 1971. 114p. (5th Printing 1976). (FAO Irrigation and Drainage Paper no. 6).	First
First	Food and Agriculture Organization. Drainage of Salty Soils. Rome; FAO, 1973. 87p. (3d Printing 1976). (FAO Irrigation and Drainage Paper no. 16).	First
	Food and Agriculture Organization. Employment of Draught Animals in Agriculture. Rome; Food and Agriculture, and Organization for Centre d'Etudes et d'Experimentation du Machinisme Agricole Tropical, 1972. 249p.	First
	Food and Agriculture Organization. Energy Cropping versus Food Production; FAO expert consultation, Rome, 2–6 June 1980. Rome; Food and Agriculture Organization of the United Nations, 1981. 59p. (FAO Agricultural Services Bulletin no. 46).	Third
First	Food and Agriculture Organization. Equipment for Rice Production. Rome; FAO, 1966. 169p. (FAO Agricultural Development Paper no. 84).	Second
Second	Food and Agriculture Organization. Groundwater Models. Rome; FAO, 1973. 198p. (4th Printing 1983. (FAO Irrigation and Drainage Paper no. 21).	First
Second	Food and Agriculture Organization. Groundwater Pollution—Technology, Economics and Management. Rome; FAO, 1979. 149p. (FAO Irrigation and Drainage Paper no. 31).	Second
Second	Food and Agriculture Organization. Hydrological Techniques for Upstream Conservation. Rome; FAO, 1976. 145p. (2d printing 1979). (Conservation Guide no. 2).	First
First	Food and Agriculture Organization. Irrigation, Drainage and Salinity. London; Hutchinson, FAO with Unesco, 1973. 510p.	First
	Food and Agriculture Organization. Irrigation in Africa South of the Sahara. Rome; Food and Agriculture Organization, 1986. (FAO Investment Centre Technical Paper no. 5).	First
Second	Food and Agriculture Organization. Localized Irrigation: Design, Installation, Operation, Evaluation. Rome; FAO, 1980. 203p. (2d Printing 1984). (FAO Irrigation and Drainage Paper no. 36).	First

Third	Food and Agriculture Organization. Mathematical Models in Hydrology. Rome; FAO, 1973. 282p. (5th Printing 1984). (FAO Irrigation and Drainage Paper no. 19).	
First	Food and Agriculture Organization. Mechanization of Irrigated Crop Production. Rome; FAO, 1977. 413p. (2d Printing 1979.) (FAO Agricultural Services Bulletin no. 28).	First
First	Food and Agriculture Organization. Mechanized Sprinkler Irrigation. Rome; FAO, 1982. 438p. (FAO Irrigation and Drainage Paper no. 35). (Previous ed., 1980 by Lionel Rolland in English and French. 465p.)	First
Second	Food and Agriculture Organization. Multifarm Use of Agricultural Machinery. Rome; FAO, 1985. 63p. (FAO Agricultural Services Bulletin no. 17). (Available from UNIPUB). (Previous edition by H. Lonnemark, 1967. 113p., as FAO Agricultural Development Paper no. 85).	First
Second	Food and Agriculture Organization. Pesticide Application Equipment and Techniques. Rome; FAO, 1979. 261p. (FAO Agricultural Services Bulletin no. 38).	First
	Food and Agriculture Organization. Replacement Parts for Agricultural Machinery. Rome; FAO, 1981. 33p. (FAO Agricultural Services Bulletins no. 44). (Available from Unipub).	Second
Third	Food and Agriculture Organization. Residue Utilization; Management of Agricultural and Agro-Industrial Residues. Rome; FAO, 1977. 2 Vols. (Available from UNIPUB).	Second
Third	Food and Agriculture Organization. Rice Milling in Developing Countries: Case Studies and Some Aspects of Economic Policies. Rome; Food and Agriculture Organization, 1969. 33p. (Commodity Bulletin Series no. 45).	
Second	Food and Agriculture Organization. Selected Terms of Irrigation. Rome; Food and Agriculture Organization, 1978. 94p. (FAO Terminology Bulletin no. 34). (Available from UNIPUB).	Second
Third	Food and Agriculture Organization. Simulation Methods in Water Development. Rome; FAO, 1974. 62p. (2d Printing 1984). (FAO Irrigation and Drainage Paper no. 23).	First
	Food and Agriculture Organization. Small Hydraulic Structures, Vol. 1 and 2. Rome; FAO, 1975. 730p. (2d printing 1982.) (FAO Irrigation and Drainage Paper no. 26).	First
Second	Food and Agriculture Organization. Soils Bulletins; no. 33: Soil Conservation and Management in Developing Countries; 1977. no. 30: Soil Conservation in Developing Countries; 1976. no. 24: Shifting Cultivation and Soil Conservation in Africa; 1974. Rome; FAO, 1974–77.	First
Second	Food and Agriculture Organization. Some Aspects of Earth Moving Machines as Used in Agriculture. Rome; FAO, 1975. 56p. (2d Printing 1976.) (FAO Agricultural Services Bulletin no. 27).	First
Second	Food and Agriculture Organization. The Technology Applications Gap: Overcoming Constraints to Small-Farm Development. Rome; FAO, 1986. 144p. (FAO Research and Technology Paper no. 1).	First

Second	Food and Agriculture Organization. Water-Lifting Devices; Proceedings of the FAO/DANIDA Workshop on Water Lifting Devices in Asia and the Near East. Rome; FAO, Water Resources, Development and Management Service, 1986. 295p. (FAO Irrigation and Drainage Paper no. 43)	First
Third	Food and Agriculture Organization. World Directory of Institutions Concerned with Residues of Agriculture, Fisheries, Forestry and Related Industries. Repertoire Mondial des Institutions Interessees aux Residus de L'agriculture des Peches, des Forets, et des Industries Connexes. 3d ed. Rome; Food and Agriculture Organization, 1982. 219p. (FAO Agricultural Services Bulletin, no. 21, rev. 2).	
Third	Forest Transportation Symposium, Casper, Wyom., 1984. Proceedings . . . Denver, Colo.; United States Forest Service, Rocky Mountain Region, Engineering Staff, 1985. 238p.	
Second	Framji, K. K., B. C. Garg and S. D. L. Luthra. Irrigation and Drainage in the World; A Global Review. 3d ed. New Delhi; International Commission on Irrigation and Drainage, 1981–83. 3 vols. (1st ed. by Niranjan Das Gulhati, published under title: Irrigation in the World, 1955; 2d ed. by Framji et al., 1978–81, 2 vols.)	Second
Third	Framji, K. K., ed., and J. Shalhevet, and J. Kamburov, compilers. Irrigation and Salinity: A World Wide Survey. New Delhi; International Commission on Irrigation and Drainage, 1976. 106p.	Second
Second	France. Ministere de la Cooperation. Manuel de Motorisation des Cultures Tropicales. Antony; Centre D'Estudes et D'Experimentation du Machinisme Agricole Tropical, 1977. 2 vols.	Third
	Frederick, Kenneth D. Water Management and Agriculture Development: A Case Study of the Cuyo Region of Argentina. Washington, D.C.; Resources for the Future, Inc., 1975. 204p.	Third
Third	Fuentes-Yague, Jose-Luis. Construcciones para la Agricultura. 4th ed. Madrid; Publicaciones de Extension Agraria, 1980. 478p. (1st ed., 1970, 351p.)	Third
Third	Fukuda, Hitoshi., ed. Irrigation in the World: Comparative Development. Japan; University of Tokyo Press, 1976. 329p.	Second

G

Third	Gaffney, J. J. Quality Detection in Foods. Compiled in cooperation with ASAE Technical Committee FE-74 on Food Handling. St. Joseph, Mich.; American Society of Agricultural Engineers, 1976. 240p.	
	Garcie-Fernandez, J. Arados, Explanadoras, Traillas y Niveladoras. Madrid; Dosset, 1961. 146p.	Third
Second	Gasser, J. K. R., ed. Composting of Agricultural and Other Wastes. Proceedings of a Seminar Organised by the Commission of the European Communities at Brasenose College, Oxford, UK, 1984. London and New York; Elsevier Applied Science Publishers, 1985. 320p.	
Third	Gasser, J. K. R., J. C. Hawkins, J. R. O'Callaghan, and	Third

DC	B. F. Pain. eds. Effluents from Livestock. London; Applied Science, 1980. 712p.	TW
	Gava, Altanir J. Principios de Tecnologia de Alimentos. Sao Paulo; Nobel, 1978. 284p.	Third
Third	Geankoplis, Christie J. Transport Processes and Unit Operations. 2d. ed. Boston; Allyn and Bacon, 1983. 862p.	
Second	Gensler, William G. Advanced Agricultural Instrumentation: Design and Use. NATO Advanced Study Institute on Advanced Agricultural Instrumentation, Pisa, Italy, 1984. Dordrecht and Boston; M. Nijhof, 1986. 480p.	
	Ghai, Dharam and Samir Radwan, eds. Agrarian Policies and Rural Poverty in Africa. Geneva; International Labour Office, 1983. 311p.	Third
Third	Ghali, A. Circular Storage Tanks and Silos; Analysis and Design. London; Chapman and Hall, 1979. 210p.	
Second	Gieck, K. Engineering Formulas. Technische Formelsammlung. 5th ed. New York; McGraw-Hill Book Co., 1986. 544p. (2d ed., 1974. 382p.)	
	Gifford, R. C. Agricultural Mechanization in Development: Guidelines for Strategy Formulation. Rome; Food and Agriculture Organization, 1981. 77p. (FAO Agricultural Services Bulletin no. 45).	Third
	Gifford, R. C. and A. G. Rijk. Guidelines for Agricultural Mechanization Strategy in Development. Bangkok; Economic and Social Commission for Asia and the Pacific, Regional Network for Agricultural Machinery, 1980.	Second
	Giles, G. W. The Reorientation of Agricultural Mechanization for the Developing Countries: Policies and Attitudes for Action Programmes. Rome; Food and Agriculture Organization, 1975.	Third
	Gill, Gerard J. Farm Power in Bangladesh; Vol. 1, A Comparative Analysis of Animal and Mechanical Farm Power in Bangladesh. Reading; University of Reading, Dept. Agricultural Economics and Management, 1981. 248p. (Development Study no. 19).	Second
First	Gill, William R. and Glen E. VandenBerg. Soil Dynamics in Tillage and Traction. Washington, D.C.; U.S. Department of Agriculture, Agricultural Research Service, 1967. 511p. (USDA Agricultural Handbook 316).	First
	Gillies, M. T. Dehydration of Natural and Simulated Dairy Products. Park Ridge, New Jersey; Noyes Data Corp., 1974. 328p. (Food Technology Review no. 15).	Third
	Gillies, M. T. Fish and Shellfish Processing. Park Ridge, New Jersey; Noyes Data Corp., 1975. 338p. (Food Technology Review no. 22).	Third
	Gleck, K. Engineering Formulas. 5th ed. McGraw-Hill Book Co., 1986. 544p.	Second
	Gleizes, Claude. Evaluation des Quantites D'Eau Necessaires Aux Irrigations. Paris; Ministere de la Cooperation, 1964. 170p.	Third
	Glennie, Colin. Village Water Supply in the Decade: Lessons	Second

	From Field Experience. New York; John Wiley and Sons, Inc., 1983. 152p.	
Third	Gloyna, Earnest F. and W. Wesley Eckenfelder, Jr. Advances in Water Quality Improvement. Austin; University of Texas Press, 1968. 513p. (Water Resources Symposium no. 1).	
Third	Glysson, E. A., ed. Innovations in the Water and Wastewater Fields. London; Butterworth's, 1984. 240p.	
	Goense, D. Mechanized Farming in the Humid Tropics with Special Reference to Soil Tillage, Workability and Timeliness of Farm Operations: A Case Study for the Zanderij Area of Suriname. Wageningen, Netherlands; Landbouwuniversiteit, 1987. 136p. (Phd. Thesis).	Second
First	Goering, Carroll E. Engine and Tractor Power. Boston, Mass; Breton Publishers, 1986. 404p. (Available from the American Society of Agricultural Engineers).	First
Second	Goldberg, Dan, Baruch Gornat, and Daniel Rimon. Drip Irrigation: Principles, Design and Agricultural Practices. Kfar Shmaryahu, Israel; Drip Irrigation Scientific Publications, 1976. 296p.	Second
Third	Goldblith, Samuel A., M. A. Joslyn, and J. T. R. Nickerson. Introduction to Thermal Processing of Foods. Westport, Conn.; AVI Pub. Co., 1961. 1128p.	
Second	Goldemberg, Jose, et al. Energy for a Sustainable World. New York; Wiley, 1988. 517p.	Third
Second	Golding, E. W. The Generation of Electricity by Wind Power. Rev. ed. London; Methuen, Inc., 1976. (1st ed., New York; Philosophical Library, 1956. 318p.)	Second
Third	Golubev, Genady N. and Asit K. Biswas, eds. Large Scale Water Transfers: Emerging Environmental and Social Issues. Oxford, England; Published for U.N. by Tycooly Publishing, 1985. 158p.	Second
Third	Goodman, Louis J., John N. Hawkins and Ralph N. Love, eds. Small Hydroelectric Projects for Rural Development: Planning and Management. New York; Pergamon Press, 1981. 200p. (Published in Cooperation with the East-West Center, Hawaii).	Third
Second	Goryachkin, V. Sobranie Sochinenii v Trekh Tomakh. Collected works in Three Volumes. Translated from Russian by E. Vilim. Jerusalem; Israel Program for Scientific Translations, 1972. (United States National Technical Information Service TT 71-50087). (Original 3 Vols., Moscow, Kolos, 1968).	Third
Second	Goswami, D. Yogi, ed. Alternative Energy in Agriculture. Boca Raton, Flor.; CRC Press, 1986. 2 Vols.	Third
	Gracia, C. and E. Palau. Mecanización de los Cultivos Hortícolas. Madrid; Mundi Presna, 1981. 190p.	Third
Third	Graham, I., and P. L. Jones. Expert Systems: Knowledge, Uncertainty and Decision. London and New York; Chapman and Hall, 1988. 363p.	
Second	Grassi, G., D. Pirrwitz and H. Zibetta, eds. Energy from Biomass 4; Proceedings of the Third Contractors' Meeting,	

DC	Paestum, Italy, 1988. London and New York; Elsevier Applied Science, 1989. 627p.	TW
Second	Gray, Roy B. The Agricultural Tractor: 1855–1950. St. Joseph, Mich.; American Society of Agricultural Engineers, 1974. 160p.	
Third	Greacen, E. L., ed. Soil Water Assessment by the Neutron Method. East Melbourne, Victoria; CSIRO, 1981. 140p.	
	Great Britain. Ministry of Agriculture, Fisheries and Food. Farm Buildings Pocketbook. Amended Metric Edition. London; Her Majesty's Stationery Office, 1981. 95p.	Second
Second	Great Britain. Ministry of Agriculture, Fisheries and Food. Irrigation. 5th ed. London; Her Majesty's Stationery Office, 1982. 122p.	Second
Second	Great Britain. Ministry of Agriculture, Fisheries and Food. Water for Irrigation: Supply and Storage. 2d ed. London; Her Majesty's Stationery Office, 1977. 97p. (Bulletin 202)	Second
Second	Great Britain. Ministry of Overseas Development. Tropical Stored Products Centre. Food Storage Manual, 3 parts. Part 1, Storage Theory; Part 2, Food and Commodities; Part 3, Storage Practice. Rome; FAO, 1970. 799p.	Second
Second	Green, John H. and Amihud Kramer. Food Processing Waste Management. Westport, Conn.; AVI Pub. Co., 1979. 629p.	
Third	Greenland, D. J. and R. Lal, eds. Soil Conservation and Management in the Humid Tropics. International Conference on Soil Conservation and Management in the Humid Tropics, Ibadan, Nigeria, 1975. Sponsored by the International Institute of Tropical Agriculture, and the Agricultural Research Council of Nigeria. Chichester, Eng. and New York; Wiley, 1977. 283p.	Third
Second	Gregg, Billy R., Alvin G. Law, Sher S. Virdi, and John S. Balis. Seed Processing. Aram Nagar and New Delhi; U.S. AID and Avion Printers, 1970. 396p.	Second
Third	Gregory, Kenneth J. and D. E. Walling. Drainage Basin Form and Process; A Geomorphological Approach. London; E. Arnold, and New York; Wiley, 1973. 458p.	
Second	Griffin, George A. Combine Harvesting. 3d ed. Moline, Ill.; Deere and Co., 1987. (1st ed., 1973. 196p.)	Second
Second	Gurfinkel, German. Wood Engineering. New Orleans, Louisiana; Southern Forest Products Association, 1972.	
FIrst	Gustafson, Robert J. Fundamentals of Electricity for Agriculture. 2d ed. St. Joseph, Mich.; American Society of Agricultural Engineers, 1988. 411p. (ASAE Textbook no. 2). (1st ed., Westport, Conn.; AVI Pub. Co., 1980. 293p.).	First
Third	Gutcho, Marcia H. Freeze Drying Processes for the Food Industry. Park Ridge, New Jersey; Noyes Data Corp., 1977. 401p. (Food Technology Review no. 41).	Second

H

First	Haan, Charles T., H. P. Johnson, and D. L. Brakensiek, eds. Hydrologic Modeling of Small Watersheds. St. Joseph, Mich.;	Second

Citation Analysis and the Core Monographs 269

	American Society of Agricultural Engineers, 1982. 533p. (ASAE Monograph no. 5).	
Second	Haan, Charles. T. Statistical Methods in Hydrology. Ames; Iowa State University Press, 1977. 378p.	
First	Hagan, Robert M., Howard R. Haise, and T. W. Edminster, eds. Irrigation of Agricultural Lands. Madison, Wisc.; American Society of Agronomy, 1967. 1180p.	First
Third	Hahn, Gerald J. and Samuel S. Shapiro. Statistical Models in Engineering. New York; Wiley, 1967. 355p.	Second
Third	Haidan, M. and Erhardt Thum. Maschinen und Anlagen fur die Tie Produktion. (Machinery and Equipment for Animal Production.) Berlin; VEB Deutscher Landwirtschaftsverlag, 1985. 426p.	
	Haiduk, V. M. and M. F. Sahach. Elektroteplovi Sil'S'kohospodars'ki ustanovky. (Thermoelectric Systems in Agriculture.) Kiev; Derzhsil'hospvydav URSR, 1961. 138p.	Third
	Hall, A. Drought and Irrigation in North-East Brazil. Cambridge; New York; Cambridge University Press, 1978. 152p. (Latin American Studies no. 29).	Third
First	Hall, Carl W., A. W. Farrall, and A. L. Rippen. 2d ed. Encyclopedia of Food Engineering. Westport, Conn.; AVI Publishing Co., 1986. 882p. (1st ed., 1971. 755p.)	Second
Second	Hall, Carl W. and G. Malcolm Trout. Milk Pasteurization. Westport, Conn.; AVI Publishing Co., 1968. 234p.	Second
Second	Hall, Carl W. and T. I. Hedrick. Drying of Milk and Milk Products. Westport, Conn.; AVI Publishing Co., 1971. 338p. (1st ed., 1966. 338p.)	Second
First	Hall, Carl W. Bibliography of Agricultural Engineering Books. St. Joseph, Mich.; American Society of Agricultural Engineers, 1976. 82p. (ASAE Special Publication 8–76).#2	Third
First	Hall, Carl W. Bibliography of Bibliographies of Agricultural Engineering and Related Subjects. St. Joseph, Mich.; American Society of Agricultural Engineers, 1976. 62p.	Third
Second	Hall, Carl W. Biomass as an Alternative Fuel. Rockville, Md.; Government Institutes, 1981. 267p.	
Second	Hall, Carl W. Dictionary of Drying. New York; Marcel Dekker, 1979. 350p.	Second
Second	Hall, Carl W. Drying and Storage of Agricultural Crops. Westport, Conn.; AVI Pub. Co., 1980. 381p.	Second
Second	Hall, Carl W. Processing Equipment for Agricultural Products. 2d. ed. Westport, Conn.; AVI Pub. Co., 1979. 294p. (1st ed.: Ann Arbor, Mich.; Edwards Bro., 1963. 272p.)	First
Second	Hall, D. W. Handling and Storage of Food Grains in Tropical and Subtropical Areas. Rome; Food and Agriculture Organization, 1970. 350p. (FAO Agricultural Development Paper no. 90).	First
	Hall, David. O., G. W. Barnard and P. A. Moss. Biomass for Energy in the Developing Countries: Current Role, Potential Problems, Prospects. 1st ed. Oxford and New York; Pergamon Press, 1982. 220p.	Third
	Hall, H. S. Standardized Pilot Milk Plants. Rome; Food and	Second

DC	Agriculture Organization, 1976. 104p. (FAO Animal Production and Health Series no. 3).	TW
	Hall, H. S., Y. Rosen, and H. Blombergsson. Milk Plant Layout. Rome; Food and Agriculture Organization, 1963. 156p. (FAO Agricultural Studies no. 59).	Second
First	Hansen, Vaughn E., Glen E. Stringham, and Orson W. Israelsen. Irrigation Principles and Practices. 5th ed. New York; John Wiley and Sons, Inc., 1984. 450p. (1st ed. by Israelsen and Stringham, 1932. 442p.)	First
	Hanson, L. P. Commercial Processing of Vegetables. Park Ridge, New Jersey; Noyes Data Corp., 1975. 449p. (Food Technology Review, no. 27)	Second
	Hardcastle, J. E. Y. Mechanizing Rice Cultivation in Nigeria. Bamako; Inter-African Committee on the Mechanization of Agriculture, 1961.	Third
	Harms, R. Greg, editor. Ventilation Handbook: Livestock and Poultry. Victoria, British columbia; The Ministry, Agricultural Engineering Branch, 1985. 1 Vol. (various pagings)	Second
Third	Harper, John C. Elements of Food Engineering. Westport, Conn.; AVI, 1976. 282p.	Second
Second	Harper, Judson M. Extrusion of Foods. Boca Raton, Flor.; CRC Press, 1981. 2 Vols.	Second
Second	Harris, Anthony G., T. B. Muckle, and J. A. Shaw. Farm Machinery. 2d ed. London; Oxford University Press, 1974. 280p. (1st ed., 1965. 240p.)	
Second	Harrold, L. L., G. O. Schwab, and B. L. Bondurant. Agricultural and Forest Hydrology. Columbus, Ohio; Agricultural Engineering Dept., Ohio State University, 1974. 273p.	Second
Third	Hatfield, Jerry L. and Ivan J. Thomason, eds. Conference on Biometeorology and Integrated Pest Management; Biometeorology in Integrated Pest Management: Proceedings of a Conference held at the University of California, Davis, 1980. New York; Academic Press, 1982. 491p.	
Second	Hathaway, Louis. Preventive Maintenance. 4th ed. Moline, Ill.; Deere and Co., 1990. (1st ed., 1973. 238p.)	Second
Second	Hathway, Gordon. Appropriate Technology in Rural Development: Vehicles Designed for On and Off Farm Operations. Washington, D.C.; World Bank, 1978.	
Second	Hathway, Gordon. Low-Cost Vehicles: Options for Moving People and Goods. London; Intermediate Technology Publications, 1985. 106p.	
Third	Hawkins, J. C. Engineering Problems with Effluents from Livestock. Luxembourg; Commission of the European Communities, 1979. 538p.	Third
Third	Hayami, Yujiro, and Vernon W. Ruttan. Agricultural Development: An International Perspective. 2d ed. Baltimore: Johns Hopkins University Press, 1986. 506p. (1st ed., 1971. 367p.)	
Second	Hayward, Alan. Flowmeters: A Basic Guide and Source-Book for Users. London; Macmillan, 1979. 197p.	
	Hazlewood, A. and I. Livingstone. Irrigation Economics in Poor Countries: Illustrated by the Usango Plains of Tanzania. Oxford and New York; Pergamon Press, Inc., 1982. 150p.	Second

	Heid, J. L. and Maynard A. Joslyn. Fundamentals of Food Processing Operations: Ingredients, Methods, and Preparation. Westport, Conn.; AVI Pub. Co., 1967. 730p.	Second
Third	Heinrich, Rudolf. FAO/Austria Training Course on Mountain Forest Roads and Harvesting, 3d, 1981, Ossiach and Ort, Austria. Logging of Mountain Forests; Report of the Third . . . Rome; Food and Agriculture Organization of the United Nations, 1982. 285p. (FAO Forestry Paper no. 33)	
Third	Held, R. Burnell and Marion Clawson. Soil Conservation in Perspective. Baltimore, MD; Johns Hopkins Press, published for Resources for the Future Inc., 1965. 344p.	
Second	Heldman, Dennis R. and R. Paul Singh. Food Process Engineering. Westport, Conn.; AVI Publishing Co., 1981. 415p. (1st ed., 1975. 401p.)	First
Second	Hellickson, Mylo A. and John N. Walker, eds. Ventilation of Agricultural Structures. St. Joseph, Mich.; American Society of Agricultural Engineers, 1983. 372p. (ASAE Monograph no. 6.)	Second
Third	Helsel, Zane R., ed. Energy in Plant Nutrition and Pest Control. Amsterdam and New York; Elsevier, 1987. 293p. (Distributors for the United States and Canada, Elsevier Science Pub. Co.)	
Second	Henderson, Silas M. and R. L. Perry. Agricultural Process Engineering. 3d ed. Westport, Conn.; AVI Publishing, 1976. 442p. (1st ed., New York; Wiley, 1955. 402p.; 2d ed., Ann Arbor; Edwards, 1966. 434p.)	First
Second	Hendrick, James G. An Annotated Bibliography on Rotary Tillage Tools. Auburn, Alabama; U.S. Department of Agriculture, Agricultural Research Service, National Tillage Machinery Laboratory, 1970. Various pagings.	
Second	Hendrick, James G., ed. An Annotated Bibliography on Vibratory Soil Dynamics (1969–1979). Auburn, Alabama; U.S. Department of Agriculture, Agricultural Research Service, National Tillage Machinery Laboratory, 1979. 59p.	Second
Third	Henrickson, Robert Lee. Meat, poultry, and seafood technology. Englewood Cliffs, New Jersey; Prentice-Hall, 1978. 276p.	Second
Third	Hepher, Balfour and Yoel Pruginin. Commercial Fish Farming: With Special Reference to Fish Culture in Israel. New York; Wiley, 1981. 261p.	
Second	Hess, Oleen. Small Farms Appropriate Technology: Research, Production, Hardware, and Animal Traction. Beltsville, Md.; Four Six Zero Five Brandon Lane Press, 1985. 175p.	Second
Second	Hexem, Roger W. and Earl O. Heady. Water Production Functions for Irrigated Agriculture. Ames; Iowa State University Press, 1978. 215p.(2)	Second
Second	Heyde, Heinrich. Landmaschinenlehr. Berlin; Leitfaden fur Studierende der Landwirtschaft, 1967.	
Second	Hicks, Tyler G. and S. David Hicks, eds., Standard Handbook of Engineering Calculations. 2d ed. McGraw-Hill Book Co., 1985. 1600p. (1st ed., Hicks, 1972)	
Second	Hignett, Travis P. Fertilizer Manual. Dordrecht, Germany and	Third

DC	Boston; Nijhoff/W. Junk Publishers for the International Fertilizer Development Center, Hingham, Mass., 1985. 363p. (Developments in Plant and Soil Science, Vol. 15)	TW
First	Hiler, Edward A. and Bill A. Stout, eds. Biomass Energy: A Monograph. Texas; Texas Engineering Experiment Station, Texas AandM University Press, 1985. 313p.	Third
First	Hillel, Daniel, ed. Advances in Irrigation. New York; Academic Press, Inc., 1982 87. 4 vols.	First
Third	Hillel, Daniel. Applications of Soil Physics. New York; Academic Press, 1980. 385p.	Third
Second	Hillel, Daniel. Fundamentals of Soil Physics. New York; Academic Press, 1980. 413p.	First
Second	Hillel, Daniel. Soil and Water: Physical Principles and Processes. New York; Academic Press, 1971. 288p.	First
	Hine, Howard J. Dictionary of Agricultural Engineering. Cambridge, England; W. Heffer and Sons, 1961. 252p.	Second
	Hinton, William. Iron Oxen: A Documentary of Revolution in Chinese Farming. New York; Monthly Review Press, 1970. 225p.	Third
Third	Ho, Chester S. Biotechnology Processes: Scale-up and Mixing. New York; American Institute of Chemical Engineers, 1987. 267p.	
Second	Hobson, P. N. and A. M. Robertson. Waste Treatment in Agriculture. London; Applied Science Publishers, 1977. 257p.	
Second	Hobson, P. N., R. Summers and S. Bousfield. Methane Production from Agricultural and Domestic Wastes. London; Applied Science Pub.; New York; Halsted Press Division, Wiley, 1981. 269p. #5	Second
	Hoki, M., ed. Studies on the Conventional Farm Tools and Evolution of Farming Systems in Southeast Asia. Mie, Japan; Mie University, Faculty of Bioresources, 1988. 160p.	Third
Third	Holman, Jack P. and W. J. Gajda, Jr., Experimental Methods for Engineers. 5th ed. New York; McGraw-Hill, 1989. 549p. (1st ed.: 1966. 412p.)	
	Holtkamp, Rudolf. Small Four-Wheeled Tractors for the Tropics and Sub-Tropics: Their Role in Agricultural and Industrial Development. CTZ Weikersheim; Margraf, 1990. 248p.	Third
Third	Ho-pei Nung Yeh ta Hsueh. Nung Yeh She Ch'an Chi Hsieh Hua: Hua Pei Pen. (Mechanization of Agricultural Production.) Peiking; Chung-kuo Nung Yeh Chi Hsieh Ch'u Pan She, 1987. 421p.	
First	Hopfen, Hans J., and E. Biesalski. Small Farm Implements. Rome; Food and Agriculture Organization, 1953. 79p. Reprinted in 1978. (FAO Agricultural Development Paper no. 32; FAO Agricultural Series no. 5).	First
	Hopfen, Hans J. Farm Implements for Arid and Tropical Regions. Revised ed. Rome; Food and Agriculture Organization, 1969. 159p. (Reprinted 1981). (1st ed., 1960). (FAO Agricultural Development Paper no. 67, FAO Agricultural Development Paper no. 91).	First

Citation Analysis and the Core Monographs 273

Second	Hord, R. Michael. Remote Sensing Methods and Applications. New York; John Wiley and Sons, Inc., 1986.	
Third	Hua Nan Nung Hsueh Yuan. Nung Chi Chiao Yen Shih. Shui Tao Lien Ho Shou Ko Chi Yuan Li Yu She Chi. (The Principle and Design of Rice Harvesters.) Pei-ching; Chung-kuo Nung Yeh Chi Hsieh Ch'u Pan She, Hsin Hua Shu Tien Pei-ching Fa Hsing So Fa Hsing, 1981. 193p.	Third
	Hudson, A. W., H. G. Hopewell, D. G. Bowler, and M. W. Cross. The Draining of Farm Lands. 2d ed. New Zealand; Massey College, 1962. (1st ed., 1950).	Third
Second	Hudson, Norman W. Field Engineering for Agricultural Development. Oxford, England; Clarendon Press, 1983. 240p.	Second
First	Hudson, Norman W. Soil Conservation. 2d ed. Ithaca, New York; Cornell University Press, 1981. 324p. (1st ed., 1971, 320p.)	First
Second	Hughes, Harold A. Crop Chemicals. 2d ed. Moline, Ill.; Deere and Co., 1982. 229p. (Fundamentals of Machine Operation series). (1st ed., 1976, 229p.)	Second
Second	Huisman, Willem. Optimum Cereal Combine Harvester Operation by Means of Automatic Machine and Threshing Speed Control. Wageningen, The Netherlands; Dept. of Agricultural Engineering, Agricultural University, 1983. 293p.	
First	Hunt, Donnell. Engineering Models for Agricultural Production. Westport, Conn.; AVI Pub. Co., 1986. 260p.	First
First	Hunt, Donnell. Farm Power and Machinery Management. 8th ed. Ames, Iowa; Iowa State University Press, 1983. 352p. (2d ed., 1956. 149p.)	First
Third	Hurd, Clarence J. Sprinkler Irrigation Guidebook. Washington, D.C.; USAID, Office of War on Hunger, Agriculture and Rural Dev. Service, 1969. 103p.	

I

Second	Imholte, Thomas J. Engineering for Food Safety and Sanitation: A Guide to the Sanitary Design of Food Plants and Food Plant Equipment. Crystal, Minn.; Technical Institute of Food Safety, 1984. 283p.	Second
Third	Improving Irrigation Water Management on Farms. Annual Technical Reports . . . Submitted to U.S. Agency for International Development. Fort Collins; Colorado State University, Water Management Research Project, 1978–1980. 3 vols.	
Indian	Council of Agricultural Research. Indigenous Agricultural Implements of India; An All-India Survey. New Delhi; Indian Council of Agricultural Research, 1960. 401p.	Second
Second	Indian Society of Agricultural Engineers. Convention 23d, 1987. Role of Agricultural Engineering in Dryland Agriculture: Proceedings . . . (Jawaharlal Nehru Krishi Vishwa Vidyalaya). New Delhi; The Society, 1987. 273p.	Second
Second	Inglett, George E., ed. Symposium on Processing Agricultural and Municipal Wastes, New York, 1972. Westport, Conn.; AVI Publishing Co., 1973. 221p.	Second
	Institut fur Tropische und Subtropische Landwirtschaft, Karl-	Third

DC	Marx-Universitat. Recent Stage and Prospects of Agricultural Mechanization in Developing Countries: Proceedings of a Scientific Conference of the Institute of Tropical Agriculture and Veterinary Science . . . Berlin; Akademie-Verlag, 1976. 261p. (Studien Uber Asien, Afrika und Lateinamerika no. 22). (In English)	TW
Third	Institut Pertanian Bogor. Kokusai Kyoryoku Jigyo Dan. Text Book on Agricultural Products Processing: Agricultural Products Processing Pilot Plant Project (AP4 Project), Indonesia. Bogor; Bogor Agricultural University; Japan International Cooperation Agency, 1979. 197p.	Third
Third	Institute of Gas Technology. Energy from Biomass and Wastes, VII, Lake Buena Vista, Florida, 1983. Symposium Papers . . . Chicago, Ill.; Institute of Gas Technology, 1983. 1417p.	
Second	Institute of Gas Technology. Symposium on Clean Fuels from Biomass and Wastes, Orlando, Florida, 1977. Chicago; Institute of Gas Technology, 1977. 521p.	Second
Third	Instituut Voor Mechanisatie, Arbeid en Gebouwen. Automatisering Ten Behoeve van het Bedrijfsbeheer op Melkveebedrijven: Toepassingsmogelijkheden, Wensen en Normen = Automation of Dairy Farm Management. Wageningen; IMAG, 1981. 76p. (Instituut Voor Mechanisatie, Arbeid en Gebouwen. Publikaie, no. 160)	
Third	International Commission of Agricultural Engineering. Latest Developments in Livestock Housing. 2d Technical Section Seminar. St. Joseph, Mich; American Society of Agricultural Engineers, 1987. 417p.	
Third	International Commission of Agricultural Engineering. Section III Hungarian National Committee. Magyar Agrartudomanyi Egyesulet. Proceedings of the II International Conference on Physical Properties of Agricultural Materials and their Influence on Technological Process. Godollo, Hungary; 1980. 5 Vols.	Second
	International Commission of Agricultural Engineering. Section IV. Magyar Elektrotechnikai Egyesulet. "Rural Electrification". 7th International Conference to be held from 6th to 11th October 1975: Reports = Section IV, l'electrification Rurale" : 7e Journees d'etude Internationales du 6 au 11 Octobre 1975 : Rapports. Budapest; Magyar Elektrotechnikai Egyesulet, 1975. 451p.	Second
	International Commission of Agricultural Engineering. Section V. Symposium, 1979. (Darmstadt, Germany). Planning Elements for Agriculture = Planungsunterlagen fur die Landwirtschaft = Elements de Planification en Agriculture. Darmstadt; KTBL, 1979. 3 vols.	Third
	International Commission on Irrigation and Drainage. International Cooperation in the Development of Water Resources for Agriculture. New Delhi; ICID, 1970. 318p.	Second
Second	International Commission on Irrigation and Drainage. The Application of Systems Analysis to Problems of Irrigation, Drain-	Second

Citation Analysis and the Core Monographs 275

	age and Flood Control: A Manual for Water and Agricultural Engineers. Prepared by the Permanent Committee on Application of Systems Analysis to Irrigation, Drainage and Flood Control. 1st ed. Oxford; New York; Published for the ICID by Pergamon Press, 1980. 211p.#6.	
Third	International Conference and Exhibition on Aseptic Packaging, 3d, 1985, Princeton, N.J. Aseptipak '85 : Proceedings . . . Princeton, N.J.; Schotland Business Research, 1985. 544p.	
Second	International Conference on Biomass. Energy from Biomass: Proceedings of the Conference . . . Brighton Centre, England, 1980. London; Applied Science Pub., 1981. 982p.	
Second	International Conference on Plant and Vegetable Oils as Fuels, 1982, Fargo, N. D. Vegetable Oil Fuels: Proceedings of the . . . St. Joseph, Mich.; American Society of Agricultural Engineers, 1982. 400p.	
Third	International Conference on Robotics and Intelligent Machines in Agriculture; Tampa, Florida, 1983. Robotics and Intelligent Machines in Agriculture: Proceedings . . . St. Joseph, Mich.; American Society of Agricultural Engineers, 1984. 155p. (ASAE Pub.: 4–84)	
Second	International Conference on Soil Dynamics, Auburn, Alabama, 1985. Proceedings . . . Jointly Sponsored by National Tillage Machinery Laboratory and Alabama Experiment Station. Auburn; Auburn University, Office of Continuing Education, ICSD Conference, 1985. 5 vols. 1157p.	
Third	International Congress on Agricultural Engineering, 11th, Dublin, 1989. Proceedings . . . Agricultural Engineering, edited by Vincent A. Dodd and Patrick M. Grace. Rotterdam and Brookfield, Vt.; A.A. Balkema, 1989. 4 vols. (3,114 pp) Vol. 1: Land and Water Use; Vol. 2: Agricultural Buildings; Vol. 3: Agricultural Mechanisation; Vol. 4: Power, Processing and Systems.	Third
Second	International Directory of New and Renewable Energy Information Sources and Research Centres. 2d ed. Paris; UNESCO, 1986. 661p. (1st ed., 1982.)	
Second	International DLG-Congress for Computer Technology, Hanover, 1986. Microelectronics in Agriculture, Facts and Trends: Papers . . . Frankfurt am Main; Deutsche Landwirtschafts-Gesselschaft, 1986. 230p.	
First	International Drip/Trickle Irrigation Congress, 3d, 1985, Fresno, California. Drip/Trickle Irrigation in Action: Proceedings . . . St. Joseph, Mich.; American Society of Agricultural Engineers, 1985. 2 vols. (931p.) (ASAE Publication 10-85)	First
First	International Institute for Land Reclamation and Improvement. Drainage Principles and Applications. 4 vols. Wageningen; IILRI, 1972–80. Vol. 1, Introductory Subjects, 1972. 241p.; vol. 2, Theories of Field Drainage and Watershed Runoff, 1973. 374p.; Vol. 3, Surveys and Investigations, 2d ed., 1980. 364p.; Vol. 4, Design and Management of Drainage Systems, 2d ed., 1980. 470p.	First
Second	International Institute of Tropical Agriculture, and Food and	Second

DC	Agriculture Organization of the United Nations. Report on the Expert Consultation Meeting on the Mechanization of Rice Production. Ibadan, Nigeria; International Institute of Tropical Agriculture, 1975. 280p.	TW
	International Labour Office. Excess Manufacturing Capacity and Farm Equipment Needs in Kenya. Geneva; International Labour Office, 1986. 117p.	Second
	International Labour Office. Mechanisation and Employment in Agriculture: Case Studies from Four Continents. Geneva; ILO, 1973. 192p.	First
Second	International Labour Organisation. Appropriate Farm Equipment Technology for the Small-Scale Traditional Sector. Geneva; International Labor Organisation, Employment and Technology Branch, 1983–1985. 6 vols.	First
Second	International Livestock Environment Symposium, 2d, Iowa State University, 1982. Livestock Environment II: Proceedings . . . St. Joseph, Mich.; American Society of Agricultural Engineers, 1982. 624p. (ASAE Pub. no. 3-82.)	
Second	International Livestock Environment Symposium, 3d, Toronto, 1988. Livestock Environment III: Proceedings . . . St. Joseph, Mich.; American Society of Agricultural Engineers, 1988. 470p. (ASAE Pub. no. 1-88).(2)	Second
	International Rice Commission. Report of the Working Party on the Agricultural Engineering Aspects of Rice Production, Storage and Processing. 3 vols. Rome; Food and Agriculture Organization, 1962–64.	Third
First	International Rice Research Institute and the Agricultural Development Council. Consequences of Small-Farm Mechanization. Los Banos, Philippines; International Rice Research Institute, 1983. 184p.	Second
Second	International Rice Research Institute. Small Farm Equipment for Developing Countries: Past Experiences and Future Priorities. Proceedings of the International Conference, 1985. Manila; International Rice Research Institute, 1986. 629p.	First
Third	International Soil Tillage Research Organization: Proceedings of the International Soil Tillage Research Organization, ISTRO (8th Conference), University of Hohenheim, 1979. Stuttgart; German Federal Republic; University of Hohenheim, 1979. 2 vols. (439p.)	Third
Second	International Symposium on Agricultural Wastes, 5th, Chicago, 1985. Agricultural Waste Utilization and Management; Proceedings of the . . . St. Joseph, Mich.; American Society of Agricultural Engineers, 1985. 770p.	
Second	International Symposium on Livestock Wastes, 1st, Ohio State University, 1970. Livestock Waste Management and Pollution Abatement; Proceedings . . . St. Joseph, Mich.; American Society of Agricultural Engineers, 1971. 360p.	
Second	International Symposium on Livestock Wastes, 3d, University of Illinois, 1975. Managing Livestock Wastes; Proceedings . . . St. Joseph, Mich.; American Society of Agricultural Engineers, 1975. 631p.	
First	International Symposium on Livestock Wastes, 4th, Amarillo,	

	Texas, 1980. Livestock Waste; A Renewable Resource; Proceedings . . . St. Joseph, Mich.; American Society of Agricultural Engineers, 1981. 430p.	
Second	International Symposium on Mechanization of Forage, Horticultural, Industrial and Fruit-Tree Crops Harvesting, Bologna, Italy, 1981. Consiglio Nazionale Delle Richerche (Italy). Atti del Simposio Internazionale su Meccanizzazione della Raccolta delle Produzioni Foraggere, Ortive, Industriali e Arboree: Risultati Ottenuti e Possibilita di Trasferimento: Esposizione delle Macchine e dei Prototipi Messi a Punto Nel Corso della Ricerca. Proceedings . . . Bologna; Consiglio Nazionale delle Richerche, 1981. 336p.	Third
	International Wind Energy Symposium. Papers presented . . . New York; American Society of Mechanical Engineers, 1982. 1 Vol.	Third
Third	Internationales Symposium Mechanisierung in der Waldarbeit, 19th, 1985, Diemelstadt, Germany (International Symposium on Mechanization in Forestry.). Hamburg; Kommissionverlag Wiedebusch, 1985. 230p. (Mitteilungen der Bundesforschungsanstalt fur Forst- und Holzwirtschaft, no. 51). (Hamburg-Reinbek, 0368–8798; no. 151)	
Second	Irrigation and Drainage Specialty Conference, Lincoln, Neb., 1971. Optimization of Irrigation and Drainage Systems: Proceedings . . . New York; American Society of Civil Engineers, 1971. 626p. (Available from Books on Demand)	
	Irrigation Association. Irrigation Technical Conference, 1983, Proceedings: Irrigation Today and Tomorrow . . . Priorities, Conflicts and Opportunities. Silver Spring, Maryland; Irrigation Association, 1984. 252p.	Third
Third	Irrigation Association. Irrigation: The Hope and the Promise. Silver Spring, Maryland; Irrigation Association, 1981. 191p.	
Second	Irrigation Association. Water, Energy and Economic Alternatives: Proceedings of the 1982 Technical Conference. Silver Springs, Maryland; Irrigation Association, 1982. 337p.	
First	Irrigation Scheduling Conference, 1981, Chicago. Irrigation Scheduling for Water and Energy Conservation in the 80's. St. Joseph, Mich.; American Society of Agricultural Engineers, 1981. 231p. (ASAE Pub.: 23–81).	
Second	Irrigation Technical Manual: Engineering Data. Lafayette, Calif.; Irrigation Association, Technical Services, Pub. Div., 1969. 72p.	Second
Third	Ivanov, V. P. Biologicheskie i Agrotekhnicheskie Osnovy Oroshaemogo Zemledeliia (Biological and Agrotechnological Fundamentals of Irrigation Farming). Moscow; Akademiia Nauk SSSR, Ordena Trudovogo Krasnogo Znameni Inst. Fiziologii Rastenii im K. A. Timiriazeva, 1983. 269p.	Third

J

Second	Jackson, Ian J. Climate, Water and Agriculture in the Tropics. Harlow, Essex; Longman Group, 1989. 377p. (1st ed., 1977, 248p.)	Second

Second	Jacobs, Clinton O., William R. Harrell and Glen C. Shinn. Agricultural Power and Machinery. New York; Gregg Division, McGraw-Hill, 1982. 472p.	
	Jaeger, William K. Agricultural Mechanization: The Economics of Animal Draft Power in West Africa. Boulder, Colo.; Westview Press, 1986. 199p.	Third
Third	Jain, B. K. S. Sprinkler Irrigation Techniques. Bombay, India; Voltas, Ltd., 1962.	
Second	James, David W., R. J. Hanks, and J. J. Jurinak. Modern Irrigated Soils. New York; John Wiley and Sons, Inc., 1982. 235p.	
First	James, Larry G. Principles of Farm Irrigation System Design. New York; John Wiley and Sons, Inc., 1988. 543p.	First
	Jaquier, N. and G. Blanc. The World of Appropriate Technology: A Quantitative Analysis. Paris; Development Centre of the Organization for Economic Cooperation and Development, 1983. 210p.	Second
	Jegatheeswaran, P. et al. Energy Needs for Food Production in Developing Countries. Witzenhausen; University of Kassel, Faculty of International Agriculture, 1985. 240p. (Tropenlandwirt, no. 22).	Third
Third	Jenkins, S. H., ed. The Agricultural Industry and its Effects on Water Quality: Proceedings . . . New Zealand, 1979. New York; Pergamon Press, Inc., 1979. 727p.	
First	Jensen, Marvin E., ed. Consumptive Use of Water and Irrigation Water Requirements: A Report prepared by the Technical Committee on Irrigation Water Requirements of the Irrigation and Drainage Division of the American Society of Civil Engineers. New York; American Society of Civil Engineers, 1974. 215p.	
First	Jensen, Marvin E., ed. Design and Operation of Farm Irrigation Systems. St. Joseph, Mich.; American Society of Agricultural Engineers, 1981. 829p. (ASAE Monograph no. 3.)	First
Second	Jewell, William J., ed. Energy, Agriculture and Waste Management; Proceedings of the 1975 Cornell Agricultural Waste Management Conference, 7th, Syracuse, 1975. Ann Arbor, Mich.; Ann Arbor Science Publishers, 1975. 540p.	
	Jha, Divakar. Evaluation of Benefits of Irrigation: Tribenji Canal Report. Bombay; Orient Longmans, 1967. 601p.	Second
Third	Jobling, G. A. Trickle Irrigation Design Manual. Canterbury, New Zealand; New Zealand Agricultural Engineering Institute, 1974. 2 vols. (New Zealand Agricultural Engineering Institute. Miscellaneous Publication, no. 6–7)	
	Joglekar, G. D. Irrigation Engineering. Poona, India; United Book Corporation, 1963. 506p.	Second
Third	Johannsen, Chris J. and James L. Sanders, eds. Remote Sensing for Resource Management. Ankeny, Iowa; Soil Conservation Society of America, 1982. 665p.	
Second	Johl, S. S., ed. Irrigation and Agricultural Development. Oxford and New York; Pergamon Press, 1980. 370p.	Second
First	John Deere and Co. The Operation, Care and Repair of Farm Machinery. 28th ed. Moline, Ill.; John Deere, 1957. 279p.	First

Third	Johnson, Arnold H. and Martin S. Peterson. Encyclopedia of Food Technology. Westport, Conn.; AVI Pub. Co., 1974. 993p.	Second
Second	Johnson, Gary L. Wind Energy Systems. Englewood Cliffs, N.J.; Prentice-Hall, Inc., 1985. 360p.	
	Johnston, Bruce F. and W. C. Clark. Redesigning Rural Development: A Strategic Perspective. Baltimore, Md., Johns Hopkins University Press, 1982. 311p.	Third
FIrst	Jones, Fred R. and William H. Aldred. Farm Power and Tractors. 5th ed. New York; McGraw-Hill Book Company, 1980. 466p. (1st ed. as Farm Gas Engines and Tractors. 1938)	Second
Third	Jones, K. R., O. Berney, D. P. Carr, and E. C. Barrett. Arid Zone Hydrology for Agricultural Development. Rome; Food and Agriculture Organization, 1981. 383p. (FAO Irrigation and Drainage Paper no. 37)	
	Jordan, Wayne R., ed. Water and Water Policy in World Food Supplies: Proceedings of the Conference, 1985, Texas A&M University. College Station; Texas A&M University Press, 1987. 444p.	Third
Second	Jowitt, Ronald, ed. Hygienic Design and Operation of Food Plant. Westport, Conn.; AVI, 1980. 292p.	Third
Second	Juergenson, Elwood M. Handbook of Livestock Equipment. 2d ed. Danville, Ill.; Interstate, 1979. 371p. (1st ed., 1971. 266p.)	Second
	Jurion, F., and J. Henry. Can Primitive Farming Be Modernised? (Trans. from French.) Bruxelles, Belgium; Republique Democratique du Congo, Office National de la Recherche et du Developpement, 1969. 457p.	Second

K

	Kaburaki, H., A. Hosokawa, and K. Maeda, eds. Farm Mechanization in Japan. Tokyo Association of Agricultural Relation in Asia. Japan; 1982. 244p.	Third
Third	Kader, Adel A. et al., ed. Postharvest Technology of Horticultural Crops. Berkeley; College of Agriculture and Natural Resources, Univ. of California, 1985. 192p.	
Second	Kallen, Howard P. Handbook of Instrumentation and Controls; A Practical Design and Applications Manual for the Mechanical Services Covering Steam Plants, Power Plants, Heating Systems, Air-conditioning Systems, Ventilation Systems, Diesel Plants, Refrigeration, and Water Treatment. New York; McGraw-Hill, 1961.	
First	Kalman, Robert E. and J. Martinez. Computer Applications in Food Production and Agricultural Engineering; Proceedings of the IFIP TC 5 Working Conference on Food Production and Agricultural Engineering, Havana, 1981. Amsterdam and New York; North-Holland, 1982. 334p. (Sole Distributor for the U.S. and Canada, Elsevier Science Publishing Company, Inc.)	
Second	Kamarck, Andrew M. The Tropics and Economic Development. Baltimore; Johns Hopkins University Press, for the World Bank, 1976. 113p.	Second
Second	Kane, L. A., ed. Handbook of Advanced Process Control Sys-	

DC	tems and Instrumentation. Houston, Texas; Gulf Pub. Co., 1987. 356p.	TW
Second	Karafiath, Leslie L. and Edward A. Nowatzki. Soil Mechanics for Off-Road Vehicle Engineering. Clausthal, Germany; Trans Tech Publs., 1978. 515p.	
Second	Karassik, I., et al., eds. Pump Handbook. New York; McGraw-Hill, 1976. 1102p.	Second
Third	Karel, Marcus, Owen R. Fennema and Daryl B. Lund. Physical Principles of Food Preservation. New York, NY; Marcel Dekker, Inc., 1975. 474p. (Food Science Vol 4). (Principles of Food Science, pt. 2).	Second
	Karmas, Endel. Processed Meat Technology. Park Ridge, New Jersey; Noyes Data Corp., 1976. 368p. (Food Technology Review no. 33).	Second
	Karpenko, A. N. and A. A. Zelenev. Agricultural Machines. Translation from Russian by E. Vilim. Springfield, Virg.; National Technical Information Service, 1968. 452p.	Second
	Kato, Yoshikazu. Keisha: Kigu no Sekkei. (Design of Poultry Houses and its Equipment). Tokyo; Nosan Gyoson Bunka Kyokai, 1971. 283p.	Third
Second	Katzan Associates, International. Grain Motor Fuel Alcohol: Technical and Economic Assessment Study. Washington, D. C.; U.S. Dept. of Energy, Assistant Secretary for Policy and Evaluation, Office of Technical Programs Evaluation, 1979. 343p.	
Third	Katzman, Martin T. Solar and Wind Energy: An Economic Evaluation of Current and Future Technologies. Totowa, N.J.; Rowman and Allenheld, Publishers, 1984. 187p.	
Third	Kay, Melvyn. Surface Irrigation: Systems and Practice. Cranfield, Bedford.; Cranfield Press, 1986. 142p.	
	Kelly, William W. Water Control in Tokugawa Japan: Irrigation Organization in a Japanese River Basin, 1600–1870. Ithaca, New York; Cornell University China-Japan Program, 1982. 260p. (East Asia Papers no. 31).	Second
	Kennedy, W. K. and T. A. Rogers. Human and Animal Powered Lifting Devices. London; Intermediate Technology Publications, 1985. 111p.	Third
Third	Kent, Norman L. Technology of Cereals: An Introduction for Students of Food Science and Agriculture. 3d ed. Oxford and New York; Pergamon, 1983. 221p.	
First	Kepner, Robert A., Roy Bainer and E. L. Barger. Principles of Farm Machinery. 3d ed. Westport, Conn.; AVI Publishing Company, 1978. 527p. (1st ed. with Bainer as first author, 1955, 571p.; 2d ed., 1972. 486p.)	First
	Keterlaars, E. H. and Iwema, S. Boer., eds. Animals as Waste Converters: Proceedings of an International Symposium, Wageningen, Netherlands, 1983. Wageningen; Pudoc, 1984. 153p.	Second
Second	Kezdi, Arpad. Soil Physics. Amsterdam; Elsevier Science Publs., 1974. 256p.	
Third	King, Alexander and Harlan Cleveland, eds. Bioresources for Development: The Renewable Way of Life; Proceedings of the	

	International Conference on Bioresources for Development, University of Houston, 1978. New York; Pergamon Press, 1980. 345p.	
Second	Kinori, B. Z. Manual of Surface Drainage Engineering, Vol. I. Amsterdam and New York; Elsevier Pub. Co., 1970. 224p. (Transl. from Hebrew by Amihoud Nov)	
Second	Kirkby, M. J. and R. P. C. Morgan, eds. Soil Erosion. Chichester, Eng. and New York; John Wiley, 1980. 312p.	
Second	Kirkham, Don and W. L. Powers. Advanced Soil Physics. New York; Wiley Interscience, 1972. 534p. (Reissued, Malabar, Flor.; R. E. Krieger Pubs., 1984)	
	Kjaerby, F. Problems and Contradictions in the Development of Ox-Cultivation in Tanzania. Copenhagen; Scandinavian Institute of African Studies and the Centre for Development Research, 1983. 163p. (Scandinavian Institute of African Studies Research Report no. 66).	Second
Second	Klenin, N. I., I. F. Popov, and V. A. Sakun. Agricultural Machines: Theory of Operation, Computation of Controlling Parameters and the Condititons of Operation. (Translation of Sel'skokhozyaistvennye Mashiny: E'lementy Teorii Rabochikh Protsessov, Raschet Regulirovochnykh Parametrov i Rezhimov Raboty; Kolos Publishers, 1970.) Translated by A. Jaganmohan. New Delhi; Amerind Pub. Co. for U.S. Department of Agriculture and the National Science Foundation, 1985. 633p.	Second
Second	Kline, C. K., D. A. G. Green, Roy L. Donahue, and B. A. Stout. Agricultural Mechanization in Equitorial Africa. E. Lansing, Mich.; Institute of International Agriculture, Michigan State University, 1969. 465p.	First
Second	Knorr, Dietrich W. Sustainable Food Systems. Westport, Conn.; AVI Pub. Co., 1983. 416p.	
	Koegel, R. G. Self-Help Wells. Rome, Food and Agriculture Organization, 1977. 78 p. (FAO Irrigation and Drainage Paper no. 30)	Third
Third	Kollar, L. Automatisierung in der Landwirtschaft. (Automation in Agriculture.) Berlin; Verlag Technik, 1975. 352p.	
Second	Koolen, A. J. and H. Kuipers. Agricultural Soil Mechanics. Berlin and New York; Springer-Verlag Inc., 1983. 241p. (Advances in Agricultural Science Ser. Vol. 13).	First
	Koval'skaia, L. P. Tekhnologiia Pishchevykh Proizvodstv. Uchebniki i Uchebnye Posobiia Dlia Uchashchikhsia Tekhnikumov. (Technology of Foodstuff Production.) Moscow; Agropromizdat, 1988. 286p.	Third
Third	Kozlowski, T. T., ed. Water Deficits and Plant Growth. New York; Academic Press, 1968–73. 7 vols.	Third
Second	Kraatz, D. B. Irrigation Canal Lining. 2d ed. Rome; Food and Agriculture Organization of the U.N., 1977. 199p. (1st ed., 1971. 170p.) (FAO Land and Water Development Series no. 1).	Second
Third	Kramer, J. Drip System Watering for Bigger and Better Plants. New York; W. W. Norton and Co., 1980. 140p.	

Third	Kreider, Jan F. and Frank Kreith. Solar Heating and Cooling: Engineering, Practical Design, and Economics. Washington, D.C.; Scripta Book Co., 1975. 342p.	
	Kreyger, J. Drying and Storing Grains, Seeds and Pulses in Temperate Climates. Wageningen; Institute for Storage and Processing of Agricultural Produce, 1972. 333p. (Wageningen Instituut voor Bewaring en Verwerking van Landbouwprodukten, Publikatie no. 205).	Second
Third	Krishnawsami. Irrigation. 4th ed. Madras; Prabha Pub., 1963. 199p.	
Third	Kriukov, Aleksei I. Proektirovanie Potochnogo Stroitel'stva Zhivotnovodcheskikh Kompleksov. (Planning Construction of Livestock Production Complexes by Means of Linear Analysis.) Moscow; Stroiizdat, 1977. 261p.	
	Kristoferson, Lars A. and V. Bokalders. Renewable Energy Technologies: Their Applications in Developing Countries. 1st ed. Oxford and New York; Pergamon Press, 1986. 319p.	Third
Third	Krol, B., Pieter S. von Roon, and J. H. Houben, compilers. International Symposium: Trends in Modern Meat Technology (1984: Wageningen, Netherlands). Trends in Modern Meat Technology; Proceedings . . . Wageningen; Pudoc, 1985. 125p.	
Third	Kruse, E. Gordon, Chuck R. Burdick, and Yousef A. Yousef, eds. Environmentally Sound Water and Soil Management. Proceedings . . . of a Conference, Orlando, Florida, 1982. New York; American Society of Civil Engineers, 1982. 524p.	
	Krutz, Gary W., Lester Thompson and Paul Claar. Design of Agricultural Machinery. New York; John Wiley and Sons, Inc., 1984. 472p.	Second
Third	Kubyshev, V. A. Kompleksnaia Tekhnologiia Pervichnoi Pererabotki Sakharnoi Svekly. (Complex Technology for the Initial Processing of Sugar Beets.) Moscow; VIM, 1983. 111p. (Vsesoiuznyi Nauchno-issledovatel'skii Institut Mekhanizatsii sel'skogo Khoziaistva. Sbornik Nauchnykh Trudov, no. 98)	
Third	Kubyshev, V. A. Mekhaniko-tekhnologicheskie Osnovy Zashchity Pochv ot Erozii. (Mechanical and Technological Bases for the Protection of Soil from Erosion.) Moscow; VIM, 1983. 179p. (Sbornik Nauchnykh Trudov. Vsesoiuznyi Nauchno-issledovatel'skii Institut Mekhanizatsii Se;'skogo Khoziaistva, no. 96).	
Third	Kubyshev, V. A. Nauchnye Osnovy Ekspluatatsii Mashinno-Traktornogo Parka. (Scientific Bases of Utilizing the Machine and Tractor Fleet.) Moscow; VIM, 1982. 215p. (Sbornik Nauchnykh Trudov. Vsesoiuznyi Nauchno-Issledovatel'skii Institut Mekhanizatsii Sel'skogo Khoziaistva, no. 95).	
	Kugler, Klaus and Gerd Bernhardt. Maschinen und Gerate fur die Pflanzenproduktion: Lehrbuch fur die Sozialistische. Berlin; VEB Deutscher Landwirtschaftsverlag, 1981. 228p.	Third
Third	Kunze, George W., J. S. White, and Richard H. Rust. Mineralogy in Soil Science and Engineering. Madison, Wisc.; Soil Science Society of America, 1968. 106p. (SSSA Special Publication No.3.)	

Third	Kunze, Robert F. Lexikon der Landtechnik: Futterernte und-konservierung. (Lexicon of Agricultural Engineering: Feeds and Storage). Wurzburg; Vogel, 1985. 221p.	Third
Third	Kunze, Robert F. Lexikon der Landtechnik: Getreide- und Hackfruchternte. (Lexicon of Agricultural Engineering: Grain and Root Crops.) Wurzburg; Vogel, 1987. 261p.	Third
Third	Kuprits, I. N. Technology of Grain Processing and Provender Milling. Translation from Russian of Tekhnologiia Pererabotki Zerna. Jerusalem; Israel Program for Scientific Translations, 1967. 555p.	
Second	Kurtz, Gary W. Lester Thompson and Paul Claar. Design of Agricultural Machinery. John Wiley and Sons, 1984. 472p.	

L

Second	L'Hermite, P., editor. Processing and Use of Organic Sludge and Liquid Agricultural Wastes. Proceedings of the Fourth International Symposium, Rome, 1985. Dordrecht and Boston; D. Reidel Pub. Co., 1986. 576p.	
	Lachenmaier, Fritz. 50 Jahre KTBL 50 Jahre Fortschritt in der Agrarechnik. (Fifty Years of KTBL: Fifty-year Progress in Agricultural Engineering). Darmstadt; Kuratorium fur Technik und Bauwessen in Derlandwirtschaft, 1973. 140p.	Third
	Lal, D. Wells and Welfare: An Exploratory Cost-Benefit Study of the Economics of Small-Scale Irrigation in Maharashtra. Paris; Organization for Economic Co-operation and Development, 1972. 162p. (Series on Cost-Benefit Analysis, Case Study no. 1).	Second
	Landon, J. R., ed. Booker Tropical Soil Manual: A Handbook for Soil Survey and Agricultural Land Evaluation in the Tropics and Subtropics. London; Booker Agriculture International Ltd., and New York; Longman, 1984. 320p.	First
	Langford-Smith, Trevor and John Rutherford. Water and Land: Two Case Studies in Irrigation. Canberra, Australia; Australia National University Press, 1966. 270p.	Third
	Laurent, Claude. Conservation des Produits d'Origine Animale en Pays Chauds. Paris; Presses Universitaires de France, 1974. 154p.	Third
Third	Laxminarayan, H. Impact of Harvest Combines on Labour-Use, Crop Pattern, and Productivity. New Delhi; Agricole Pub. Academy, 1981. 177p.	Third
	Leach, G. Energy and Food Production. London; International Institute for Environment and Development, 1975. 151p. (Later republ. by IPC Science and Tech. Press, Guildford, 1976).	Second
Second	Lehr, Jay H., Tyler E. Gass, and Wayne E. Pettyjohn. Domestic Water Treatment. McGraw-Hill Book Company, 1979.	Second
Second	Leliavsky, Serge. Irrigation and Hydraulic Design. London; Chapman and Hall, 1955–1960. 3 vols. (782p.)	
	Lenard, Jan Z. Podstawy Budownictwa Rolniczego: Podr Ecznik dla Studentow Akademii Rolniczych. Wyd. 1. (Foundation of Agricultural Architecture.) Warszawa; Panstwowe Wydawn, Rolnicze i Lesne, 1983. 351p.	Third

	Lepajoe, Jaan. Taimekasvatussaaduste Sailitamine ja Umbertootamine. (The Storage and Processing of Agricultural Products.) Tallinn; Valqus, 1979. 239p.	Third
Third	Levy, Maurice and John L. Robinson, eds. Energy and Agriculture: Their Interacting Futures: Policy Implications of Global Models; United Nations University Symposium on Energy and Agriculture Futures; Policy Implications of Global Models, 1982, Paris. Chur, Switzerland and New York; Harwood Academic Publishers for the United Nations University, 1983. 370p.	
First	Liljedahl, John B., Paul K. Turnquist, David W. Smith, and Makoto Hoki. Tractors and Their Power Units. 4th ed. New York; Van Nostrand Reinhold, 1989. 463p. (1st and 2d eds. under principle authorship of E. L. Barger. 2d ed., 1963.)	First
Second	Lindblad, C. and Laurel Druben. Small Farm Grain Storage. Washington, D.C.; U.S. Government Printing Office, 1976. 640p. (Volunteers in Technical Assistance Manual Series no. 35E). (Reprinted in 3 vols. in 1977: Preparing Grain for Storage; Enemies of Stored Grain; Storage Methods.)	First
Second	Linsley, Ray K., Jr., Max A. Kohler and Joseph L. H. Paulhus. Hydrology for Engineers. 3d ed. New York; McGraw Hill, 1982. 508p. (1st ed., 1958. 340p.; 2d ed., 1975. 482p.)	
Second	Lockeretz, William, ed. Agriculture and Energy; Proceedings of a Conference, Washington University, St. Louis, 1976. New York; Academic Press, 1977. 750p.	First
Third	Loehr, Raymond C. Agricultural Waste Management: Problems, Processes and Approaches. New York; Academic Press, 1974. 576p.	
Second	Loehr, Raymond C. Pollution Control for Agriculture. 2d ed. Orlando; Academic Press, 1984. 467p. (1st ed., New York, 1977. 383p.)	
	Loncin, Marcel and J. Carballo. Ingenieria Alimentaria. Madrid; Edial Dossat, 1965. 781p.	Third
Second	Loncin, Marcel and Richard L. Merson. Food Engineering, Principles and Selected Applications. New York; Academic Press, 1979. 494p.	Second
Third	Loncin, Marcel. The Nature of the Food Industry. Paris; Masson, 1976. 286p.	Third
Second	Lopez, Anthony. A Complete Course in Canning. 11th ed. Books I and II. Baltimore, Md.; The Canning Trade, Inc., 1981. 1000p.	
Second	Luikov, Aleksei Vasilevich. Heat and Mass Transfer in Capillary-Porous Bodies. Translated from Russian by P.W.B. Harrison. 1st English ed. Oxford and New York; Pergamon, 1966. 523p.	
	Lovegrove, H. T. Crop Production Equipment. London; Hutchinson, 1968. 406p. (3d printing in 1973.) (Reprinted in paperback as Crop Production Equipment: A Practical Guide for Farmers, Operators, and Trainees; Brookfield Publishing Co., 1981.)	Second
	Lucas, Barbara G. and Stephen Freedman, eds. Technology	Third

DC	Choice and Change in Developing Countries: Internal and External Constraints. Dublin; Tycooly International Pub., 1983. 155p.	TW
First	Luthin, James N. Drainage Engineering. 2d ed. Huntington, New York; Robert E. Krieger Publishing Co., 1973. 250p. (1st ed., New York; John Wiley, 1966. 250p.)	First
First	Luthin, James N., ed. Drainage of Agricultural Lands. Madison, Wisc.; American Society of Agronomy, 1957. 620p. (Monograph no. 7).	First

M

	MacDonald, Neil B., William F. Barnicke, Francis W. Judge, and K. E. Hansen. Farm Tractor Production Costs. Ottawa, Canada; Queen's Printer for Canada, 1970. 286p.	Third
Third	Maeda, Koichi. Kokusai Shokuryo Nogyo Kyokai. (Mechanization of Japanese Agriculture.) Tokyo; Japan FAO Assoc., 1985. 108p.	Third
Third	Magnien, E. and D. de Nettancourt, editors. Genetic Engineering of Plants and Micro-Organisms Important for Agriculture. Sponsored by the Commission of the European Communities. Dordrecht and Boston; Nijhoff/Junk for the Commission of the European Communities; Hingham, Mass.; Distributors for the U.S. and Canada, Kluwer Academic, 1985. 197p.	
Second	Majumdar, S. K. Irrigation Engineering. New York; McGraw-Hill, 1984. 350p. #6	First
	Makeham, J. Thailand's Harvest: Man or Machine. Bangkok; Economic and Social Commission for Asia and the Pacific, Regional Network for Agricultural Machinery, 1979.	Third
Second	Makhijani, Arjun and Alan Poole. Energy and Agriculture in the Third World. Cambridge, Mass; Ballinger Pub. Co., 1975. 168p.	Third
	Mann, H. T. and D. Williamson. Water Treatment and Sanitation: Simple Methods for Rural Areas. Volunteers in Technical Assistance, 1982. 92p.	Second
	Mann, Robert D. How to Build a "Cretan Sail" Wind Pump for Use in Low-Speed Wind Conditions. Repr. of 1979 ed. London; Intermediate Technology Publications, 1983. 79p.	Second
	Manoiu, Gheorghe. Aplicatiile Ergonomice in Procesul Muncii Agricole. (Ergonomic Applications in Agricultural Processing.) Bucuresti; "Ceres", 1979. 165p.	Third
First	Mark's Standard Handbook for Mechanical Engineers. 8th ed. Edited by Theodore Baumeister. New York; McGraw-Hill, 1978. 1,952p. (1–3d., 1916–1941, ed. by L. S. Mark).	Second
Second	Marley, Stephen J., Wesley F. Buchele, editors. Proceedings of the Fourth International Conference on the Mechanization of Field Experiments; July 1976, Iowa State University. Ames, Iowa; Iowa State University, 1976. 420p.	Second
Second	Marshall, B. and F. I. Woodward. Instrumentation for Environmental Physiology. Cambridge and New York; Cambridge	

DC	University Press, 1985. 241p. (Society for Experimental Biology Seminar Series no. 22)	TW
Second	Marshall, Theo J., and J. W. Holmes. Soil Physics. 2d ed. Cambridge and New York; Cambridge University Press, 1988, 374p. (1st ed., 1979. 345p.)	Second
Third	Martem'ianov, Askol'd I. and Anatolii D. Ternovskii. Spravochnik Sel'skogo Stroitelia. (Handbook for Rural Builders.) Moscow; Stroiizdat, 1975. 2 vols.	
	Martinusen, Donald and Bernard Barry. Revenues, Costs and Profits in the Farm Machinery Industry. Ottawa, Canada; Royal Commission on Farm Machinery, 1970. 219p.	Third
	Massey-Ferguson, Ltd. The Pace and Form of Farm Mechanization in the Developing Countries. Toronto, Calif.; 1975. 38p.	Second
Third	Maton, Andre, J. Daelemans, and J. Lambrecht. Housing of Animals: Construction and Equipment of Animal Houses. Amsterdam and New York; Elsevier Science Publishing, 1985. 458p. (Developments in Agricultural Engineering no. 6).	Second
Second	Matthews, G. A. Pesticide Application Methods. London and New York; Longman, 1979. 334p.	Second
Second	Matthews, J. and A. A. Knight. Ergonomics in Agricultural Equipment Design. Silsoe; National Institute of Agricultural Engineering, 1971. 61p.	
Second	Matz, Samuel A., ed. Bakery Technology and Engineering. 2d ed. Westport, Conn.; AVI Publishing Co., 1972. 598p. (1st ed., 1960. 669p.)	Second
Third	Matz, Samuel A. Snack Food Technology. 2d ed. Westport, Conn.; AVI Pub. Co., 1984. 415p. (1st ed., 1976. 349p.)	
Second	May, Brian. How to Make the Most of Your Tractor. Intermediate Technology Development Group of North America, 1985. 184p.	Second
	Mayeux, Mansel M. Retailing Farm and Light Industrial Equipment. 2d ed. Westport, Conn.; AVI Publishing Co., 1983.	Second
Second	McColly, H. F. and J. W. Martin. Introduction to Agricultural Engineering. New York; McGraw Hill, 1955. 553p.	First
Third	McGinnes, William G. and Bram J. Goldman, eds. Arid Lands in Perspective. Tucson, AR; University of Arizona Press, and Washington, D.C.; American Association for the Advancement of Science, 1969. 421p.	
Third	McGuigan, D. Small Scale Water Power. Dorchester, England; Prism Press, 1978. 113p.	
Third	McGuigan, D. Small Scale Wind Power. Dorchester, England; Prism Press, 1979. 148p.	
	McInerney, John P. and G. F. Donaldson. The Consequences of Farm Tractors in Pakistan. Washington, D.C.; International Bank for Reconstruction and Development, 1975. 94p. (World Bank Staff Working Paper no. 210).	Second
Third	McKyes, Edward. Agricultural Engineering Soil Mechanics. Amsterdam and New York; Elsevier Science Pub., 1989. 292p.	
First	McKyes, Edward. Soil Cutting and Tillage. Amsterdam and	First

	New York; Elsevier, 1985. 217p. (Developments in Agricultural Engineering 7).	
Second	McLean, K. A. Drying and Storing Combinable Crops. 2d ed. Ipswich; Farming Press, 1989. 257p. (1st ed., 1980. 281p.)	Second
	Meij, J. L., ed. Mechanization in Agriculture. Amsterdam; North-Holland, 1960.	Second
Second	Merriam, John L. and Jack Keller. Farm Irrigation System Evaluation. Silver Spring; Irrigation Association, 1978. 300p. (Originally published by Utah State Univ., Agricultural Irrigation and Engineering Dept., 1978).	First
Second	Merkel, James A. Basic Engineering Principles. 2d ed. Westport, Conn.; AVI Publishing Co., 1983. 223p. (1st ed., 1974. 204p.)	
Third	Merkel, James A. Managing Livestock Wastes. Westport, Conn.; AVI Publishing Co., 1981. 419p.	
Second	Merva, George E. Physioengineering Principles. Westport, Conn.; AVI Publishing Co., 1975. 353p.	Third
Third	Meske, Christoph. Fish Aquaculture: Technology and Experiments. Translation of AquaKultur von Warmwasser-Nutzfischen. Edited and translated from German by Frederick Vogt. Oxford and New York; Pergamon Press, 1985. 237p.	
	Mettrick, H. M. and D. P. James. Farm Power in Bangladesh; Vol. 2; Part 1, Some Aspects of the Economics of Animal Power; Part 2. Mechanisation and Institutions in Noakhabi. Reading; University of Reading, Dept. Agricultural Economics and Management, 1981. 156p. (Development Study no. 21).	Second
	Mettrick, Hal, S. Roy and D. S. Thornton. Agricultural Mechanisation in Southern Asia: A Report for the Ministry of Overseas Development. Reading, England; University of Reading, 1976. 150 leaves.	Third
Third	Mialhe, Luiz Geraldo. Maquinas Motoras na Agricultura. (Mechanized Agricultural Machinery.) Sao Paulo; Dom Jose Gaspar, E. P. U., 1980. 2 vols.	
	Michael, A. M. Irrigation: Theory and Practice. New York; State Mutual Book and Periodical Service, 1982. 801p. (1st ed., New Delhi; Vikas Pub., 1978).	First
Second	Midwest Plan Service. Beef Housing and Equipment Handbook. 4th ed. Ames; Iowa State University, Midwest Plan Service, 1987. 84p. (1st ed., 1968. 64p. 3d ed., 1975)	
Second	Midwest Plan Service. Dairy Housing and Equipment Handbook. 4th ed. Ames; Iowa State University, Midwest Plan Service, 1985. 110p. (2d ed., 1971. 86p.)	
Second	Midwest Plan Service. Grain Drying, Handling and Storage Handbook. 2d ed. Ames; Iowa State University, Midwest Plan Service, 1987. 86p.	
Second	Midwest Plan Service. Horse Handbook: Housing and Equipment. Ames, Iowa; Midwest Plan Service, 1971. 60p.	
First	Midwest Plan Service. Livestock Waste Facilities Handbook. 2d ed. Ames; Iowa State University, Midwest Plan Service, 1985. 112p. (1st ed., 1975. 94p.)	
Second	Midwest Plan Service. Planning Grain-Feed Handling for	

DC	Livestock and Cash-Grain Farms: Unloading, Drying, Elevating, Storing, Processing. Ames; Iowa State University, Midwest Plan Service, 1974. 70p. (1st ed., 1968. 39p.)	TW
Second	Midwest Plan Service. Sheep Housing and Equipment Handbook. 3d ed. Ames; Iowa State University, Midwest Plan Service, 1982. 116p. (2d ed., 1974. 80p.)	
First	Midwest Plan Service. Structures and Environment Handbook. 11th ed. Ames; Iowa State University, Midwest Plan Service, 1983. 658p. (7th ed., 1975. 412p.; 9th ed., 1977. 422p.)	
Second	Midwest Plan Service. Swine Housing and Equipment Handbook. 4th ed. Ames; Iowa State University, Midwest Plan Service, 1983. 112p. (2d ed., 1968. 74p.; 3d ed., 1972. 84p.)	
Third	Mikhal'chuk, Al'bert N. Elektroavtomatizatsiia v Sel'skokhoziaistvennom Proizvodstve. (Automatization of Electricity in Agricultural Production.) Moscow; Rossel'khozizdat, 1973. 81p.	
	Miller, Duncan. Self-Help and Popular Participation in Rural Water Systems. Paris; Centre of the Organisation for Economic Co-operation and Development, 1979. 149p.	Second
Third	Miner, J. R. and R. J. Smith, eds. Livestock Waste Management with Pollution Control. Ames, Iowa; Midwest Plan Service, 1975. 89p. (North Central Regional Research Publication no. 222)	
	Mishra, P. K. Water in the Land of Sand Dunes. Jaipur; Pragya Prakashan Publication, 1970. 155p.	Second
First	Mitchell, Bailey W., ed. Instrumentation and Measurement for Environmental Sciences. 2d ed. St. Joseph, Michigan; American Society of Agricultural Engineers, 1983. various pagings. (1st ed., Zachary A. Henry, Gerald C. Zoerb and Gerald S. Birth, eds., 1975. 220p.)	
Second	Moens, A. and A. H. J. Siepman, compilers. Development of the Agricultural Machinery Industry in Developing Countries: Proceedings of the Second International Conference, Amsterdam, 1984. Wageningen; Pudoc, 1984. 400p.	Second
Third	Mohsenin, Nuri N. Electromagnetic Radiation Properties of Foods and Agricultural Products. New York, New York; Gordon and Breach Science Publishers, Inc., 1984. 673p.	Second
First	Mohsenin, Nuri N. Physical Properties of Plant and Animal Materials: Structure, Physical Characteristics and Mechanical Properties. 2d ed., revised and updated. New York; Gordon and Breach Science Publishers, 1986. 891p. (Predecessor ed. with briefer title issued by Dept. of Agricultural Eng., Penn. State University, 1966–68.)	First
Third	Mohsenin, Nuri N. Thermal Properties of Foods and Agricultural Materials. London; Gordon and Breach, 1980. 407p.	Second
Second	Molenaar, Aldert. Irrigation by Sprinkling. Rome; Food and Agriculture Organization, 1960. 90p. (FAO Agricultural Development Paper no. 65).	Second
	Molenaar, Aldert. Water Lifting Devices for Irrigation. Rome; Food and Agriculture Organization of the United Nations, 1956. 75p. (FAO Agricultural Development Paper no. 60).	First
Second	Møller, Arne. Applications of Microprocessors within Agri-	

culture. Translation from Danish of Anvendelse af Mikroprocessorer i Landbruget II. Hundested, Denmark; Agri. Contact, 1985. 191p.

Second — Monteith, J. L., ed. Vegetation and the Atmosphere. London and New York; Academic Press, 1975–76. 2 vols.

Monti, Jacques R., Freddy Detraux, and Otto Oestges. La Mecanisation de L'Agriculture. Gembloux; J. Duculot, 1970. 2 Vols. — Third

Third — Montieth, J. L. and L. E. Mount, eds. Heat Loss from Animals and Man: Assessment and Control; Proceedings of the 20th Easter School in Agricultural Science, Univ. of Nottingham, 1973. London; Butterworths, 1974. 457p.

Moore, Vic. Farm Workshop & Maintenance: The Book of the *Farmers Weekly*. 3d ed. London; New York; Granada, 1984. 224p. — Second

Third — Morgan, Bryan A. Keyguide to Information Sources in Agricultural Engineering. London and New York; Mansell Publishing, 1985. 209p.

Third — Morgan, R. P. C., ed. Soil Erosion and its Control. New York; Van Nostrand Reinhold, 1986. 311p.

Third — Morgan, R. P. C. Soil Erosion and Conservation. 2d ed. Edited by D. A. Davidson. Harlow, Essex; Longman, New York; Wiley, 1986. 298p. (1st ed., 1979, 113p.)

Moris, Jon R. and Derrick J. Thom. African Irrigation Overview: Main Report. Logan; Utah State University, Agricultural and Irrigation Engineering, 1987. 635p. (Water Management Synthesis Report no. 37). — Third

Morris, J. and S. Pollard. The Place of Small Tractors in the Mechanisation Process. International Agricultural Development, 1981. — Second

Mothander, Bjorn, Finn Kjaerby, and Kjell Havnevik. Farm Implements for Small- Scale Farmers in Tanzania. Uppsala, Sweden; Scandinanianv Institute of African Studies, 1989. 214 p. — Third

Second — Moverley J. Microcomputers in Agriculture. London; Collins, 1986. 219p.

Muckle, T. B., C. P. Crossley, and J. Kilgour. Low Cost Primary Cultivation. Silsoe, England; National College of Agricultural Engineering, 1973. (Its Occasional Paper no. 1) — Second

Mudra, Stanislav. English, Czech, Slovak, Russian Dictionary of Agricultural Mechanization. Prague; Ustav Vedeckotechnickych Informaci, 1975. 160p. — Third

Second — Mujumdar, Arun S., ed. Drying '86. 2 vols. (919p.); Drying '87. 237p.; Drying '89. 550p. Washington; Hemisphere Pub. Corp., 1986–1989.

Multilingual Dictionary of Remote Sensing and Photogrammetry: English Glossary and Dictionary; Equivalent Terms in French, German, Italian, Portuguese, Spanish and Russian, George A. Rabchevsky, Editor-in-Chief. Falls Church, Virg.; American Society of Photogrammetry and Remote Sensing, 1984. 343p. — Second

Second — Multilingual Technical Dictionary on Irrigation and Drainage — Second

DC	= Dictionnaire Technique Multilingue des Irrigations et du Drainage. New Delhi; International Commission on Irrigation and Drainage, 1967. 805p.	TW
Second	Multilingual Technical Dictionary on Irrigation and Drainage. (Rus. and Eng.). United States Committee on Irrigation and Drainage, 1978.	Third
Second	Multilingual Technical Dictionary on Irrigation and Drainage. (Sp. and Eng.). United States Committee on Irrigation and Drainage, 1977.	Second
Second	Multon, J. L., ed. Preservation and Storage of Grains, Seeds and Their By-Products. Translation of: Conservation et Stockage des Grains et Graines et Produits Derives, by D. Marsh. New York; Lavoisier Pub. Inc., 1988. 1095p.	Second
Second	Munasinghe, Mohan. Rural Electrification for Development: Policy Analysis and Applications. Boulder, Colo.; Westview Press, 1987. 440p.	Second
Second	Munzinger, P. Animal Traction in Africa. Eschborn; Deutsche Gesellschaft fur Technische Zusammenarbeit; GmbH (German Agency for Technical Co-operation), 1982. 490p.	Second

N

Second	Nakayama, F. S., and D. A. Bucks, eds. Trickle Irrigation for Crop Production: Design, Operation, and Management. Amsterdam and New York; Elsevier, 1986. 383p. (Available from St. Joseph, Mich.; American Society of Agricultural Engineers.)	First
Third	Nash, Michael J. Crop Conservation and Storage in Cool Temperate Climates. 2d ed. Oxford and New York; Pergamon Press, 1985. 286p. (1st ed., 1978. 393p.)	
Second	Nassar, Mohammed, compiler. Technical Dictionary: Agricultural Engineering. Edited by El W. Abd and Mahmoud Anwar. Leipzig; Edition Leipzig, 1977. 320p.	Third
Second	National Academy of Sciences (U.S.). Panel on Methane Generation. Methane Generation from Human, Animal and Agricultural Wastes: Report of an Ad Hoc Panel of the Advisory Committee on Technology Innovation, Board on Science and Technology for International Development. Washington; National Academy of Sciences, 1977. 131p. (Available from National Technical Information Service, PB-276 469)	
First	National Academy of Sciences (U.S.). Panel on Renewable Energy Resources. Energy for Rural Development: Renewable Resources and Alternative Technologies for Developing Countries; Report of an ad hoc panel of the Advisory Committee on Technology Innovation, Board on Science and Technology for International Development, Commission on International Relations. Washington, D.C.; National Academy of Sciences, 1976. 306p. (Suppl., 1981) (BOSTID Advisory Studies and Special Reports no. 18).	First
Second	National Academy of Sciences (U.S.). Transportation Research Board. Small Drainage Structures. Washington, D.C.,	Second

	Transportation Research Board, 1978. 297p. (Publications for Developing Countries: Compendium 3).	
Third	National Conference on Advances in Evapotranspiration, Chicago, 1985. Advances in Evapotranspiration; Proceedings . . . St. Joseph, Mich.; American Society of Agricultural Engineers, 1985. 453p.	
Third	National Conference on Advances in Infiltration, Chicago, 1983. Advances in Infiltration; Proceedings of the . . . St. Joseph, Mich.; American Society of Agricultural Engineers, 1983. 385p.	Second
	National Conference on Agricultural Electronics Applications, Chicago, 1983. Agricultural Electronics—1983 and Beyond; Proceedings of the . . . St. Joseph, Mich.; American Society of Agricultural Engineers, 1984. 2 vols. (798p.)	Second
Third	National Conservation Tillage Conference, Des Moines, 1973. Proceedings . . . Soil Conservation Society of America in cooperation with the American Society of Agronomy and others. Ankeny, Iowa; Soil Conservation Society of America, 1973. 241p.	
Second	National Dairy Housing Conference, 2d, Madison, Wisconsin, 1983. Dairy Housing II; Proceedings . . . St. Joseph, Mich.; American Society of Agricultural Engineers, 1983. 350p.	
Second	National Drainage Symposium, Proceedings of the Third . . . Chicago, 1976. Proceedings of the Fourth . . . Chicago, 1982. Advances in Drainage; 177p. 5th, 1987, Drainage Design and Management. St. Joseph, Mich.; American Society of Agricultural Engineers, 1976–1988.	First
Second	National Food Processors Association. Laboratory Manual for Food Canners and Processors. 2 Vols. Westport, Conn.; AVI/VNR, Inc., 1980. 780p. (3d ed., 1968)	Second
Third	National Irrigation Symposium, 2d, University of Nebraska, Lincoln, 1980. Irrigation Challenges the 80s: The Proceedings . . . St. Joseph, Mich.; American Society of Agricultural Engineers, 1981. 252p. (ASAE Publication 6–81)	
	National Research Council (U.S.). Diffusion of Biomass Energy Technologies in Developing Countries. Washington, D.C.; National Academy Press, 1982. 120p.	Third
Second	National Symposium on Animal Waste Management, East Lansing, Mich., 1966. Management of Farm Animal Wastes; Proceedings . . . St. Joseph, Mich.; American Society of Agricultural Engineers, 1966. 160p. (ASAE Pub.: SP-0366).	
Third	National Symposium on Conservation Systems, Chicago, 1987. Optimum Erosion Control at Least Cost; Proceedings of the . . . St. Joseph, Mich.; American Society of Agricultural Engineers, 1987. 418p. (ASAE Publication: 08-87).	
Third	National Symposium on Erosion and Soil Productivity, New Orleans, 1984. Erosion and Soil Productivity; Proceedings of the . . . St. Joseph, Mich.; American Society of Agricultural Engineers, 1985. 289p. (ASAE Publication: 08-85).	
Second	National Symposium on Individual and Small Community Sewage Treatment; Proceedings of the . . . 3d, 1981, Chicago,	

DC	1982. 345p.; 4th, 1984, New Orleans, 1985. 381p.; 5th, 1988: On-site Sewage Treatment; 1988. 411p. St. Joseph, Mich.; American Society of Agricultural Engineers, 1982–1988.	TW
Third	National Symposium on Water Resources Law, Chicago, 1986. Water Resources Law; Proceedings of the . . . St. Joseph, Mich.; American Society of Agricultural Engineers, 1986. 243p.	
	Neelov, Vadim A. Promyshlennye i Sel'skokhoziaistvennye Zdaniia. (Industrial and Agricultural Buildings.) Moscow; Stroiizdat, 1980. 222p.	Third
Second	Nelson, Gordon L., H. B. Manbeck, and N. F. Meador. Light Agricultural and Industrial Structures: Analysis and Design. New York; Van Nostrand Reinhold Co., 1988. 558p.	
Third	Nelson, Philip E., James V. Chambers and Judy H. Rodriguez, eds. Principles of Aseptic Processing and Packaging. Washington, D. C.; Food Processors Institute, 1987. 120p.	
	Nerli, Nerlo. Meccanica Agraria. 7th ed. Bologna; Soc. Edit. Internazional R. Patron, 1961. 557p.	Third
	New Zealand Agricultural Engineering Institute, Lincoln College. A Simulation Model of Hill and High Country Pastoral Systems for Evaluation of Irrigation Investment. Canterbury, NZ; New Zealand Agricultural Engineering Institute, Lincoln College, 1985. 161p.	Third
Third	Nilsson, Lennart. Laster i Ensilagesilor: Genomgang av Litteratur och Normer = Loads in Silos for Forage Crops: Review of Literature and Standards. Lund; Swedish University of Agricultural Sciences, Dept. of Farm Buildings, Division of Farm Building Constructions, 1982. 96p. (Sveriges Lantbruksuniversitet. Institutionen for Lantbrukets Byggnadsteknik. Rapport; 23.)	
Third	Nilsson, Milton. The Farm Tractor in the Forest. Vernamo, Sweden; Swedish National Board of Forestry, 1982. 100p. (Translated to English by Forest Extension Service, New Brunswick, Canada, Dept. of Natural Resources)	
Third	Nobe, Kenneth C. and Rajan K. Sampath, eds. Irrigation Management in Developing Countries: Current Issues and Approaches. Boulder; Westview Press, 1986. 547p.	First
Third	Nolle, Jean. Machines Modernes a Traction Animale: Itineraire d'un Inventeur au Service des Petits Paysans. Paris; Harmattan, 1986. 478p.	Third
Third	Noton, N. H. Farm Buildings. Reading: College of Estate Management, 1982. 359p.	

O

First	O'Brien, Michael, Burton F. Cargill, and Robert B. Fridley, eds. Principles and Practices for Harvesting and Handling Fruits and Nuts. Westport, Conn.; AVI Publishing Co., 1983. 636p.	First
Second	O'Connor, Colin. Extrusion Technology for the Food Industry. New York; Elsevier Science Pub. Co. Inc., 1987. 177p.	Second

Third	O'Loughlin, E. M., ed. Irrigation and Water Use in Australia; Papers Presented to a Meeting of the Science and Industry Forum of the Australian Academy of Science, 1979. Canberra; Australian Academy of Science, 1980. 87p.	
Third	O'Mara, Gerald T., ed. Efficiency in Irrigation—The Conjunctive Use of Surface and Groundwater Resources. Washington, D.C.; World Bank, 1988. 196p.	First
Third	Olesen, H. T. Grain Drying. Lyngby, Denmark; Polyteknisk Boghandel og Forlag, 1987. 218p.	
	Olson, G. W. Soil Survey Interpretation for Engineering Purposes. (Eng., Fr. and Span.). Rome; Food and Agriculture Organization, 1973. 30p. (Available from Unipub). (FAO Soils Bulletins no. 19).	Second
Third	Organisation for Economic Cooperation and Development. OECD Standard Codes for the Official Testing of Agricultural Tractors. Paris; Organisation for Economic Cooperation and Development, 1988. 194p.	Third
Third	Orr, Clyde, Jr., and J. M. Dallavalle. Fine Particle Measurement: Size, Surface, and Pore Volume. New York; Macmillan, 1959. 353p.	
Third	Ortiz-Canavate, Jaime. Las Maquinas Agricolas y su Aplicación. 3d ed., revised. Madrid; Mundi-Presna, 1989. 488p. (1st ed.: Madrid; Ediciones Mundi-Prensa, 1980.)	
	Ortiz-Cañavate, Jaime and J. L. Hernanz. Técnica de la Mecanización Agraria. 3d Rev. ed. Madrid; Ediciones Mundi-Presna, 1989. 642p. (Coleccion Biblioteca Mecanizacion Agraria no. 1) (1st ed. as Tecnica de la Mecanizacion Agraria. Vol 1: Tractores y Aperos de Labranza y de Cultivo; Vol. 2: Maquinaria de Recoleccion y Granja. Madrid; Instituto Nacional de Investigaciones Agrarias, 1972. (Coleccion Monografias I.N.I.A., no. 14). (First volume is revised edition of: Tecnica de la Mecanizacion Agraria: Tractores y Aperos de Cultivo y Labranza. London and Madrid; Editorial Garsi, 1968.)	Third
Third	Ostrovskii, E. V. and Evgenii V. Eidel'man. Kratkii spravochnik konstruktora prodovol'stvennykh mashin. (Short handbook for the design of food processing machines.) Moscow; Agropromizdat, 1986. 623p.	
Third	Ouellett, Robert P., Norman W. Lord and Paul N. Cheremisinoff. Food Industry Energy Alternatives. Westport, Conn.; Food and Nutrition Press, 1980. 135p.	Second
Third	Overcash, Michael R., Frank James Humenik and J. Ronald Miner. Livestock Waste Management. Boca Raton, Flor.; CRC Press, 1983. 2 vols.	
Second	Ower, Ernest and R. C. Pankhurst. The Measurement of Air Flow. 5th ed. Oxford and New York; Pergamon Press, 1977. 367p. (4th ed., 1966).	

P

Third	Page, A. L., R. H. Miller and D. R. Keeney, eds. Methods of Soil Analysis. 2d ed. Madison, Wisc.; American Society of	First

DC	Agronomy and Soil Science Society of America, 1982. 2 vols. (1st ed. by Charles A. Black, et al., 1965. 2 vols. (1572p.)	TW
Third	Paine, Frank A. Modern Processing, Packaging and Distribution Systems for Food. Glasgow; Blackie, and New York; AVI, 1987. 163p.	Second
First	Pair, C. H., W. W. Hinz, K. R. Forst, et al. Irrigation. 5th ed. Silver Spring, Maryland; Irrigation Association, 1983. 686p. (Previous title: Sprinkler Irrigation). (1st ed. edited by A.W. McCulloch and J.F. Schrunk, 1955; 2d ed. edited by G.O. Woodward; 3d—5th eds. edited by C. H. Pair with others).	First
Third	Palz, W. and P. Chartier. Energy from Biomass in Europe. Barking, U.K.; Applied Science, 1980. 234p.	
Third	Palz, W., J. Coombs, and O. Hall, eds. Energy from Biomass; 3d International Conference on Biomass, Venice, 1985. London and New York; Elsevier Applied Sciences Publishers, 1985. 1211p. (Distributed in the U.S. and Canada by Elsevier Science Pub. Co.)	
	Pande, G. N. Elements of Irrigation Engineering. Allahabad, India; Soroj Prakashan, 1962. 214p.	Second
Third	Parr, J. F., W. R. Gardner, and L. F. Elliott, eds. Water Potential Relations in Soil Microbiology: Proceedings of a Symposium. Madison, Wisc.; Soil Science Society of America, 1981. 151p. (SSSA Special Publication no. 9.)	
Second	Parry, R. H., ed. Stress-Strain Behaviour of Soils; Proceedings of the Roscoe Memorial Symposium, Cambridge Univ., 1971. Henley-on-Thames, England; G.T. Foulis, 1973. 752p.	
	Pasfield, D. H. Farm Building Design and Construction. London; Temple Press, 1965. 294p.	Second
	Patel, S. M. and K. U. Patel. Economics of Tubewell Irrigation. Ahmedabad; Indian Institute of Management, 1971.	Second
	Patel, S. M., P. P. Madappa, and B. M. Desai. Management of Lift Irrigation: Report on a Pilot Research Project in Gujarat. Ahmedabad: Faculty for Management in Agriculture and Cooperative, Indian Institute of Management, 1969. 153p.	Second
Second	Patrick, J. Michael, Rex L. Clark and L. Harlan Davis. Agro-Mechanical Technologies in Latin America; A Survey of Applications in Selected Countries. Washington, D.C.; Inter-American Development Bank, Economic and Social Development Dept., General Studies Division, Agricultural Economics Section, 1978. 79p.	
	Paul, J. K. Fruit and Vegetable Juice Processing. Park Ridge, New Jersey; Noyes Data Corp., 1975. 227p. (Food Technology Review no. 21).	Second
Third	Pearce, John, George Stenzel and Thomas A. Walbridge. Logging and Pulpwood Production. 2d ed. New York; Wiley, 1985. 358p.	
Third	Pedersen, Jorgen and Soren Pedersen. Staldklimateknik : Isolering, Ventilering, Opvarmning. (Barn-Climate Technique : Isolation, Ventilation, Heating.) Copenhagen : DSR, 1979. 228p.	

Third	Peleg, Micha and Edward B. Bagley. Physical Properties of Foods. Westport, Conn.; AVI Pub. Co., Inc., 1983. 532p.	
Third	Pelley, Lee. In One Barn: Efficient Livestock Housing and Management. Woodstock, Vermont; Countryman Press, Inc., 1984. 152p.	
Second	Pellizzi, G. A., Guidobono Cavalchini and M. Lazzari, eds. Energy Savings in Agricultural Machinery and Mechanization. London and New York; Elsevier Applied Science, 1988. 143p.	Third
Second	Pellizzi, G. A. Meccanica e Meccanizzazione Agricola. (Mechanics and Agricultural Mechanization.) Bologna; Edagricole, 1987. 484p.	
Third	Penning-Rowsell, Edmund C., Dennis J. Parker and Donald M. Harding. Floods and Drainage: British Policies for Hazard Reduction, Agricultural Improvement and Wetland Conservation. London, Boston; Allen and Unwin, Inc., 1985. 199p.	
Second	Pereira, Charles; M. J. Hamlin, and M. Mansell-Moullin, eds. Scientific Aspects of Irrigation Schemes. London; Scholium International, Inc., 1986. 181p.	
First	Perry, John H., Don W. Green and James O. Maloney, eds. Perry's Chemical Engineers' Handbook. 6th ed. New York; McGraw-Hill, 1984. 1 vol., various pagings. (2d ed., 1941 3029p.)	Second
	Persson, Sverker. Mechanics of Cutting Plant Material. St. Joseph, Mich.; American Society of Agricultural Engineers, 1987. 288p. (ASAE Monograph no. 7).	Second
Third	Peterson, Arnold P. G. and Erwin E. Gross, Jr. Handbook of Noise Measurement. 9th ed. West Concord, Mass.; Genrad, Inc., 1980. 384p. (1st ed., with Leo Beranek. Cambridge, Mass.; General Radio; 1953. 102p.; 4th–8th eds., 1960–1978.)	
Third	Phillips, Richard E. Farm Buildings: From Planning to Completion. St. Louis; Doane Western Publishing, 1981. 417p.	
Second	Phillips, Ronald E. and Shirley H. Phillips, eds. No-Tillage Agriculture: Principles and Practices. New York; Van Nostrand Reinhold, 1984. 306p.	Third
Second	Phillips, Shirley H. and H. M. Young, Jr. No-Tillage Farming. Milwaukee; Reiman Associates, 1973. 224p.	
Third	Phipps, Lloyd J. and Carl L. Reynolds. Mechanics in Agriculture. 4th ed. Danville, Ill.; Interstate Publishers, 1990. 689p. (1st ed., 1967. 808p.; 2d ed., 1977. 835p.)	Second
Third	Pierotti, Anne, Andrew G. Keeler and Albert Fritsch. Energy and Food: Energy Used in Production, Processing, Delivery and Marketing of Selected Food Items. Washington; Center for Science in the Public Interest, 1977. 76p.	
Third	Pierre, William H., et al., eds. Plant Environment and Efficient Water Use. Madison, Wisc.; American Society of Agronomy, 1966. 295p.	
Second	Pillsbury, Arthur F. Sprinkler Irrigation. Rome; Food and Agriculture Organization, 1968. 179p. (FAO Agriculture Development Paper no. 88).	Second

Second	Pimentel, David, and Carl W. Hall. Food and Natural Resources. San Diego; Academic Press, 1989. 512p.	Second
Second	Pimentel, David, ed. Handbook of Energy: Utilization in Agriculture. Boca Raton; CRC Press, Inc., 1980. 475p.	First
Third	Pingale, S. V. Handling and Storage of Food Grains. New Delhi; Indian Council of Agricultural Research, 1976. 186p.	Second
Second	Pingali, Prabhul, Yves Bigot, and Hans P. Binswanger. Agricultural Mechanization and the Evolution of Farming Systems in Sub-Saharan Africa. Baltimore; Johns Hopkins University Press, 1987. 216p.	First
Third	Pintauro, Nicholas D. Food Packaging. Park Ridge, New Jersey; Noyes Data Corp., 1978. 415p. (Food Technology Review no. 47).	Second
Second	Planner, J. H., ed. Workshop on Grain Handling: Challenges, Shortcomings, Improvements, Solutions; Melbourne, 1984; Preprints of Papers. Organized by Panel on Bulk Materials Handling, College of Mechanical Engineers, Institution of Engineers, Australia. Barton, A.C.T; The Institution, 1984. 13 pieces (246p.)	
Second	Poiree, Maurice A. A. and Charles Ollier. Irrigation; les Reseaux d'Irrigation Theorie. 7th ed. Paris; Eyrolles, 1986. 503p. (3d ed., 1966, 424p.)	
Second	Pollard, S. J. A Select Bibliography of the Economics of Farm Mechanisation in Developing Countries. Silsoe, England; Overseas Department, National Institute of Agricultural Engineering, 1978. 52p.	
Second	Portland Cement Association. Concrete Culverts and Conduits. Skokie, Ill.; Portland Cement Association, 1975. 73p.	Second
Third	Potter, Norman N. Food Science. 4th ed. Westport, Conn.; AVI Pub. Co., 1986. 735p. (1st ed., 1968. 653p.; 2d ed., 1973. 706p.; 3d ed., 1978. 780p.)	Second
Third	Price, Barry L. The Political Economy of Mechanization in U. S. Agriculture. Boulder, Colo.; Westview Press, 1983. 108p.	
Third	Pritchett, William L. and Richard F. Fisher. Properties and Management of Forest Soils. 2d ed. New York; John Wiley and Sons, Inc., 1987. 494p. (1st ed., Pritchett, 1979. 500p.)	
Second	Priyani, V. B. The Fundamental Principles of Irrigation Engineering. 6th ed. Anand, India; Charotar Book Stall, 1964. 540p.	Second
Second	Promersberger, William J. and Frank E. Bishop. Modern Farm Power. 3d ed. Reston, Virg.; Reston Publishing Company, 1979. 270p. (Earlier ed., Englewood Cliffs, New Jersey; Prentice-Hall, 1962. 280p.)	Second

Q

Third	Quantick, H. R. Aviation in Crop Protection, Pollution and Insect Control. London; Collins, 1985. 428p.	
Second	Quick, Graeme R., and George F. Montgomery, eds. Bibliography on Combines and Grain Harvesting; Citations from the International Literature on the Engineering, Biological and	

Citation Analysis and the Core Monographs 297

DC	Economic Aspects of the Harvesting of Crops for Grain and Seed. St. Joseph, Mich.; American Society of Agricultural Engineers, 1974. 71p. (ASAE Special Publication SP-0274)	TW
First	Quick, Graeme R., and Wesley F. Buchele. The Grain Harvesters. St. Joseph, Mich.; American Society of Agricultural Engineers, 1978. 269p.	First

R

	Ramaswamy, N. S., and C. L. Narasimban. India's Animal-drawn Vehicles. Bangalore; Indian Institute of Management, 1984.	Second
Second	Ranken, M. D. Food Industries Manual. 22d ed. Glasgow; Blackie; and New York; AVI Pub. Co., 1988. 599p. (20th ed., Anthony Wollen, New York; Chemical Pub. Co., 1970. 509p.; 21st ed., Glasgow, L. Hill; and Washington, D.C.; Kapitan Szabo Publishers, 1984. 530p.)	
Second	Rasmussen, William O. et al. Computer Applications in Agriculture. Boulder, Colo.; Westview Press, 1985. 143p.	
Second	Rastovski, A., A. Van Es, et al. Storage of Potatoes: Post-Harvest Behavior, Store Design, Storage Practice, Handling. 2d ed. Wageningen; Pudoc, Centre for Agricultural Publishing and Documentation, 1987. 453p. (1st ed., 1981. 462p.)	
	Rauscher, Maurice. Agricultural Mechanization: Harvesting and Storage of the More Common Cereals. Geneva; United Nations, 1957. 66p.	Third
Third	Raymond, Frank, Gordon Shepperson and Richard Waltham. Forage Conservation and Feeding. 4th ed. Ipswich, Suffolk; Farming Press, 1986. 188p. (1st ed., 1972. 175p.; 2d ed., 1975. 204p.)	
Third	Rechcigl, Jr., Miloslav, ed. CRC Handbook of Agricultural Productivity. Boca Raton, Flor.; CRC Press, 1982. 2 vols.	
	Regional Network for Agricultural Machinery (U.N. Development Programme). RNAM Test Codes and Procedures for Farm Machinery. Pasay City, Philippines; United Nations D. P.; 1983. 297p.	Third
Second	Reimbert, Mariel L. and Andre M. Reimbert. Silos: Theory and Practice: Vertical Silos, Horizontal Silos (Retaining walls). 2d ed. (Transl. from French and German). New York; Lavoisier Pub., 1987. 564p. (1st ed., Aedermannsdorf, Switzerland; Trans Tech Publications, 1976. 251p.)	Second
	Research Issues in Irrigation Systems in Developing Countries: Proceedings. Lincoln Institute of Land Policy, 1978. (Lincoln Institute Monograph no. 78-6).	Second
Third	Reybold, W. U., and G. W. Petersen, eds. Soil Survey Techniques: Proceedings of a Symposium, Chicago, 1985. Madison, Wisc.; Soil Science Society of America, 1987. 98p. (SSSA Special Publication no. 20).	Second
	Reynolds, Lloyd G., ed. Agriculture in Development Theory. New Haven; Yale University Press, 1975. 510p.	Second
Second	Reznicek, Rados, editor. Physical Properties of Agricultural Materials and Products. International Conference on Physical	

DC	Properties of Agricultural Materials and Their Influence on Technological Processes; 3d; Prague; 1985. Washington, D.C.; Hemisphere Pub. Corp., 1988. 1224p.	TW
Third	Ricco, Guido Di. L' Irrigazione dei Terreni: Bas: Techiche e Realizzazioni. 6th ed. Bologna; Edagricole, 1967. 582p.	
Second	Richards, Adrian F., ed. Vane Shear Strength Testing in Soils: Field and Laboratory Studies. Philadelphia; American Society for Testing Materials, 1988. 378p.	
Second	Richards, J. A. Remote Sensing: Digital Image Analysis: An Introduction. Berlin and New York; Springer-Verlag, Inc., 1986. 281p.	
Third	Richardson, Daryl G. and Michael Meheriuk, editors. National Controlled Atmosphere Research Conference on Controlled Atmospheres for Storage and Transport of Perishable Agricultural Commodities, 3d; 1981; Oregon State University. Controlled Atmospheres for Storage and Transport of Perishable Agricultural Commodities. Beaverton, Ore.; Timber Press in cooperation with School of Agriculture, Oregon State University, 1982. 390p. (Oregon State University. School of Agriculture, Symposium Series, no. 1.)	
	Richey, C. B., Paul Jacobson, and Carl W. Hall, eds. Agricultural Engineers' Handbook. New York; McGraw-Hill, 1961. 880p.	First
Second	Riedijk, W., ed. Appropriate Technology for Developing Countries. 2d ed. Delft, the Netherlands; Delft University Press, 1984. 356p.	First
	Rijk, A. G. Agricultural Mechanization Policy and Strategy: The Case of Thailand. Tokyo; Asian Productivity Organization, 1990. 283p.	Third
Third	Ripp, B. E., H. J. Banks, et al., eds. Controlled Atmosphere and Fumigation in Grain Storages: Proceedings of an International Symposium. Amsterdam and New York; Elsevier Science Publishing Co., Inc., 1984. 798p.	
Second	Ritchie, James D. Sourcebook for Farm Energy Alternatives. New York; McGraw-Hill Book Co., 1983. 384p.	
Third	Roberts, David V. Enzyme Kinetics. Cambridge and New York; Cambridge University Press, 1977. 326p.	
Third	Roberts, Eric Hywel. Viability of Seeds. London; Chapman and Hall, 1972. 448p.	Second
Third	Robertson, Doris, ed. The Small Tree Resource: A Materials Handling Challenge. Madison, Wis.; Forest Products Research Society and Dubuque, Iowa; Kendall/Hunt, 1983. 135p.	
Second	Robertson, John. Mechanising Vegetable Production. 2d ed. Ipswich; Farming Press, 1978. 195p. (1st ed., 1974)	
	Rogin, Leo. The Introduction of Farm Machinery in Its Relation to the Productivity of Labor in the Agriculture of the U. S. During the Nineteenth Century. Berkeley; University of California Press, 1931. 260p. (Reprinted by New York; Johnson Reprint Corp., 1966).	Third
Second	Rose, C. W. Agricultural Physics. New York and Oxford; Pergamon, 1966. 226p. (Reprinted in 1969, 230p.)	

	Roth, Lawrence O., Franklin R. Crow, and George W. A. Mahoney. An Introduction to Agricultural Engineering. Westport, Conn.; AVI Publishing Company, 1975. 356p.	
Second	Ruigu, George M. and Mandivamba Rukuni, eds. Irrigation Policy in Kenya and Zimbabwe. Proceedings of seminars and workshops on agricultural irrigation policy, 1987. Nairobi, Kenya; Institute for Development Studies, University of Nairobi, 1989. 245p.r	Third
	Ruthenberg, H. Farming Systems in the Tropics. 3d ed. Oxford; Clarendon Press, 1980. 424p. (Previous ed., 1971. 313p.)	Second
Second	Ruttan, Vernon W. The Economic Demand for Irrigated Acreage: New Methodology and Some Preliminary Projections, 1954–1980. Baltimore, Maryland; Johns Hopkins University Press, 1965. 139p. (Resources for the Future Series).	Second
Second	Ryall, A. Lloyd and Werner J. Lipton. Handling, Transportation and Storage of Fruits and Vegetables. 2d ed. Westport, Conn.; AVI Publishing Co., 1979. 2 Vols. (1st ed., 1972. Vol. 1, 472p., Vol. 2, 545p.)	Second
	Ryan, J. G. and H. L. Thompson, eds. International Workshop on Socioeconomic Constraints to Development of Semi-arid Tropical Agriculture, 1979, Hyderabad, India. Andhra Pradesh, India; International Crops Research Institute for the Semi-Arid Tropics, 1980. 435p.	Third

S

Third	Sadhu, A. N. and R. K. Mahajan. Technological Change and Agricultural Development in India. Bombay; Himalaya Pub. House, 1985. 215p.	Third
Second	Sagardoy, J. A., A. Bottrall, and G. O. Uittenbogaard. Organization, Operation and Maintenance of Irrigation Schemes. Rome; Food and Agriculture Organization, 1982. 174p. (FAO Irrigation and Drainage Paper no. 40.)	
Third	Sainsbury, David and Peter Sainsbury. Livestock Health and Housing. 3d ed. London; Bailliere Tindall, 1988. 319p. (2d ed., 1979. 338p.)	
	Sainsbury, David. Pig Housing. 5th ed. Ipswich; Farming Press, 1978. 205p. (3d ed., 1972. 212p.)	Second
	Sally, Hari Lal. Irrigation Planning for Intensive Cultivation. Bombay and New York; Asia Publishing House, 1968. 208p.	Second
	Sargen, Nicholas P. Tractorization in the United States and Its Relevance for the Developing Countries. New York; Garland Publishing, Inc., 1979. 280p.	Second
Second	Sanks, Robert L., ed. Water Treatment Plant Design for the Practicing Engineer. Ann Arbor; Ann Arbor Science Publishers, 1978. 845p.	
Third	Schepens, G. et al., eds. Solar Energy in Agriculture and Industry: Potential of Solar Heat in European Agriculture, an Assessment. Dordrecht and Boston; Reidel Pub. Co. for the Commission of the European Communities, 1986. 198p. (Sold	

DC	and distributed in the U.S. and Canada by Kluwer Academic Publishers)	TW
	Schneeweiss, R., editor. Dictionary of Cereal Processing and Cereal Chemistry in English, German, French, Latin, and Russian. Amsterdam and New York; Elsevier Scientific Pub. Co., 1982. 520p.	Second
	Schoenherr, W. H., editor. Seminar in Food Storage and Handling Practices, 1979, Mombassa, Kenya, Sponsored by United States Agency for International Development, Office of Food for Peace. Proceedings . . . Washington, D.C.; A.I.D., 1979. 162p.	Second
Second	Schuler, Michael L. The Utilization and Recycle of Agricultural Wastes and Residues. Boca Raton, Flor.; CRC Press, Inc., 1980. 298p.	Second
First	Schwab, Glenn O., Delmar D. Fangmeier, William J. Elliot, and Richard K. Frevert. Soil and Water Conservation Engineering. 4th ed. New York; John Wiley and Sons, 1992. 544p. (Early eds. with others; 3d ed. under title: Elementary Soil and Water Engineering.)	First
Second	Schwartzberg, Henry G., Daryl Lund, and John L. Bomben, eds. Food Process Engineering. New York; American Institute of Chemical Engineers, 1982. 191p. (AIChE Symposium Series no. 218 = v.78)	
	Schwartzman, David. Oligopoly in the Farm Machinery Industry. Ottawa; Royal Commission on Farm Machinery, 1970. 249p.	Third
Third	Segerlind, Larry J. Applied Finite Element Analysis. 2d ed. New York; Wiley, 1984. 427p. (1st ed., 1976. 422p.)	
	Seminar on the Improvement of Irrigation Performance at the Project Level: Proceedings . . . United Nations, 1983. 111p. (Water Resources Development Ser. no. 56).	Second
Third	Serdechnyi, A. N. Senouborochnye Mashiny. (Hay-Harvesting Machinery). Moscow; Vysshaia Shkola, 1973. 108p.	
Third	Severnev, M. M., ed. Wear of Agricultural Machine Parts. Translation of Iznos Detalei Selskokhozyaistvennykh Mashin. Leningrad; Kolos and New Delhi; Amerind Pub. Co., 1972. 259p. (Available from Springfield, Vir.; National Technical Information Service, 1984.)	Second
Second	Shainberg, I. and J. Shalhevet, eds. Soil Salinity under Irrigation: Processes and Management. Berlin and New York; Springer-Verlag New York, Inc., 1984. 349p.	First
Second	Shalhevet, J., A. Mantell, H. Bielorai, and D. Shimshi, eds. Irrigation of Field and Orchard Crops under Semi-Arid Conditions. Bet Dagan, Israel; International Irrigation Information Center, and New York; Pergamon Press, 1979. 124p. (2d ed. Bet Dagan, Israel and Ottawa; International Irrigation Information Center, 1981. 132p.)	First
	Sharan, Girja, D. P. Mathew, and M. Vishwanath. Characterization of the Process of Mechanization and Farm Power Requirements. Ahmedabad; Centre for Management in Agriculture, Indian Institute of Management, 1974. 153p. (CMA Monograph no. 45).	Third
Third	Shejbal, J. ed. Controlled Atmosphere Storage of Grains; An	Second

	International Symposium, Castelgandolfo, Italy, 1980. Amsterdam and New York; Elsevier Scientific Pub. Co., 1980. 608p. (Developments in Agricultural Engineering no. 1).	
Second	Shinn, Glen C. and Curtis R. Weston. Working in Agricultural Mechanics. New York; Gregg Division, McGraw-Hill, 1978. 280p.	
Second	Shippen, J. M., C. R. Ellin, and C. H. Clover. Basic Farm Machinery. 3d ed. Oxford; Pergamon Press, 1980. 288p. (1st ed., J. M. Shippen and J. C. Turner, 1966. 2 vols.; 2d ed., 1973. 383p.)	First
Second	Sholto, Douglas James. Advanced Guide to Hydroponics (Soilless Cultivation). 2d ed. London; Pelham Books, 1985. 368p. (1st ed., 1976. 333p.)	
Second	Shultz, Richard and Richard A. Smith. Introduction to Electric Power Engineering. New York; Harper and Row Publishers, Inc., 1985. 264p.	
	Silletto, Thomas A. and Dale O. Hull. Safe Operation of Agricultural Equipment. St. Paul; Hobar Publications, 1976.	Third
Third	Sineokov, G. N. Design of Soil Tilling Machines. Translation of Proektirovanie Pochvoobrabatyvayushchikh Mashin. New Delhi; Indian National Scientific Documentation Centre for the U.S. Dept. of Agriculture, Agricultural Research Service and the National Science Foundation, 1977. 407p.	
	Singh, Gajendra and J. H. de Goede, editors. International Conference on Rural Development Technology, Bangkok, 1977. Proceedings ... An Integrated Approach. Sponsored by Canadian International Development Agency. Bangkok; Asian Institute of Technology, 1977. 847p.	Third
Third	Singh, R. Paul and Augusto G. Medina. Food Properties and Computer-Aided Engineering of Food Processing Systems; Proceedings of the NATO Advanced Research Workshop in Porto, Portugal, 1988. Dordrecht, the Netherlands, Kluwer Academic Press, 1989. 593p. (NATO ASI Series, Series E, Applied Sciences; Vol. 168).	Second
Third	Singh, R. Paul and Dennis R. Heldman. Introduction to Food Engineering. Orlando, Flor.; Academic Press, Inc., 1984. 306p.	First
Second	Singh, R. Paul. Energy in Food Processing. Amsterdam and New York; Elsevier Science Pub. Co., Inc., 1986. 375p.	Second
Second	Sinha, R. N. and W. E. Muir. Grain Storage: Part of a System. Westport, Conn.; AVI Publishing Co., 1973. 481p.	
Third	Sistler, Fred. E. The Farm Computer. Reston, Vir.; Reston Publishing Co., 1984. 194p.	
Second	Sitkei, Gyorgy. Mechanics of Agricultural Materials. English translation by S. Bars of Mezogazdasagi Anyagok Mechanikaja. Amsterdam and New York; Elsevier, Inc., 1986. 487p. (Developments in Agricultural Engineering no. 8).	Second
Third	Skovorodin, Vasilii I. and Leonid V. Tishkin Spravochnaia Kniga Po Nadezhnosti Sel'Skokhoziaistvennoi Tekhniki. (Reference Book on the Reliability of Agricultural Equipment.) Leningrad; Lenizdat, 1985. 202p.	
Third	Small, L. E., M. S. Adriano, E. D. Martin, R. Bhatia, Y. K.	

DC	Shim and P. Pradhan. Financing Irrigation Services: A Literature Review and Selected Case Studies from Asia. Sri Lanka; International Irrigation Management Institute, 1989. 286p.	TW
Second	Smedema, Lambert K. and David W. Rycroft. Land Drainage: Planning and Design of Agricultural Drainage Systems. Ithaca, New York; Cornell University Press, 1983. 376p.	Second
	Smil, Vaclav and William E. Knowland, eds. Energy in the Developing World: The Real Energy Crisis. Oxford and New York; Oxford University Press, 1980. 366p.	Third
	Smith, C. V. Some Environmental Problems of Livestock Housing. Geneva; World Meteorological Organisation, 1972. 71p. (Technical Note Ser. no. 122).	Second
Second	Smith, Harris P. and Lambert H. Wilkes. Farm Machinery and Equipment. 6th ed. New York; McGraw-Hill, 1976. 488p. (1st–5th ed., 1929–64).	Third
Third	Smith, L. P. and B. D. Trafford. Climate and Drainage. London; Her Majesty's Stationery Office, 1976. 119p. (Ministry of Agriculture, Fisheries and Food, Technical Bulletin no. 34).	
	Smith, Peter, ed. Agricultural Project Management: Monitoring and Control of Implementation. London and New York; Elsevier Science Publishing, 1984. 190p.	Second
First	Snedecor, George W., and William G. Cochran. Statistical Methods. 8th ed. Ames; Iowa State University Press, 1989. 503p. (6th ed., 1967. 593p.; 7th ed., 1980. 507p. Previous editions as Statistical Methods Applied to Experiments in Agriculture and Biology.)	
Second	Soane, B. D., ed. Compaction by Agricultural Vehicles: A Review. Penicuik, Midlothian; Scottish Institute of Agricultural Engineering, 1982. 95p.	First
	Society of Automotive Engineers. An Historical Perspective of Farm Machinery. Warrendale, Penn.; Society of Automotive Engineers, 1980.	Second
Second	Society of Automotive Engineers. Durability by Design: Integrated Approaches to Mechanical Durability Assurance of Ground Vehicles. Warrendale, Penn.; Society of Automotive Engineers, 1988. 120p.	Second
Second	Sodha, Mahendra, et al. Solar Crop Drying, Vols. 1 and 2. Boca Raton, Flor.; CRC Press, Inc., 1987. 404p.	First
Second	Solomons, G. L. Materials and Methods in Fermentation. London and New York; Academic Press, 1969. 331p.	
Second	Sopher, Charles and Jack V. Baird. Soils and Soil Management. 2d ed. Reston, Virg.; Reston Publishing Co., Inc., 1982. 312p. (1st ed., 1978. 238p.)	
	Southgate, Douglas D. and John F. Disinger, eds. Sustainable Resource Development in the Third World. Boulder, Colo.; Westview Press, 1987. 177p.	Third
	Southworth, Herman, and M. Barnett. Experience in Farm Mechanization in South East Asia. New York; Agricultural Development Council, 1974. 345p.	Second
First	Spangler, Merlin G. and Richard L. Handy. Soil Engineering.	Second

Citation Analysis and the Core Monographs 303

	4th ed. New York; Harper and Row, 1981. 819p. (1st–2d eds., Merlin G. Spangler. Scranton, Penn.; International Textbook Co., 1951–1960. Various pagings; 3d ed., New York; Intext Educational Pub., 1973. 748p.)	
Second	Spedding, C. R. W. An Introduction to Agricultural Systems. 2d ed. Elsevier Applied Sci England). London and New York; Elsevier Science Publishing Co., Inc., 1988. 189p. (1st ed., 1979. 169p.)	First
Third	Spillman, C. K. Low Temperature and Solar Grain Drying Handbook. Ames; Iowa State University, Midwest Plan Service, 1980. 86p.	
Third	Spillman, C. K. Solar Livestock Housing Handbook. Ames; Iowa State University, Midwest Plan Service, 1983. 88p.	
Second	Standard Methods for the Examination of Water and Wastewater, Including Bottom Sediments and Sludges: Prepared and Published jointly by the American Public Health Association, the American Water Works Association and the Water Pollution Control Federation. 17th ed. New York; American Public Health Association, 1989. various pagings, near 1000p. (11th ed., 1960. 626p.)	
Second	Stanhill, G., ed. Energy and Agriculture. Berlin and New York; Springer, 1984. 192p.	
	Starkey, Paul and Fadel Ndiame, eds. Animal Power in Farming Systems: Proceedings of the 2d West African Animal Traction Networkshop, Sept. 1986, Freetown, Sierra Leone. Braunschweig; Vieweg, 1988. 363p.	Second
Second	Starkey, Paul. Animal-Drawn Wheeled Toolcarriers: Perfected Yet Rejected; A Cautionary Tale of Development. Eschborn; Deutsche Gesellschaft fur Technische Zusammenarbeit (GYZ); and Braunschweig; Vieweg, 1988. 161p.	First
Third	Starkey, Paul. Harnessing and Implements for Animal Traction: An Animal Traction Resource Book for Africa. Braunschweig/Wiesbaden, Germany; Friedr. Vieweg, 1989. 245p.	First
	Stavis, B. The Politics of Agricultural Mechanization in China. Ithaca, New York; Cornell University Press, 1978. 288p.	Second
Second	Steel, Robert George Douglas and James H. Torrie. Principles and Procedures of Statistics: A Biometrical Approach. 2d ed. New York; McGraw-Hill., 1980. 633p.	
Third	Stefferud, Alfred, ed. Power to Produce; U.S. Department of Agriculture, Yearbook of Agriculture, 1960. Washington, D.C.; United States Government Printing Office, 1960. 480p. (Yearbook of Agriculture, 1960.)	
	Stern, Peter H. Field Engineering: An Introduction to Development Work and Construction in Rural Areas. London: Intermediate Technology Publications, 1983. 251p.	Second
Second	Stern, Peter H. Small Scale Irrigation: A Manual of Low-Cost Water Technology. London; Intermediate Technology Publications; Bet Dagan, Israel; International Irrigation Information Center, 1979. 152p.	First
Third	Stevens, G. N. Equipment Testing and Evaluation. Silsoe,	

DC	Bedford, England; National Institute of Agricultural Engineering, 1982. 137p.	TW
	Stewart, George F. and Maynard A. Amerine. Introduction to Food Science and Technology. 2d ed. New York; Academic Press, 1982. 289p. (1st ed., 1973. 294p.)	Second
Third	Stoate, David. The Miller's Manual. The Author, 1981.	
First	Stone, Archie A. and Harold E. Gulvin. Machines for Power Farming. 3d ed. New York; John Wiley, 1977. 533p. (1st ed., 1952. 616p.; 2d ed., 1967. 559p.)	First
	Stout, B. A. Energy for World Agriculture. Rome; Food and Agriculture Organisation, 1979. 286p. (FAO Agricultural Series no. 7).	First
	Stout, B. A. Energy Use and Management in Agriculture. North Scituate, MA; Breton Publishers, 1984. 318p.	Second
	Stout, B. A. Equipment for Rice Production. Rome; Food and Agricultural Organization, 1966. 169p.	Second
First	Stout, B. A. Handbook of Energy for World Agriculture. London and New York; Elsevier, Applied Science, 1990. 506p.	First
	Strauch, D. Animal Production and Environmental Health. Amsterdam and New York; Elsevier Science Pub. Co., Inc., 1987. 324p.	Second
Third	Strub, A., P. Chartier, and G. Schleser, eds. Energy from Biomass; 2d E. C. International Conference on Biomass, Germany, 1982. Proceedings . . . London and New York; Applied Science Publishers, 1983. 1148p. (Distributed in the U.S. and Canada by Elsevier Science Pub. Co.)	
Second	Sweeten, John M. and Frank J. Humenik, eds. Agriculture and the Environment. Report Developed by Environmental Quality Coordinating Committee (T-9) of the A.S.A.E. St. Joseph, Mich.; American Society of Agricultural Engineers, 1984. 117p.	
	Symposium on Computers, Electronics and Control Engineering in Agriculture, 1986; Are, Sweden. Proceedings . . . Uppsala; Sveriges Lantbruksuniversitet, Institutionen for Lantbruksteknik, 1987. 1 Vol. (Various Pagings) (Sveriges Lantbruksuniversitet, Institutionen for Lantbruksteknik. Rapport 17).	Third
Third	Symposium on Cow Identification Systems and their Applications, Wageningen, 1976, Proceedings . . . Wageningen; IMAG, 1976. 97p.	
Third	Symposium on Recent Developments in Food Packaging, Mysore, India, 1985. Proceedings of the Symposium on Recent Developments in Food Packaging, sponsored by the Association of Food Scientists and Technologists; Central Food Technological Research Institute. Mysore, India; The Association, 1986. 192p.	Second

T

	Tam, OnKit. China's Agricultural Modernization: The Socialist Mechanization. London; Dover, New Hampshire; Croom Helm, 1985. 241p.	Third

First	Tapley, Bryon D. Handbook of Engineering Fundamentals. 4th ed. New York; Wiley, 1989. 2,368p. (1st–3d ed., 1936–1975, by O. W. Eshbach and M. Souders. Various pagings).	Second
Second	Teixeira, Arthur A. and Charles F. Shoemaker. Computerized Food Processing Operations. New York; Van Nostrand Reinhold, 1989. 202p.	
Third	Terrängmaskinen (Forest Terrain Machines). Stockholm; Forskningsstiftelsen skogsarbeten, 1970–1981. 3 vols.	
First	Terzaghi, Karl and R. B. Peck. Soil Mechanics in Engineering Practice. 2d ed. New York; John Wiley, and Sons, 1967. 729p. (1st ed., 1948. 566p.)	First
	Teter, Norman. Paddy Drying Manual. Rome; Food and Agriculture Organization, 1987. 123p. (FAO Agricultural Services Bulletin no. 70).	Third
	Theobald, G. H. Methods and Machines for Tile and Other Tube Drainage. Rome; Food and Agriculture Organization, 1963. 104p. (Agricultural Development Paper no. 78).	Third
	Thorne, D. Wynne, and Marlowe D. Thorne, eds. Soil, Water and Crop Production. Westport, Conn.; AVI Publishing Company, Inc., 1979. 252p.	Second
	Thurm, Richard. Technologie der Landwirtschaftlichen Produktion. Berlin; Deutscher Landwirtschaftwverlag, 1970. 381p.	Third
Second	Tools for Agriculture; A Guide to Appropriate Equipment for Smallholder Farmers. 4th ed. London; Intermediate Technology Publications, 1992. 240p. (1st ed.: Tools for Progress. Anthony Gater, Tim Meller, and John Warsanny. London; ITDG, 1968. 192p.)	First
Second	Transnational Corporations in the Agricultural Machinery and Equipment Industry. New York; United Nations, United Nations Centre on Transnational Corporations, 1983. 134p.	Third
	Tresemer, David. The Scythe Book: Mowing Hay, Cutting Weeds, and Harvesting Small Grains with Hand Tools. London; By Hand and Foot, Ltd., 1981. 128p.	Second
Third	Tressler, Donald K., Wallace B. Van Arsdel, Michael J. Copley and Willis R. Woolrich. The Freezing Preservation of Foods. Vol. 1. Principles of Refrigeration. 4th ed. Westport, Conn.; AVI Publishing Co., 1968. 325p. (1st and 2d eds. by Tressler and Clifford F. Evers. New York; AVI Pub. Co., 1943–1947. Various pagings; 3d ed., 1957. 2 vols.)	Second
Third	Troeh, Frederick R., J. Arthur Hobbs, and Roy L. Donahue. Soil and Water Conservation for Productivity and Environmental Protection. Englewood Cliffs, New Jersey; Prentice-Hall, 1980. 718p.	Second
Third	Troller, John A. Sanitation in Food Processing. New York; Academic Press, 1983. 456p.	Third
Third	Trow-Smith, Robert. Power on the Land: A Centenary History of the Agricultural Engineers Association, 1875–1975. London; Agripress Publicity Ltd for the Association, 1975. 93p.	
	Tschiersch, J. E. Appropriate Mechanization for Small	Second

DC	Farmers in Developing Countries. Heidelberg; Verlag Breitenbach, 1978. 106p. (Research Centre for International Agrarian Development no. 8).	TW
Third	Tyler, Colin and John Haining. Ploughing by Steam: A History of Steam Cultivation over the Years. Hemel, Hempstead; Model and Allied Publications, 1970. 360p.	

U

Third	Unger, P. W. and D. M. Van Doren, Jr., eds. Predicting Tillage Effects on Soil Physical Properties and Processes. Madison, Wisc.; American Society of Agronomy, 1982. 198p. (ASA Special Publication 44)	
	United Nations Industrial Development Organization. Appropriate Industrial Technology for Sugar. New York; United Nations, 1980. 88p. (Monographs on Appropriate Industrial Technology no. 8).	Third
Second	United Nations Industrial Development Organization. Appropriate Industrial Technology for Agricultural Machinery and Implements. New York; United Nations, 1979. 159p. (Monographs on Appropriate Industrial Technology no. 4).	Third
Third	United Nations Industrial Development Organization. Appropriate Industrial Technology for Food Storage and Processing. New York; United Nations, 1979. 120p. (Monographs on Appropriate Industrial Technology no. 7).	Third
Third	United Nations Industrial Development Organization. Appropriate Industrial Technology for Energy for Rural Requirements. New York; United Nations, 1979. 169p. (Monographs on Appropriate Industrial Technology, no. 5)	
Third	United Nations Industrial Development Organization. Appropriate Industrial Technology for Oils and Fats. New York; United Nations, 1979. 50p. (Monographs on Appropriate Industrial Technology, no. 9)	
Second	United Nations Industrial Development Organization. Information Sources on Agricultural Wastes. New York; United Nations, 1982. (UNIDO Guides to Information Sources, no. 34) #7	
Third	United Nations Industrial Development Organization. Information Sources on the Agricultural Implements and Machinery Industry. Revised. New York; United Nations, 1982. 108p. (UNIDO Guides to Information Sources, no. 8, rev.)	
Second	United Nations Industrial Development Organization. Small-Scale Maize Milling. Prepared under the auspices of the International Labour Office and the U.N. Industrial Development Organisation. Geneva; ILO, 1984. 155p.	
Second	United Nations Industrial Development Organization. The Agricultural Machinery Industry: An Appraisal of the Current Global Situation: Production and Market Outlook. 2 vols. Vienna; UNIDO, 1983. Vol. 1, 87p. Vol. 2, 89p. UNIDO/IS.408; UNIDO/IS.408 Add. 1. (Sectoral Studies Series no. 5).	Second
Second	United Nations. Development of Airborne Equipment to Inten-	Third

	sify World Food Production. New York; United Nations, 1981. 281p.	
Second	United States Committee on Irrigation and Drainage. Design Practice of Open Drainage Channels in an Agricultural Land Drainage System: A World Wide Survey, 1984. 343p.	First
Second	United States. Agricultural Research Service. Field Manual for Research in Agricultural Hydrology. Washington, D.C.; Soil and Water Conservation Research Division, 1962. 215p. (U.S.D.A. Agriculture Handbook 224)	
Second	United States. Bureau of Reclamation. Design of Small Dams. 3d ed. Denver, U.S. Bureau of Reclamation, 1987. 860p. (1st ed., 1960. 611p.)	Second
First	United States. Bureau of Reclamation. Drainage Manual: A Guide to Integrating Plant, Soil, and Water Relationships for Drainage of Irrigated Lands. Washington, D.C., 1978. 286p.	First
Second	United States. Bureau of Reclamation. Ground-Water Manual; A Guide for the Investigation, Development, and Management of Ground-Water Resources. Washington, D.C.; U.S. Dept. of the Interior, Government Printing Office, 1981. 480p. (1st ed., 1977. 477p.)	Second
Third	United States. Department of Agriculture. Soil Survey Staff. Soil Taxonomy: A Basic System of Soil Classification for Making and Interpreting Soil Surveys. Washington, D.C.; U.S.D.A., Soil Conservation Service, U.S. Govt. Printing Office, 1975. 754p.	Third
Third	United States. Department of Agriculture. Soil Survey Staff. Soil Survey Manual. Washington, D.C.; U.S. Government Printing Office, 1951. 503p. (U.S.D.A. Agriculture Handbook no. 18)(Revision and enlargement of U.S. Dept. of Agriculture Misc. Pub. 274, 1937.)	
Second	United States. Office of Technology Assessment. Energy from Biological Processes. Washington, D.C.; Congress of the United States, Office of Technology Assessment, 1980. 195p. (for sale by the Superintendent of Docs., U.S. Government Printing Office).	
Second	United States. Soil Conservation Service. Agricultural Waste Management Field Manual. Washington. D.C.; U.S. Dept. of Agriculture, Soil Conservation Service, 1975. 366p.	
First	United States. Soil Conservation Service. Drainage of Agricultural Land; A Practical Handbook for Planning, Design, Construction and Maintenance of Agricultural Drainage Systems. Port Washington, New York; Water Information Center, 1973. 430p.	First
First	United States. Soil Conservation Service. Engineering Field Manual for Conservation Practices. Washington, D.C.; U.S. Dept. of Commerce, National Technical Information Service, 1984. 1074p. (1st ed., 1969; 2d ed., 1975. 1024p.)	First
Second	United States. Soil Conservation Service. Engineering Practice Standards: Part 1, Engineering Conservation Practices (National Engineering Handbook Section 2, looseleaf); Irrigation (Soil Conservation Service National Engineering Handbook	First

DC	Section 15, 476p.) Section 3, Sedimentation, 1971. Section 4, Hydrology, 1954. Section 5, Hydraulics, 1954. Section 18, Ground Water, 1968. Washington, D.C.; U.S. Government Printing Office.	TW
	Uphoff, Norman. Getting the Process Right: Improving Irrigation Management with Farmer Participation. Boulder, Colo.; Westview Press, 1986. 200p.	First
V		
	Valdes, Alberto, Grant M. Scobie, and John L. Dillon, eds. International Conference on Economic Analysis in the Design of New Technology, for Small Farmers. Economics and the Design of Small-Farmer Technology. Ames, Iowa; Iowa State University Press and Centro Internacional de Agricultura Tropical, 1979. 211p. (Available from Books on Demand.)	Second
Third	Valentine, Fredrick A., ed. Forest and Crop Biotechnology: Progress and Prospects; Papers from a Colloquium, SUNY College of Environmental Science and Forestry, Syracuse, 1985. New York; Springer-Verlag, 1988. 466p.	
Second	Van Elderen, E. Scheduling Farm Operations: A Simulation Model. Wageningen, Netherlands; Pudoc, 1987. 226p.	
Third	Van Hemert, P. A., H. L. M. Lelieveld and J. W. M. la Riviere. Biotechnology in Developing Countries: A Symposium, Delft, October, 1982. Delft; Delft University Press, 1983. 158p.	Third
Second	Van Hoorn, J. W., ed. Agrohydrology—Recent Developments; Proceedings of the Symposium Agrohydrology at the International Agricultural Centre IAC Wageningen, the Netherlands, 1987. Amsterdam and New York; Elsevier, 1988. 550p. (Distributors for the U.S. and Canada, Elsevier Science Pub.)	Second
First	Van Schilfgaarde, Jan, ed. Drainage for Agriculture. Madison, Wisc.; American Society of Agronomy, 1974. 700p.	First
Second	VandenBerg, Glen E. Agricultural Sensors. St. Joseph, Mich.; American Society of Agricultural Engineers, 1988. 81p. (ASAE Publication 09-88)	
Second	Vanoni, Vito A., ed. Sedimentation Engineering. New York; American Society of Civil Engineers, Task Committee for the Preparation of the Manual on Sedimentation, 1975. 745p. (Manuals and Reports on Engineering Practice, no. 54)	
Second	Vogt, Frederick, ed. Energy Conservation and Use of Renewable Energies in the Bio industries; Proceedings of the International Seminar on Energy Conservation and the Use of Solar and other Renewable Energies in Agriculture, Horticulture, and Fishculture, at the Polytechnic of Central London, Sept. 1980. Oxford and New York; Pergamon Press, 1981. 574p.	
Third	Volunteers in Technical Assistance. Manual for Rural Water Supply. Arlington, Virg.; Volunteers in Technical Assistance. 175p.	First

W

	Wade, J. E. and E. W. Hewson. Wind Power Prospecting: A Guide to Biological Indicators. WindBooks, 1984. 110p. (Repr. of 1981 ed.)	Second
	Wade, Robert. Irrigation and Agricultural Politics in South Korea. Boulder, Colo; Westview Press, 1982. 160p.	Third
	Wakeman, Truman J. and V. L. McCoy. The Farm Shop. New York; MacMillan Co., 1960. 597p.	Second
Second	Wakeman, Truman J. Modern Agricultural Mechanics. Danville, Ill.; Interstate Printers and Publs., Inc., 1977. 503p.	Second
Second	Walker, Wynn R. and Gaylord V. Skogerboe. Surface Irrigation: Theory and Practice. Englewood Cliffs, New Jersey; Prentice-Hall, Inc., 1987. 386p.	First
Second	Wang, Jaw-Kai and Ross E. Hagan. Irrigated Rice Production Systems: Design Procedures. Boulder, Colo.; Westview Press, 1981. 300p.	Second
Third	Ward, Shane M. Computer Modelling in Agricultural Mechanization. Dublin; Book Press, 1985.	
Second	Warne, D. F. Wind Power Equipment. London; Methuen, Inc., 1983. 220p.	
Second	Warnecke, H. J., ed. Agricultural Electronics—1983 and Beyond. Vol. I: Field Equipment, Irrigation and Drainage. Vol. II: Controlled Environments, Livestock Production Systems, Materials Handling and Processing. Proceedings of a National Conference, Chicago, 1983. St. Joseph, Mich.; American Society of Agricultural Engineers, 1984. 2 vols., 798p. (ASAE Pub. 8-84 and 9-84)	
Third	Warnecke, H. J., ed. Knowledge Based Systems in Agriculture: Prospects for Application. Proceedings of the 2d International DLG-Congress, Frankfurt, 1988. Frankfurt am Main; Deutsche Landwirtschafts-Gesellschaft (DLG), 1988. 546p.	
Second	Weber, Edward J., et al. Cassava Harvesting and Processing ... ICTA Conference ... Colombia, 1978. Ottawa, Canada; International Development Research Centre, 1978. 84p.	Second
Second	Weisz, Paul B. and John F. Marshall. Fuels from Biomass: A Critical Analysis of Technology and Economics. New York; M. Dekker, 1980. 119p.	
Third	Weller, John B. Farm Buildings. Vol. 1: Techniques—Design—Profit. Vol. 2: Structural Techniques and Materials. London; Crosby Lockwood, 1965, 1972. 2 vols.	
Third	Wells, G. D. et al. Small Farms—Livestock Buildings and Equipment. Ames, Iowa; Midwest Plan Service, and Ithaca, New York; Northeast Regional Agricultural Engineering Service, 1984. 85p.	Second
	Welte, E., ed. Water and Fertilizer Use for Food Production in Arid and Semi-Arid Zones. Symposium Proceedings. Gottingen, West Germany; Centre International des Engrais Chimiques, 1979. 357p.	Second
Second	Wesseling, J., editor. Proceedings of the International Drainage Workshop, Wageningen, The Netherlands, 1978. Wag-	

DC	eningen; International Institute for Land Reclamation and Improvement, 1979. 731p. (International Institute for Land Reclamation and Improvement Publication no. 25)	TW
Second	Westerman, R.L., ed. Soil Testing and Plant Analysis. 3d ed. Madison, Wis.; Soil Science Society of America, 1990. 784p. (SSSA Book Series no. 3 (1st ed., 1967)	
Third	Wexler, Arnold, ed. Humidity and Moisture: Measurement and Control in Science and Industry; Proceedings of an International Symposium on Humidity and Moisture. New York; Reinhold Publishing Corp., 1965. 4 vols.	
Second	Wheaton, Frederick W. Aquacultural Engineering. New York; John Wiley & Sons, Inc., 1977. 708p.	Third
Third	Whitaker, James H. Agricultural Buildings and Structures. Reston, Virg.; Reston Pub. Co., 1979. 530p.	Second
Second	White, G. F., ed. Environmental Effects of Arid Land Irrigation in Developing Countries. Paris; Unesco, 1978. 67p.	First
Third	White, L. P. Aerial Photography and Remote Sensing for Soil Survey. Oxford; Clarendon Press, 1977. 104p.	Second
Third	Whyte, William F., and Damon Boynton, eds. Higher-Yielding Human Systems for Agriculture. Ithaca, New York; Cornell University Press, 1983. 342p.	Second
	Wickes, J. A., ed. Consequences of Small-Farm Mechanization. Papers from a Joint Agricultural Development Council/ International Rice Research Institute Conference. Los Banos, Philippines; International Rice Research Institute, 1983. 184p.	Second
Second	Wickham, T., ed. Irrigation Management: Research from Southeast Asia. Bangkok, Thailand; Agricultural Development Council, 1985. 248p.	Second
Second	Wilkinson, Robert H. and Oscar A. Braunbeck. Elements of Agricultural Machinery, Vol. 1. (In English and Spanish). Rome; Food and Agriculture Organization, 1977. 306p. (FAO Agricultural Services Bulletins: no. 12, Supp. 2) (Available from Unipub)	
	Wilkinson, Robert H. and Oscar A. Braunbeck. Elements of Agricultural Machinery, Vol. 1. (In English and Spanish). Rome; Food and Agriculture Organization, 1977. 306p. (FAO Agricultural Services Bulletins no. 12, Supp. 2). (Available from Unipub)	Second
Third	Williams, J. Richard. Solar Energy: Technology and Applications. 2d ed. Ann Arbor; Ann Arbor Science Publishers, 1977. 176p. (1st ed., 1974. 120p.)	
Second	Williams, Phil, and Karl H. Norris, eds. Near-Infrared Technology in the Agricultural and Food Industries. St. Paul, Minn.; American Association of Cereal Chemists, 1987. 330p.	
Second	Wills, R. H. H., T. H. Lee, D. Graham, W. B. McGlasson, and E. G. Hall. Postharvest: An Introduction to the Physiology and Handling of Fruit and Vegetables. 3d ed. New York; Van Nostrand, 1989. 174p. (1st and 2d eds., Westport, Conn.; St. Albans.)	Second
Second	Winter, E. J. Water, Soil and the Plant. London; Macmillan, 1974. 141p.	

Third	Wise, Donald L., ed. Fuel Gas Production from Biomass. Boca Raton, Flor.; CRC Press, 1981. 2 vols.	
	Wit, Theodorus. P. M. de. The Wageningen Rice Project in Surinam; A Study on the Development of a Mechanized Rice Farming Project in the Wet Tropics. Gravenhage; Mouton, 1960. 293p.	Third
Second	Withers, Bruce and Stanley Vipond. Irrigation: Design and Practice. 2d ed. Ithaca, New York; Cornell University Press, 1980. 306p.	First
Second	Witney, Brian. Choosing and Using Farm Machines. Harlow, Essex, England; Longman Scientific and Technical; New York; Wiley, 1988. 412p.	Second
Second	Wittmuss, Howard and Wesley F. Buchele. Conservation Tillage and Water Quality Control. Lincoln, Neb.; Water Quality Dept., University of Nebraska, 1972.	
Second	Woolrich, Willis R. Handbook of Refrigerating Engineering. 4th ed. Westport, Conn.; AVI Publishing Co., 1965–1966. 2 vols. (460p., 434p.)	
	Workshop on Appropriate Agricultural Technology, Dacca, Bangladesh, 1975. Proceedings . . . Dacca; Bangladesh Agricultural Research Council, 1976. 312p.	Second
	Workshop on Irrigated Agriculture in Africa, Harare, April 1988. Atelier de Travail sur l'Agriculture Irriquee en Afrique. Harare, 1988. 12 pieces.	Third
Second	Worthington, E. Barton, ed. Arid Land Irrigation in Developing Countries: Environmental Problems and Effects. Based on the International Symposium, Egypt, 1976. Oxford and New York; Pergamon Press, 1977. 463p.	Second
Second	Wray, Warren K. Measuring Engineering Properties of Soil. Englewood Cliffs, N.J.; Prentice-Hall, Inc., 1986. 276p.	
Second	Wright, Forrest B. Rural Water Supply and Sanitation. 3d ed. Huntington, New York; Robert E. Krieger Pub. Co., Inc., 1977. 305p. (1st ed., New York; Wiley; London; Chapman and Hall, 1939. 288p.; 2d ed., New York; Wiley, 1956. 347p.)	Second
Second	Wright, P. A. Old Farm Tractors. Newton Abbot, England; David and Charles, 1972. 77p.	
	Wu, Shou I. Nung Yeh Chi Hsieh Hsueh. (Study of Agricultural Engineering.) Chung-kuo Nung Yeh Chi Hsieh Ch'u Pan She: Hsin Hua Shu Tien Pei-ching Fa Hsing So Fa Hsing, 1987. 2 Vols.	Third
	Wu, T. H. Soil Mechanics. 2d ed. Boston, Mass.; Allyn and Bacon, 1976. 440p. (Reprinted 1982).	Third
	Wyatt, Alan and Sam Baldwin. Wind Energy Activities in Africa. Volunteers in Technical Assistance, 1982. 56p.	Second

Y

Third	Yamawaki, Sanpei. Ringyo Kikaigaku (Forest Machinery). Tokyo; Asakura Shoten, 1980. 253p. (In Japanese)	
	Yamazaki, Fujio. Agricultural Engineering (In Japanese). 1971–72. 2 vols.	Third

	Yamazaki, Fujio. Paddy Field Engineering. Translated from the Japanese. Bangkok; Asian Institute of Technology, 1988. 425p.	Third
Second	Yaron, Bruno, E. Danfors, and Y. Vaadia. Arid Zone Irrigation. Berlin and New York; Springer Verlag, 1973. 434p.	Second
Third	Yen, T. F. Recycling and Disposal of Solid Wastes—Industrial, Agricultural, Domestic. Ann Arbor, Mich.; Ann Arbor Science, 1974. 372p.	
	Yeomans, P. A. Water for Every Farm. Sydney and Melbourne; Murray, 1965. 223p.	Second
	Yokata, Hiroshi. Agricultural Mechanization in Selected Asian Countries. Unipub, 1985. 176p.	Second
First	Yong, Raymond N., Ezzat A. Fattah, and Nicolas Skiadas, eds. Vehicle Traction Mechanics. Amsterdam and New York; Elsevier Science Publishing Company, Inc., 1984. 307p. (Developments in Agricultural Engineering Ser. Vol. 3).	First
	Yudelman, M., G. Butler, and R. Banerji. Technological Change in Agriculture and Employment in Developing Countries. Paris; Development Centre of the Organisation for Economic Co-operation and Development, 1971. 204p.	Second

Z

Second	Zachar, D. Soil Erosion. Translation of Erozia Pody. Amsterdam and New York; Elsevier, 1982. 547p.	First
Zhalnin,	Eduard V., A. S. Mnatsakanov, and A. I. Filippov. Harvesting of Grain and Pulse Crops. Translated from Russian. Arlington, VA; Joint Publications Research Service, 1978. 82p.	Second
Second	Zhang, Dejun and Chunhui He, editors. International Conference on Agricultural Systems Engineering; 1st, 1987, Chang-Chun, Shih, China. Proceedings . . . Beijing, China; China Machine Press, 1987. 624p.	Second
Second	Zimmerman, Josef D. Irrigation. New York; Wiley, 1966. 516p.	First
Third	Zot'ev, Aleksei I. Sovremennye Sredstva Razmola Zerna. (Modern Means of Grinding Grain.) Moscow; Kolos, 1982. 135p.	Third

D. The Nature of the Core Lists

The 1950–90 titles in the Core monographic lists are characterized in Table 14.1.

The strong showing of the commercial presses in agricultural engineering does not follow the pattern of citations in the literature analysis. This means, of course, that the commercial publications are the most valuable in academic instruction and research, with the citations in the analysis spread very broadly. It must be noted in this regard the top two publishers are not commercial presses; this distinction belongs to the American Society of Agricultural Engineerings and the Food and Agricultural Organization.

Table 14.1. Characteristics of monographs

	Developed countries list (n = 795)	Third world list (n = 623)	Titles common to both lists (n = 386)
Types of publishers			
Commercial presses	50.6%	46.5%	51.5%
Governments (incl. FAO & UN)	22.1	27.2	23.8
Independent organizations (societies, etc.)	20.0	15.3	16.1
Universities (presses, dept., institutes)	7.3	11.0	8.6
Place of publication			
United States	56.1%	45.8%	54.0%
United Kingdom	15.1	12.1	11.7
Italy	7.5	10.7	11.4
Netherlands	4.5	3.8	4.4
Germanies	3.9	4.1	3.4
India	2.4	5.0	4.2
Non-English publications	5.5%	5.4%	2.9%
Non-English publications and translations into English	8.5	7.4	5.5
Median year of publications	1980.9	1980.1	1980.4
Primary publishers			
American Society of Agricultural Engineers	7.9%	4.8%	5.5%
Food and Agricultural Organization	7.2	10.9	11.4
AVI Publishers	5.1	5.6	8.1
Wiley	4.1	3.6	4.2
Elsevier	3.6	3.0	3.6

The place of publication is of limited significance with the 50% commercial press titles because of the internationalization of so many publishing houses. Source of publication was determined as the first cited city or country in the imprint of the publication. For example, Elsevier generally uses its Netherlands home first but not always. Oxford University Press publishes from Oxford to Kuala Lumpur.

E. The Best Monographs

The following two lists of the top-ranking monographs reaches through the top seventeen in the developed countries list and the top twenty-two in the Third World list. They are provided here for insights into the closeness and diversity of the two communities in agricultural engineering.

Top Developed Country Agricultural Engineering Monographs

Rank

1   Donnell Hunt. *Farm Power and Machinery Management*. 8th ed. Ames, Iowa; Iowa State University Press, 1983. 352p.

2   John B. Liljedahl, Paul K. Turnquist, David W. Smith, and Makoto Hoki. *Tractors and Their Power Units*. 4th ed. New York; Van Nostrand Reinhold, 1989. 463p. (1st and 2d eds. under principal authorship of E. L. Barger.)

3–4   Donald B. Brooker, Fred W. Bakker-Arkema, and Carl W. Hall. *Drying Cereal Grains*. Westport, Conn.; AVI Publishing Co., 1974. 265p.
Robert A. Kepner, Roy Bainer and E. L. Barger. *Principles of Farm Machinery*. 3d ed. Westport, Conn.; AVI Publishing Company, 1978. 527p. (1st ed. with Bainer as primary author, 1955.)

5   Nuri N. Mohsenin. *Physical Properties of Plant and Animal Materials: Structure, Physical Characteristics and Mechanical Properties*. 2d ed., revised and updated. New York; Gordon and Breach Science Publishers, 1986. 891p. (The predecessor ed. has a briefer title and was issued by the Department of Agricultural Engineering, Pennsylvania State University, 1966–68.)

6   John Deere and Co. *The Operation, Care and Repair of Farm Machinery*. 28th ed. Moline, Ill.; John Deere, 1957. 279p.

7   Robert J. Gustafson. *Fundamentals of Electricity for Agriculture*. 2d ed. St. Joseph, Mich.; American Society of Agricultural Engineers, 1988. 411p. (*ASAE Textbook* no. 2). (1st ed. by AVI Pub. Co., Westport, Conn., 1980.)

8–9–10   Claude Culpin. *Farm Machinery*. 12th ed. Oxford; Blackwell Scientific, 1992. 480p. (1st–9th eds., 1938–1976 by Lockwood Publ.)
Glenn O. Schwab, Delmar D. Fangmeier, William J. Elliot and Richard K. Frevert. *Soil and Water Conservation Engineering*. 4th ed. New York; John Wiley and Sons, 1992. 544p. (Early eds. with others; 3d ed. under title: *Elementary Soil and Water Engineering*.)
B. A. Stout. *Handbook of Energy for World Agriculture*. London and New York; Elsevier, Applied Science, 1989. 506p.

11   Robert M. Hagan, Howard R. Haise, and T. W. Edminster, eds. *Irrigation of Agricultural Lands*. Madison, Wisc.; American Society of Agronomy, 1967. 1180p.

12–17   Arlen D. Brown and R. Mack Strickland. *Tractor and Small Engine Maintenance*. 5th ed. Danville, Ill.; Interstate Printers and Publishers, Inc., 1983. 350p. (1st ed. by I. G. Morrison; 2d–3d eds. by A. D. Brown and I. G. Morrison, early editions entitled: *Farm Tractor Maintenance*.)
Stanley E. Charm. *The Fundamentals of Food Engineering*. 3d ed. Westport, Conn.; AVI Publishing Co., 1978. 646p. (1st ed., 1963.)
L. L. Christianson and Roger P. Rohrbach. *Design in Agricultural Engineering*. St. Joseph, Mich.; American Society of Agricultural Engineers, 1986. 310p. (*ASAE Textbook* no. 1)
P. J. Dieleman and D. B. Trafford. *Drainage Testing*. Rome; Food and Agriculture Organization, 1976. 172p. (*FAO Irrigation and Drainage Papers* no. 28).
Vaughn E. Hansen, Glen E. Stringham, and Orson W. Israelsen. *Irrigation Principles and Practices*. 5th ed. New York; John Wiley and Sons, Inc., 1984. 450p. (1st ed. by Israelsen and Stringham, 1932.)
Claude H. Pair, editor-in-chief, with others. *Irrigation; Formerly Sprinkler Irrigation*. Compiled and edited by the Textbook Re-editing Committee, Irri-

gation Association. 5th ed. Silver Spring, Md.; Irrigation Association, 1983. 686p. (1st ed. edited by A. W. McCulloch and J. F. Schrunk, 1955; 2d ed. by G. O. Woodward and 3d–5th eds. by C. H. Pair with others).

Top Third World Agricultural Engineering Monographs

Rank

1	Donnell Hunt. *Farm Power and Machinery Management*. 8th ed. Ames, Iowa; Iowa State University Press, 1983. 352p.
2–3–4	L. J. Booher. *Surface Irrigation*. Rome; Food and Agriculture Organization, 1974. 160p. (*FAO Agricultural Development Paper* no. 95).
	Hans J. Hopfen. *Farm Implements for Arid and Tropical Regions*. Rev. ed. Rome; Food and Agriculture Organization, 1969. 159p. (Reprinted in 1981). (*FAO Agricultural Development Paper* no. 67 and *FAO Agricultural Development Paper* no. 91). (1st published in 1960.)
	John B. Liljedahl, Paul K. Turnquist, David W. Smith, and Makoto Hoki. *Tractors and Their Power Units*. 4th ed. New York; Van Nostrand Reinhold, 1989. 463p. (1st–2d eds. under principal authorship of E. L. Barger.)
5–6	C. B. Richey, Paul Jacobson, and Carl W. Hall, eds. *Agricultural Engineers' Handbook*. New York; McGraw-Hill, 1961. 880p.
	Glenn O. Schwab, Delmar D. Fangmeier, William J. Elliot and Richard K. Frevert. *Soil and Water Conservation Engineering*. 4th ed. New York; John Wiley and Sons, 1992. 544p. (Early eds. with others; 3d ed. under title: *Elementary Soil and Water Engineering*.)
7	Marvin E. Jensen, ed. *Design and Operation of Farm Irrigation Systems*. St. Joseph, Mich.; American Society of Agricultural Engineers, 1981. 829p. (*ASAE Monograph* no. 3.)
8	Donald B. Brooker, Fred W. Bakker-Arkema, and Carl W. Hall. *Drying Cereal Grains*. Westport, Conn.; AVI Publishing Co., 1974. 265p.
9–15	Food and Agriculture Organization. *Multifarm Use of Agricultural Machinery*. Rome; FAO, 1985. 63p. (*FAO Agricultural Services Bulletin* no. 17). (Earlier ed. by H. Lonnemark, 1967, as *FAO Agricultural Development Paper* no. 85.)
	International Institute for Land Reclamation and Improvement, Wageningen, Netherlands. *Drainage Principles and Applications*. Wageningen; IILRI, 1972–80. 4 vols.
	Robert A. Kepner, Roy Bainer, and E. L. Barger. *Principles of Farm Machinery*. 3d ed. Westport, Conn.; AVI Publishing Company, 1978. 527p. (1st ed., Roy Bainer was first author.)
	J. R. Landon, ed. *Booker Tropical Soil Manual: A Handbook for Soil Survey and Agricultural Land Evaluation in the Tropics and Subtropics*. London; Booker Agriculture International Ltd., and New York; Longman, 1984. 320p.
	James N. Luthin, ed. *Drainage of Agricultural Lands*. Madison, Wisc.; American Society of Agronomy, 1957. 620p. (*ASA Monograph* no. 7.)
	I. Shainberg and J. Shalhevet, eds. *Soil Salinity under Irrigation: Processes and Management*. Berlin and New York; Springer-Verlag New York, Inc., 1984. 349p.
	B. A. Stout. *Handbook of Energy for World Agriculture*. London and New York; Elsevier, Applied Science, 1989. 506p.
16–22	W. C. Burrows, ed. with others. *Proceedings of the International Conference*

on Mechanized Dryland Farming held in Moline, Ill. and Great Falls, Mont.; 1969. Moline, Ill.; Deere and Co., 1970. 344p.

Daniel Hillel, ed. *Advances in Irrigation.* New York; Academic Press, Inc., 1982–1987. 4 vols.

John Deere and Co. *The Operation, Care and Repair of Farm Machinery.* 28th ed. Moline, Ill.; John Deere, 1957. 279p.

Nuri N. Mohsenin. *Physical Properties of Plant and Animal Materials: Structure, Physical Characteristics and Mechanical Properties.* 2d ed., revised and updated. New York; Gordon and Breach Science Publishers, 1986. 891p. (The predecessor ed. has a briefer title, and was issued by the Department of Agricultural Engineering, Pennsylvania State University, 1966–68.)

A. L. Page, R. H. Miller, and D. R. Keeney, eds. *Methods of Soil Analysis.* 2d ed. Madison, Wisc.; American Society of Agronomy and Soil Science Society of America, 1982–86. 2 vols. (1st ed. in two volumes edited by Charles A. Black and others.)

Tools for Agriculture; A Buyer's Guide to Appropriate Equipment. 3d ed. Introduction by Ian Carruthers; edited by a team led by Patrick Mulvany. London; Intermediate Technology Publications, 1985. 264p. (1st ed. by Anthony Gater, Tim Meller, and John Warsanny. *Tools for Progress.* London; 1968.)

Jan van Schilfgaarde, editor. *Drainage for Agriculture.* Madison, Wisc.; American Society of Agronomy, 1974. 700p.

These rankings were made by poling two distinct groups of agricultural engineers. The groups did not see each others lists and there was no interchange of discussion among these widely dispersed individuals. Yet the similarity of their choice of the top fifteen titles is remarkable, and conclusive.

Don Hunt, long-time professor at the University of Illinois, had top ranking in both communities. The same second place distinction goes to Liljedahl and colleagues. Straight down the list the parallel is evident.

	Ranking in developed countries	Ranking in Third World
Brooker, et al.	3–4	8
Kepner, et al.	3–4	9–15
Mohsenin	5	16–22
John Deere Operations	6	16–22
Schwab and Frevert	8–10	5–6
Stout	8–10	9–15

Eight of the seventeen and twenty-two in respective lists are in both lists; a strong indication of agreement.

Two subject areas accounted for 75% of the titles: Farm Equipment and Machinery was tops in both communities, with Drainage and Irrigation a close second in both instances. The latter subject was of greater interest in the developing countries than in the developed. Only one book on energy and agriculture made both lists, a 1989 book by the well-known B. A. Stout, which because of its recentness must have been partially measured on the basis of Stout's scholarly reputation and earlier writings.

All other recently published books are newer editions of a title. In fact, thirteen of the seventeen in the developed countries' top monographs are new editions, and ten of twenty-two in the Third World. The most unique title is the John Deere publication on operation of equipment in its 28th edition in 1957. It appears that getting into a second edition in agricultural engineering may guarantee a great number of years of updating and reissue. Three of the titles which have continued through several editions have little representation of authors or editors from earlier editions. The titles go on even if the original authors don't. It should be noted that ¾ of the books are authored or edited by more than one person.

Commercial publishers account for 79% of the developed countries top seventeen titles, but a much lower 45% in the Third World group. This is spread among twelve different commercial publishers with only AVI registering more than one title. The other heavy publishers concentration is that of the Food and Agriculture Organization and the American Society of Agronomy. The American Society of Agricultural Engineers also scored well as a publisher.

15. Primary Journals and Reports

WALLACE C. OLSEN
Mann Library, Cornell University

A. Source Documents and Methodology

The twenty-four monographs and the twenty-nine journal articles used as source documents for the citation analysis (see Sources of Citations in Agricultural Engineering, Chapter 14) were also used to obtain data on journals, serials, and report series. The same methods, definitions, and caveats apply concerning the journal titles as outlined for monographs in Section B of Chapter 14. All journal titles cited in the source documents were recorded and tabulated by title and date of publication, as were the report series.

The handling of proceedings volumes requires clarification. Of the numerous proceedings identified, approximately 85% were tallied and evaluated as monographs because they had distinctive titles and different editors and concentrated on a specialized aspect of some area of agricultural engineering. Proceedings volumes were counted as journals when they represent the continuing deliberations of an organization or society, and with no varying subject focus or title other than Proceedings. An example are the *Proceedings of the American Society of Civil Engineers*, which run consistently under the same title for years and cover happenings at annual meetings, including technical papers.

B. Literature Cited in Journals

Figure 14.1 indicates that 58.1% of the references in the analyzed documents were to journal articles. This is close to the 57.4% journal items in the CABI database for agricultural engineering. As mentioned earlier, any registration under 75% of journal literature in a subject field is unusual in the sciences, including agriculture. Report series which are heavily used in engineering and which were counted as monographs account for some of this difference, along with the fact that both instructional and research literature were analyzed. The 75% rule of thumb generally applies to research literature.

C. Journal Literature Findings

The Sources of Citations in Agricultural Engineering (Chapter 14) documents yielded 8,184 citations of which 4,755 (58.1%) referred to journal articles. Those 4,700 references were scattered among 590 distinctive journals. The monographs analyzed yielded the greatest scatter, with 548 journal titles, while the journal articles identified only 299 journal or annual titles. It may come as a shock to agricultural engineers that citations in their field reach to at least 590 journals. These journals are the thousand connections of agricultural engineering to the full range of agricultural subjects mentioned at the beginning of Chapter 13. This scattering pattern is very common, particularly in the application sciences with agriculture the primary example. A listing of all 590 titles is provided in Appendix A.

Counts were made of the number of times each journal was cited; 79% (467) were cited no more than once each. Approximately 75% of the citations are concentrated in 125 journal titles; the top fifty journals account for 37.8% of the citations. Figure 15.1 demonstrates the influence of the top two titles and the singleton citings. Table 15.1 lists the fifty top ranked journals resulting from the citation analysis of 4,755 references to journals and annuals.

A quick glance at Table 15.1 confirms the fact that agricultural engineers deal with the biological, soil, water and plant journals almost as much as those devoted exclusively to agricultural engineering. These journals have information and basic data which bear directly on the problems agricultural engineers are solving; they must use this literature. A great number of the top journals classify outside basic agricultural engineering on the basis of their general contents. Those clearly not agricultural engineering classify in this manner:

- 11 General Agriculture or Science
- 10 Air and Water
- 6 Plants
- 5 Animals
- 3 Soils

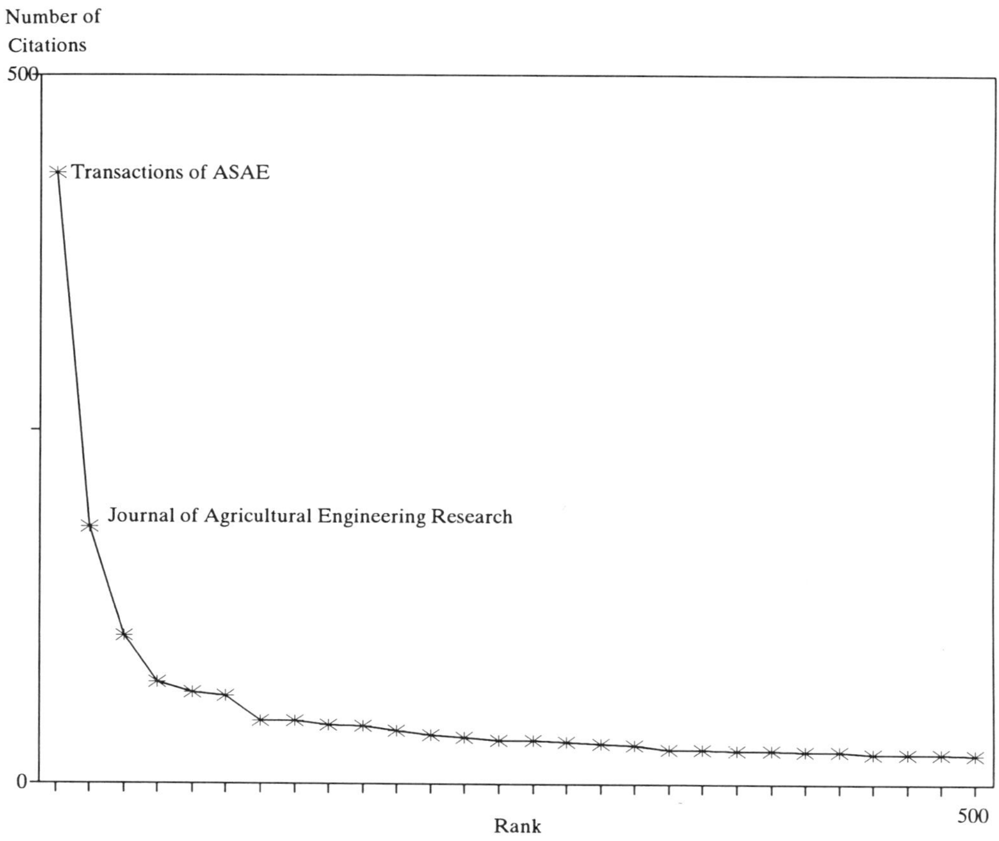

Figure 15.1. Distribution of journal titles and annuals.

This indicates that 68% of the titles are in related fields with which agricultural engineers work or which they support.

Table 15.1 provides other indicators or rankings which help explain the value of individual journals. Columns A and B identify the journals devoted exclusively to agricultural engineering. Supporting or related titles as *Agronomy Journal* and *Journal of Animal Science* (overall rank 3 and 4) are not included in these columns. Column A is a ranking of agricultural engineering titles without the supporting titles.

Column B covers the same titles in A but ranked on the basis of the number of times cited, 1980–88, in *Science Citation Index*.[1] *SCI* does not

---

1. *Science Citation Index Citation Report.* Ranking was on the average of the sums of the nine years for each title.

have citing information which is statistically valid for several titles, for example *The Agricultural Engineer* (overall ranking 7–8) is unranked in Column B. Similarly *Grundlagen der Landtechnik* has no data in *SCI*. Considering the scope of the titles included in the *SCI* database, it is rather surprising that thirteen journals had adequate data. A special case has been made for the *Journal of Soil and Water Conservation* ranked overall beyond the fiftieth journal. This important journal could be considered outside the general scope of agricultural engineering journals as was *Agronomy Journal* in Columns A and B. However, upon examination of the contents and authorship of the journal articles, 50% or more of the journal is clearly within the scope of agricultural engineering. Because of its high rank in Column B, the title is included as an illustration of two points:

(1) Titles at the periphery of agricultural engineering demonstrate the scattering across an array of agricultural subjects; this requires careful analysis;
(2) For this reason, the very high ranking of 3 in *SCI* citings (Column B) must be read as including probably no more than half of its citations to agricultural engineering. It is not easily possible to sort for this difference on this or the other important titles excluded in Columns A and B.

The same is true for *Biotechnology and Bioengineering* which is included here as a basic agricultural engineering title. Biotechnology and bioengineering are disputed territories within academia since they cover a variety of disciplines, with this title reaching heavily into biology. The title has been included here for purposes of comparison because the field is rapidly expanding and because this title ranked high in the project's citation analysis of agricultural engineering literature. Because the field is popular today and broad in scope, the journal is cited widely by disciplines close to agricultural engineering representing a new research front. This is why it outranks the *Transactions of the ASAE* in the times cited in the *SCI*.

Column C includes the same primary agricultural engineering journals (Columns A and B) ranked by the number of substantive articles published in the journal. This methodology is often used to identify weighty journals on the basis of number of articles or pages published. It is not an accurate measure of value. However, positive correlation with other indicators give credence to the value of a journal. Publishing data were gathered from the AGRICOLA and CABI databases and when these data did not agree, the higher number was used. Totals were pulled for 1980–88 or smaller subsets of years for these titles when data were not available for all years; yearly averages were computed which were used for ranking. The relative standing of the top fifteen journals in Columns A, B and C varies only slightly, a positive indication that these titles are of prime importance in agricultural engineering based partially on publishing output.

Table 15.1. Top fifty journals cited by rank

Overall ranking		Groups by rank				
		A	B	C	D	E
1	Transactions of the American Society of Agricultural Engineers	1	2	1	1	.32
2	Journal of Agricultural Engineering Research	2	4	4	13	.26
3	Agronomy Journal				2	.70
4	Journal of Animal Science				—	1.51
5	Water Resources Research				4	1.60
6	Journal of Irrigation and Drainage Engineering	3	5	10–11	3	.21
7–8	The Agricultural Engineer	4–5	—	13	18–19	.04
7–8	Agricultural Engineering	4–5	6	7	18–19	
9	Agricultural Wastes (ceased 1986)	6	7	5	8	.26
10	Grundlagen der Landtechnik	7	—	12		
11	Plant Physiology				—	2.81
12	Proceedings of the Soil Science Society of America			7	6	
13	Soil Science				—	.49
14–15	Journal of Terramechanics	8	12		12	.00
14–15	Science			25	9–10	16.46
16	Agricultural and Forest Meteorology (prior to 1984 as Agricultural Meteorology)				9–10	.60
17	Journal of Agricultural Science (Cambridge)				20	.50
18	Soil Science Society of America Journal			11	1.20	
19–20	Traktory i Sel'skhomashiny	9	—	2[a]		
19–20	Acta Horticulturae				—	—
21–22	Landtechnische Forschung (ceased 1976)	10	—		—	—
21–22	Agricultural Water Management	11	8–9	9	14	.28
23–24	Journal of Hydrology				15	.69
23–24	Journal of Soil Science				25	.74
25–27	Journal of the Water Pollution Control Federation				—	.78
25–27	Netherlands Journal of Agricultural Science				26	.39

25–27	Agricultural Mechanization in Asia, Africa and Latin America	12	—	6	16	—
28	Proceedings of the Royal Society of London—Biology				17	2.22
29–32	Biotechnology and Bioengineering	13–14	1	3	—	1.38
29–32	Soil and Tillage Research	13–14	8–9	16	—	.64
29–32	Poultry Science				—	.58
29–32	Journal of the Science of Food and Agriculture				—	.71
33	Journal of Dairy Science				—	.42
33–35	American Potato Journal				—	.62
34–35	Canadian Agricultural Engineering	15	11	10–11	—	.41
33–38	Irrigation Science	16–17	10	14–15	21	.31
36–38	Landtechnik	16–17	—	8	—	—
36–38	World Crops (ceased 1976)				22	—
39–41	Energy in Agriculture (ceased 1988)	18	13	23	23	.09
39–41	Water Pollution Control				—	.40
39–41	Journal of Hydraulic Engineering				—	.60
42–43	Chemical Engineering (N.Y.)				—	.38
42–43	Canadian Journal of Plant Science				—	.32
44–45	Nature				28	15.76
44–45	Proceedings of the National Academy of Sciences (U.S.)	19	14	24	—	10.03
46–48	Farm Building Progress				—	.06
46–48	Process Biochemistry				—	.87
46–48	World Animal Review				24	.08
49–51	Journal of the British Grasslands Society				—	—
49–51	Journal of Stored Products Research				—	.38
49–51	Power Farming	20	—	14–15	27	—
65–67	Journal of Soil and Water Conservation	28–29	3		—	.64

Column B: "—" means no data in *SCI Citation Report*; Column D: "—" means inadequate data for ranking; Column E: "—" indicates no impact factor in *SCI Citation Report*.

[a] *Traktory i . . .* is based on three years of data only from AGRICOLA and CABI.

Column D is a ranking of all the titles based on citation analysis of sixteen monograph source documents. By excluding citation counts from the analyzed journals, the influence of journals versus monographs is demonstrated. For example, the overall second-ranked *Journal of Agricultural Engineering Research* ranked 13th in Column D meaning that the analyzed journals more heavily cited the *Journal of Agricultural Engineering Research* than did the monographs. Limited conclusions can be made. For example, heavy citation of journals by journals is indicative of a research-oriented journal, and in this case *research* is in the title. Heavy journal-to-journal citing usually demonstrates more highly specialized subject matter in the sciences. The closeness of the rankings in Columns A and D also demonstrates a strong journal for both educational and research purposes.

Column E shows the impact factor of a journal as computed by *Science Citation Index* for 1988. The number is derived by dividing the number of times a journal is cited by the number of citations published in that journal. A high number indicates that a journal is cited by other journals far more times than it cites those journals; the larger the number the greater the impact the journal has. General science journals of high quality which include a diversity of science subjects will have a heavy impact as demonstrated by *Science* and *Nature* in this list. If the citations to a journal equal the number of citations in that journal, the factor is 1.0. Impact factors are influenced by several factors such as the length of time a journal has been published. The impact factor is a partial guide to the value of a journal on the basis of citation analysis.

Bradford, Goffman/Warren, and other adjustment techniques to overcome skews and influences including the use of impact factors were recently discussed by Brookes.[2] Some of these techniques were used to arrive at legitimate groupings and rankings in this chapter, but impact factors were provided in Table 15.1 only for information. The intertwining disciplines represented in journals such as *Biotechnology and Bioengineering* cause impact factors to overshadow the lower impact titles. Table 15.1 offers the opportunity to identify any glaring conflicts; other apparent inconsistencies are explainable.

D. Journals for the Third World

One aim of this study was to determine the agricultural engineering journals most valuable for the Third World as distinct from those for developed

---

2. Terrence A. Brookes, "Literature Core Zones Adjusted by Impact Factors," *Journal of Information Science* 16 (1990): 51–57.

countries. To accomplish this, a subset of journals with data was constructed from citations in the eight source monographs whose aim was to serve Third World needs. This information was then compared to the total journal information.

From comparisons, observations and accumulated data, these conclusions can be made:

(1) The Third World primary journals are an 80% match with the primary journals for the developed countries.
(2) There are few agricultural engineering journals published in the Third World, at either the continental, regional or country level. Most of the local or site specific literature is contained in two genres of publications:
   (a) The general agricultural literature of a country or region such as *Farming in Zambia* or in the engineering literature such as the *Journal of the Institute of Engineers* (India). The quantity of agricultural engineering literature is slight in either circumstance.
   (b) Agricultural engineering in many developing countries is in an early publication state often appearing in technical reports from international or consulting groups, or very limitedly from country governments or universities.

It may be some time before quality agricultural engineering literature which has application beyond a very limited locale will be published in other than these few identified journals. Additional journals may not be very swift in coming since *Agricultural Mechanization in Asia, Africa and Latin America* now is the primary vehicle for Third World agricultural engineering. The two most active areas of Third World agricultural engineering today are irrigation and mechanization. The following list includes the journals with statistically valid information.

The list distinguishes between those journals which are primarily or exclusively agricultural engineering, and those from agricultural disciplines which are used heavily by agricultural engineers in pursuit of their application or supporting efforts. Those in supporting disciplines are marked with a dagger (†).

E. Core Journal List

	Developed countries	Third World
†*Acta Horticulturae*. Vol. 1 (1963)+. The Hague; International Society for Horticultural Science.	X	
African Farming and Food Processing. Jan./Feb. 1984+. London; Alain Charles Publ. Ltd.		X
Agrartechnik: Landtechnische Fachzeitschrift. Vol. 23 (1973)+.	X	

Berlin; Verlag Technik. Monthly. (Formerly *Deutsche Agrartechink*)

†*Agricultural and Forest Meteorology.* Vol. 31 (Feb. 1984)+. Amsterdam; Elsevier Scientific Publisher. 4–5 times a year. (Formerly *Agricultural Meteorology*)	X	
The Agricultural Engineer; The Journal and Proceedings of the Institution of Agricultural Engineers. Vol. 27 (1972)+. Bedford, England; Institution of Agricultural Engineers. Quarterly.	X	X
Agricultural Engineering. Vol. 1 (Sept. 1920)+. St. Joseph, Mich.; American Society of Agricultural Engineers. 13 nos. a year since 1986.	X	X
Agricultural Engineering Australia. Vol. 1 (1970)+. Werribee, Victoria; The Agricultural Engineering Society of Australia. 2 nos. a year.	X	X
Agricultural Mechanization in Asia, Africa and Latin America: AMA. Vol. 12 (winter 1981)+. Tokyo; Farm Machinery Industrial Research Corporation. Quarterly. (Formerly *Agricultural Mechanization in Asia*)	X	X
Agricultural Water Management. Vol. 1 (Dec. 1976)+. Amsterdam; Elsevier Science Publishers. Quarterly.	X	
†*Agronomie.* Vol. 1 (1981)+. Paris; Institut National de la Recherche Agronomique. Monthly. (Formed by merger of three French journals.)	X	X
†*Agronomy Journal.* Vol. 1 (1907/09)+. Madison, Wisc.; American Society of Agronomy. Bi-monthly.	X	
†*American Potato Journal.* Vol. 3 (1926)+. Orono, Maine; Potato Association of America. Monthly.	X	
Applied Engineering in Agriculture. Vol. 1 (June 1985)+. St. Joseph, Mich.; American Society of Agricultural Engineers. 6 nos. a year.	X	
Appropriate Technology. Vol. 1 (1974)+. London; Intermediate Technology Publications Ltd. Quarterly.		X
Aquacultural Engineering. Vol. 1 (Jan. 1982)+. London; Applied Science Publishers. Quarterly.	X	
Biomass. Vol. 1 (Sept. 1981)+. Barking, England, Elsevier Applied Science Publishers Ltd. 12 nos. a year in three vols. a year since 1984.	X	
Biotechnology and Bioengineering. Vol. 4 (Mar. 1962)+. New York, J. Wiley. (Formerly *Journal of Biochemical and Microbiological Technology and Engineering*)	X	
Canadian Agricultural Engineering. Jan. 1959+. Ottawa, Ontario; Canadian Society of Agricultural Engineering. Semiannual.	X	X
†*Canadian Journal of Plant Science.* Vol. 37 (1957)+. Ottawa, Ontario; Agricultural Institute of Canada. Quarterly.	X	
†*Ceres.* Vol. 1 (1968)+. Rome; Food and Agriculture Organization. Bi-monthly.		X
†*Chemical Engineering.* Vol. 1 (1902)+. New York; McGraw-Hill.	X	
Farm and Power Equipment. St. Louis, MO, ADmore Inc. Monthly.	X	

†*Genio Rurale*. Vol. 11 (1948)+. Bologna, Italy; Gruppo Giornalistico Edagricole Svl. Monthly.	X	
Grundlagen der Landtechnik. No. 1 (1951)+. Dusseldorf, Germany; Verein Deutscher Ingenieure Verlag. Bi-monthly.	X	
ICID Bulletin. 1968/1969+. New Delhi; The International Commission on Irrigation and Drainage.		X
†*Indian Farming*. New series Vol. 1 (1951)+. New Delhi; Indian Council of Agricultural Research. Monthly.		X
Irrigation and Drainage Systems. Vol. 1 (1986)+. Dordrecht, Netherlands, and Boston; M. Nijhoff. 3 nos. a year.	X	X
Irrigation Science. Vol. 1 (1978)+. Berlin and New York; Springer International. Quarterly.	X	X
Journal of Agricultural Engineering. New Delhi; Indian Society of Agricultural Engineers. Quarterly.		X
Journal of Agricultural Engineering Research. Vol. 1 (1956)+. London; Academic Press, Inc. Quarterly. (For British Society for Agricultural Engineering.)	X	X
†*Journal of Agricultural Science*. Vol. 1 (1905/06)+. Cambridge, England; The Cambridge University Press. Bi-monthly.	X	X
†*Journal of Animal Science*. Vol. 1 (1942)+. Champaign, Illinois; American Society of Animal Science.	X	X
†*Journal of Dairy Science*. Vol. 1 (1917)+. Champaign, Illinois; American Dairy Science Association. Monthly.	X	
†*Journal of Hydraulic Engineering*. Vol. 109 (1983)+. New York, American Society of Civil Engineers. Monthly. (Formerly *Journal of the Hydraulic Division, ASCE*)	X	
†*Journal of Hydrology*. Vol. 1 (1963)+. Amsterdam; North-Holland Pub. Co. 4 nos. per vol.; 7 vols. per year.	X	X
Journal of Irrigation and Drainage Engineering. Vol. 109 (1983)+. New York, American Society of Civil Engineers. Bi-monthly since 1989. (Formerly *Journal of Irrigation and Drainage Division, ASCE*)	X	
Journal of Irrigation Engineering and Rural Planning. (Nogyo Doboku Gakkai) No. 1 (1982)+. Tokyo; Japanese Society of Irrigation, Drainage, and Reclamation Engineering. Semiannual.		X
Journal of Soil and Water Conservation. Vol. 1 (1946)+. Ankeny, Iowa; Soil Conservation Society of America. Bi-monthly.	X	
Journal of Soil Science. Vol. 1 (1949)+. Oxford, England; Clarendon Press. Quarterly.	X	
†*Journal of Stored Products Research*. Vol. 1 (1965)+. Oxford, England, Pergamon Press. Quarterly.	X	X
†*Journal of Terramechanics*. Vol. 1 (1964)+. Oxford, England and New York; Pergamon Press Ltd. Quarterly. (Published for the International Society of Terrain Vehicle Systems.)	X	
Journal of the Institution of Engineers. Dacca, Bangladesh; Institution of Engineers.		X
†*Journal of the Science of Food and Agriculture*. Vol. 1 (Jan. 1950)+. London, England; Elsevier Applied Science. Monthly. (Published for the Society for Chemical Industry.)	X	X
Landbouwmechanisatie. Wageningen, Netherlands; Instituut voor Landbouwtechniek en Rationaliste. 10 nos. a year.	X	

Die Landtechnik. Vol. 1 (1946)+. Lehrte, Germany; Verlag Eduard F. Beckmann KG. Monthly. (Published for Kuratorium fur Technik und Bauwesen in der Landwirtschaft.)   X

Mekhanizatsiia i Elektrifikatsiia Sel'skogo Khoziaistva. No. 1 (1975)+. Riga, Latvia; Zvaigzne. (Formerly *Mekhanizatsiia i Elektrifikatsiia Sotsialisticheskogo Sel'skogo Khoziaistva*)   X

Motorisation et Technique Agricole. Paris; MTA. 11 nos. a year.   X   X

†*Nature.* Vol. 247 (1974)+. London; Macmillan Journals. Weekly.   X   X

†*Netherlands Journal of Agricultural Science.* Vol. 1 (1953)+. Wageningen, Netherlands; Genootschap voor Landbouwwetenschap. Quarterly.   X

†*Norsk Landbruksforsking = Norwegian Agricultural Research.* Vol. 1 (1987)+. Ås, Norway; Statens Fagtjeneste for Landbruket = Norwegian Agricultural Advisory Centre. 4 nos. a year. (Formerly *Norsk Landbruk*)   X

†*Plant Physiology.* Vol. 1 (1926)+. Baltimore, Maryland; American Society of Plant Physiologists. Monthly; 3 vols. a year.   X   X

†*Poultry Science.* Vol. 1 (1921)+. Champaign, Illinois; Poultry Science Association. Bi-monthly.   X

Power Farming. Vol. 34 (1965)+. Surrey, England; Agriculture and Construction Press. Monthly.   X   X

†*Proceedings of the National Academy of Sciences of the United States of America.* Vol. 1 (1915)+. Washington, D.C., National Academy of Sciences. Biweekly.   X

†*Proceedings of the Royal Society of London. Series B: Biological Sciences.* Vol. 76 (1905)+. London; The Society. Monthly.   X

†*Process Biochemistry International.* Vol. 25 (1990)+. Rickmansworth, England; Turret Group. Bi-monthly. (Formerly *Process Biochemistry*)   C

†*Science.* New series, Vol. 1 (1895)+. Washington, D.C.; American Association for the Advancement of Science. Weekly.   X   X

Soil & Tillage Research. Vol. 1 (Nov. 1980)+. Amsterdam; Elsevier Scientific Publishing Co. Quarterly.   X

†*Soil Science.* Vol. 1 (1916)+. Baltimore, Maryland; Williams & Wilkins Publ. Monthly.   X   X

†*Soil Science Society of America Journal.* Vol. 40 (1976)+. Madison, Wisc.; Soil Science Society of America. Bi-monthly. (Formerly *Soil Science Society of America Proceedings*)   X

Traktory i Sel'skhomashiny. Vol. 1 (1968)+. Moscow; Ministerstvo Traktornogo i Sel'skhokhoziaistvennogo Mashinostroeniia Soiuza SSR. Monthly.   X

Transactions of the ASAE. Vol. 1 (1958)+. St. Joseph, Mich.; American Society of Agricultural Engineers. Bi-monthly.   X   X

†*Water Resources Research.* Vol. 1 (1965)+. Washington, D.C.; American Geophysical Union. Monthly.   X

Zemedelska Technika. Vol. 40 (1967)+. Prague, Czechoslovakia; Ceckoslovenska Akademie Zemedelska. Monthly.   X

† = Supporting journals; journals in related disciplines but heavily used by agricultural engineering.

These points should be noted about the titles in the list:

(1) The Core list is not a match of the overall rankings in Table 15.1. Some of the altering influential factors are mentioned earlier in this chapter. Additional influence had to be incorporated such as shifts in subject coverage in the last ten years, lack of substantive data to make correlations between the developed countries and Third World, reactions of educators in the field to the titles. Although these influences are in some cases based on subjective evaluations, they were converted to numeric tallies. Threshold levels were established with the total data which caused additional titles to fall from the list. Influential titles of the past fifteen years which recently ceased publication are not included. The aim of this list is to provide currently published titles which should be owned today by those wanting a core academic collection. *Energy in Agriculture*, for example ceased publication in 1988. For a couple of years it remained valuable, by 1992 it had but little impact.

(2) A total of sixty-three journals are identified as core for the combined developed countries and Third World. Twenty-two of these which constitute the most valuable core are under 40% of the titles common to both communities. Seven of the titles are identified as core for the Third World but not for the developed countries. This demonstrates the slightness of agricultural engineering journals devoted to Third World application. It must be emphasized again that the advancement of agricultural engineering in developing countries is largely site specific, small in quantity, and dispersed in several formats including reports. Therefore, the small number of twenty-three most valuable journals to both communities should not be surprising. The rising interest in non-mechanization in Third World countries is just beginning to impact the academic and research communities.

(3) Attention is also directed to the fact that in Third World countries much of the specialized literature such as that for agricultural engineering is included in general agricultural journals for that country or region.

(4) The dagger (†) indicating peripheral subject areas with which agricultural engineers must work is attached to thirty-one of the total sixty-three journals. This demonstrates a wide subject scope which must be accommodated. The breakout for these titles is:

    General Agriculture and Science 15
    General or Related Engineering 4
    Animals 3
    Plants 3
    Air and Water 2

Of particular interest is the relative lack of reliance of agricultural engineers on the literature of general engineering. This demonstrates a close alliance to biological and agricultural systems.

(5) This list is somewhat contrary to what might be expected. The machinery and equipment magazines do not show strongly in this list aimed at serving academic and research literature needs. In Morgan's listing of journals in 1985,[3] twenty-three of the seventy-one titles are concerned with agricultural machinery. Clearly this is not the case in this Core list, which provides a greater concentration on irrigation and drainage than does Morgan's. This results from the inclusion of Third World needs in this assessment as well as a clear swing away from the fascination with machinery and power equipment.

3. Bryan Morgan, *Keyguide to Information Sources in Agricultural Engineering* (London and New York: Mansell Pub. Ltd.). pp. 104–110.

F. Report Literature

The analyzed source documents indicated that 38.2% of all citations were to monographs (see Chapter 14, p. 233). Technical reports, whether published separately or in a series, were counted as monographs. Several individual report titles were placed in the evaluation lists and ranked along with other titles as discrete publications. Only a few of these made the final Core monograph lists. In both Third World and developing country lists, the Food and Agriculture Organization's reports were the most heavily represented in the final lists. Most other report series did not have staying power.

However, there are a few series titles or organizations which provide statistically valid data and vital publications for an academic community. Just under 150 series report titles were identified in the analysis process.

(1) *American Society of Agricultural Engineers. Papers.* This series available in microfiche far outranked the next title or publisher. The series has 200–300 technical papers a year which include informal papers, many from advanced students. The topics are as diverse as the *Transactions of the ASAE*, a peer-reviewed publication. Each paper is on a distinct subject and issued separately. This series could be considered a journal if its publication format were different. In either a monographic or journal category, it's ranking is high enough to make it a core publication.

(2) Three Food and Agricultural Organization report series constitute the next largest group of organizational report titles. They are in order:
FAO Irrigation and Drainage Paper
FAO Agricultural Development Paper
FAO Production Yearbook.
The FAO titles were ranked on a ratio of 4:1:.6. Clearly the first two would be core titles in the report series.

(3) The National Institute of Agricultural Engineers at Silsoe, England, had ten different report series cited. The reports of this important center constitute the third most valuable report publishing group. The most important title is its *Note* series followed by *Test Reports* on a ratio of 4:1. The *Report* series and the *Paper* series also were significantly cited.

(4) The agricultural experiment stations of individual states or universities in the United States were heavily represented as a group with no one state standing above another. As a composite organization, with similar or complementary agricultural engineering programs, they have strong influence in academic research. The experiment stations and agricultural engineering departments of these state land-grants were most commonly cited, usually with one report series title only:
Arkansas   Kansas
California   Michigan
Florida   Ohio
Indiana (Purdue) Washington

(5) The next most widely represented publisher of report series was the United States Department of Agriculture from which seven series were cited. The most cited was the *U.S. Soil Conservation Service Technical Publication* followed closely by the *USDA Technical Bulletin* series.

The Scottish Institute of Agricultural Engineering and the IMAG work at Wageningen were also substantially cited. Northern European countries are well represented in report and journal literature in agricultural engineering, the most prominent being Denmark, Norway, Germany, Netherlands, and Sweden. Czechoslovakia and the USSR were also well cited in the report literature of this discipline.

It should be noted that the World Bank, which is a very heavy producer of reports in agricultural economics, is only slightly involved in agricultural engineering judging from this analysis.

16. Reference Collection Update

CYNTHIA S. KAAG
Owen Science and Engineering Library
Washington State University

Two major compilations of reference tools for agricultural engineering were published in the 1980s. Bryan Morgan's *Keyguide to Information Sources in Agricultural Economics* (cited below) has at least three sections dealing with reference tools; those are headed Directories, Handbooks, and Bibliographic Sources. The other major work is the *Guide to Sources for Agricultural and Biological Research*, edited by J. Richard Blanchard and Lois Farrell (University of California Press, 1981.) Its Chapter E: Physical Sciences, by Scott Kennedy and C. Robin Burt, includes these sections:

> Agricultural Chemistry
> Soils and Fertilizers
> Water Resources and Management
> Irrigation, Drainage, and Land Reclamation
> Agricultural Engineering
> Remote Sensing and Aerial Surveying
> Meteorology
> Patents

There are 585 referenced items in Chapter E, including ninety-three in the Agricultural Engineering portion. Drawing lines around agricultural engineering is a difficult task. The update which follows includes some representation from all of the subjects in Chapter E listed above; however, the concentration is on agricultural engineering and its aspects or influences in water resources, irrigation and drainage, and soils and fertilizers.

The reader is referred to Blanchard and Farrell for the continuing and historically important reference tools. This chapter is a supplement to and an update of Chapter E: Physical Sciences, as closely related to agricultural engineering. Some titles have been updated and the successor edition noted (reference by E number is made to the related item in Chapter E).

Update of Reference Sources in Agricultural Engineering

Agricultural Machinery. Geneva; International Organization for Standardization, 1983. (ISO Standards Handbook no. 13)
Over ninety international standards dealing with tractors and machinery for agriculture and forestry. Mostly mechanical, but also including safety and noise. Graphs and drawings throughout; has 1,000+ entries. Original ed. in 1977.

Agricultural Residues, Bibliography 1975–81, and Quantitative Survey = Residues Agricoles, Bibliographie 1975–81 et Enquete Quantitative. Rome; Food and Agriculture Organization, 1982. 160p. (*FAO Agricultural Services Bulletin = Bulletin des Services Agricoles de la FAO* no. 47) English, French and Spanish.
Sample chapter titles: Animal Residues, Beverage Industry, Cereals, Fibers, Rubber Trees. First ed. 1977. In the same series: *Compendium of Residue Utilization Technologies, Quantitative Survey on Residue Availability.*

ASAE Standards: Standards, Engineering Practices and Data adopted by the American Society of Agricultural Engineers. 31st+ St. Joseph, Mich.; Society of Agricultural Engineers, 1984+ Annual. Formerly the *Agricultural Engineers Yearbook,* with only the standards continued in this publication.
ASAE intends to be a "world center for engineering standards and practices in agriculture." The 36th ed. (1989) had 189 entries, of which sixty-one were adopted, revised or reconfirmed within the year. Each annual is 550–600 pages. [E375]

Balls, R. C. *Horticultural Engineering Technology: Fixed Equipment and Buildings.* Houndmills, Basingstoke, Hampshire; Macmillan, 1986. 246p.

Balls, R. C. *Horticultural Engineering Technology: Field Machinery.* Houndmills, Basingstoke, Hampshire; Macmillan, 1985. 216p.
These two volumes are part of a *Science in Horticulture* series. Good definitions, good diagrams and outline format make them useful in a reference collection. Written in SI metric units.

Brown, R. H., ed. *CRC Handbook of Engineering in Agriculture.* Boca Raton, Fla.; CRC Press, 1988. 3 vols.
Volumes deal with crop production engineering, soil and water engineering, and environmental systems engineering. Overall reliable information, but tables and graphs tend to be United States-oriented; bibliographies tend to be old.

Coolman, Fiepko. *Who is Who: A Directory of Agricultural Engineers Available for Work in Developing Countries.* 1st ed. Paris; Commission Internationale du Génie Rural, 1985. 209p.
In two parts: specialty by country (soil and water, construction, machinery, energy, agricultural economics, industrial processing, development, support); vitae by country (giving vital statistics, addresses, specialties, languages.) Compiled from 600 questionnaire responses.

Coombs, J. *Biomass: International Directory of Companies, Products, Processes and Equipment.* New York; Stockton Press, 1986. 243p.
Covers ". . . commercial and research activities related to the production of biomass and its use as a source of fuels or bulk chemicals." Part I (161 pages) lists organizations under countries; Part II is a buyer's guide, plus company, organization, product and services indexes. Because the definition of biomass is broad, the coverage of institutions is also.

EI Vocabulary—1990 ed. New York; Engineering Information Incorporated, 1990. 512p.
The controlled vocabulary used for *Engineering Index* and its machine-readable Compendex database. Includes additions and changes from earlier editions of *SHE: Subject Headings for Engineering.* [E371]

Eshbach's Handbook of Engineering Fundamentals. 4th ed., edited and revised by Byron D. Tapley. New York; Wiley, 1989. 1,648p.

All sixteen chapters fully revised and updated; emphasis on the growing role of computers in engineering. Good for equations, tables, properties, basic laws of engineering. [E398]

Global Engineering Documents. *Directory of Engineering Document Sources*. Santa Ana, Calif.; Global Engineering Documents, 1986. 446p.

In two sections: a 302 page listing of typical document numbers representing nearly 10,000 serial publications with name of series, organization, index and source code; and 144 pages of names and addresses for the nearly 4,000 issuers of the documents in the first section. [E389]

Hall, Carl W. *Agricultural Engineering Index, 1981–1985*. St. Joseph, Mich.; American Society of Agricultural Engineers, 1987. 232p.

Hall, Carl W. and James A. Basselman. *Agricultural Engineering Index, 1971–1980*. St. Joseph, Mich.; American Society of Agricultural Engineers, 1982. 205p.

These volumes update the 1907–70 issuances. The 1971–80 volume indexes seven leading journals (one in German) and the *ASAE Papers*. Both volumes are organized by title and keyword terms; a reference may appear 1–5 times depending on the significant terms in the title. Has subject and title access only; near 10,000 references in earliest volume, 8,000 in the most recent. [E356]

Hall, Carl W. *Bibliography of Biomass Energy: Including Items Published Separately as Books, Booklets, Bulletins, Dissertations, Pamphlets, and Reports*, St. Joseph, Mich.; American Society of Agricultural Engineers, 1985. 45p.

Lists major non-journal sources of information, including bibliographies and reviews, books, bulletins, reports, and dissertations. Over 1,700 entries from the 1920s through 1983. Provides lists of abbreviations and acronyms, and an index of report numbers.

Hall, Carl W. *Comprehensive Bibliography of Drying References; Covering Bulletins, Booklets, Books, Chapters, Bibliographies*. St. Joseph, Mich.; American Society of Agricultural Engineers, 1980. 85p. (*ASAE Pub*. 3-80)

Approximately 3,400 references to world literature, circa 1950–76. Arranged by author; cross-references by keyword. Articles in magazines are not included.

Hicks, Tyler G. *Standard Handbook of Engineering Calculations*. 2d ed. New York; McGraw Hill, 1985. Various pagings. (Over 1,500 pages.)

General engineering coverage. Divided into disciplines with an index at the end. All calculation procedures use both the United States Customary System (USCS) and System International (SI) units. [E404]

Hopkins, Stephen and Douglas E. Jones. *Research Guide to the Arid Lands of the World*. Phoenix, Ariz.; Oryx Press, 1983. 391p.

Lists 3,199 bibliographies, directories, abstracting journals, statistical sources, databases, atlases and gazetteers useful to dryland researchers. Geographic arrangement; tables of desertification risk, list of bibliographies, etc. Part 2 gives access to more general materials. Subject and author indices. List of databases is useful to researchers.

International Directory of Agricultural Engineering Institutions = Répertoire International d'Institutions de Génie Rural = Reportorio Internacional de Institutiones de Ingenieria Rural. Compiled by Agricultural Engineering Service, FAO. Rome; Food and Agricultural Organization, 1983. 487p.

Information is given in the language chosen by the institution listed; includes addresses, activities and personnel. Strong Third World listings; covers sixty countries. Supersedes 1973 ed. [E384]

International Directory of Fruit, Nut and Vegetable Harvesting Researchers. Compiled by the ASAE Fruit and Vegetable Harvesting Committee (PM-48). St. Joseph, Mich.; American Society of Agricultural Engineers, 1982. 41p.
Lists 190 researchers, mostly concerned with mechanization. Has a crop/country listing; alphabetical researcher/country index, and the main section of alphabetical/country list of researchers with crops. Data from an internationally-mailed questionnaire.

International Irrigation Information Center. *Irrigation; An International Guide to Organizations and Institutions.* Oxford and New York; Pergamon Press, 1980. 153p. (*IIIC Pub.* no. 7)
Lists 864 organizations in 109 countries. Gives basic information as provided by the organizations. Name, subject, abbreviations and acronym, and country indices.
A companion to its 1980 *Irrigation Equipment Manufacturers Directory* (cf).

International Irrigation Information Center. *Irrigation Book List.* Bet Dagan, Israel, International Irrigation Information Center, 1985. 112 p. (*IIIC Pub.* no. 8) Supersedes 1st ed., 1977.
Worldwide coverage, but greatest strengths are books from Europe, North America, and international organizations. Divided into two major parts: (1) Textbooks, Handbooks and Reports (71 pages) which is sub-divided by time periods; (2) Bibliographies and Dictionaries (13 pages). Includes Author, Corporate, Subject, and Language indices; also a two page list of acronyms. A good compilation of approximately 500 titles. Bibliographic details including corporate sources are well done.

International Irrigation Information Center. *Irrigation Equipment Manufacturers Directory.* Bet Dagan, Israel; 1980. 312p. (*IIIC Pub.* no. 5)
Supersedes 1977 edition. Available from IIIC at a subsidized price for organizations in developing countries. Developed country organizations must order from Pergamon Press. [E299]

Irrigation Management. *Selected Bibliography on Irrigation Management: Documents Entered in the Irrigation Management Information Network (IMIN) Database.* Vol. 1, 1987+ Colombo, Sri Lanka; Library and Documentation Service, IIMI. Vol. 1 has title: *IMIN Bibliography.* Two issues a year.
Entries sequentially numbered from beginning of publication; total citations to date in Vol. 3, No. 1 (1989) is 4,642. That issue has subject divisions: Generalities; Social Sciences; Technology; and Irrigation Management; based on Dewey Decimal classification with expansion of the management aspect. Heavy concentration on Asian and North American literature. Includes author, geographical, keyword and titles indices. About 100 pages an issue.

Irrigation Management Network Register of Members. London; Overseas Development Institute, March 1990. 216p. Cover-title: *ODI-IIMI. Irrigation Management Network Register of Members.* IIMI is the International Irrigation Management Institute.
Entries by country from Algeria thru Zimbabwe. Individual name index is provided. Includes basic data on location, job, and contact points. Has eighteen codes for members' disciplines (e.g., Ec = Economics), and twenty-one codes for professional interests (e.g., On Farm Water Management). An impressive compilation of about 1,600 individuals.

Johnson, Jane S., compiler. *Annotated Bibliography on Development and Transfer of Agricultural Technology.* Urbana, Ill.; INTERPAKS. Office of International Agriculture, University of Illinois, 1985–86. 2 vols.
Result of US/AID research program. The 522 entries cover general agricultural development; policy and planning; technology development; technology transfer;

technology utilization. Good annotations and descriptions. Covers reports, articles, books, government publications.

Morgan, Bryan. *Keyguide to Information Sources in Agricultural Engineering.* London and New York; Mansell Pub. Inc., 1985. 209p.

Includes a survey of agricultural engineering and its literature; annotated bibliography of sources of information (journals, directories, handbooks, monographs and textbooks, bibliographic sources); list of organizations with names, addresses, interests, and publications. Includes 636 unique entries. Most work was done in late 1970s.

Munzinger, Peter. *Animal Traction in Africa.* (translation by Hilary Burgess of *Handbuch der Zugtiernutzung in Afrika).* Eschborn; Deutsche Gesellschaft für Technische Zusammenarbeit, 1982. 490p.

A handbook aimed at specialists, development planners and "decision makers within national authorities." In three parts: a brief survey of animal traction in various African countries, including a list of organizations involved; basic information on the animals, harnesses and implements; and case studies. Addresses sociological concerns such as taboos which might interfere with the successful introduction of a particular mode of animal traction.

Nursery and Greenhouse Mechanization Equipment and Manufacturers: 1979 ASAE Directory. Compiled by the ASAE Nursery and Greenhouse Mechanization Committee (PM-59). St. Joseph, Mich.; American Society of Agricultural Engineers, 1979. 30p. *(ASAE Publication 7-79).*

Half of document is a keyword subject listing (e.g., Chipping) with manufacturer's names; second half is alphabetic directory of about 1,000 manufacturers and distributors, primarily in the United States, Canada, and Europe.

Organisation for Economic Co-operation and Development. *OECD Standard Codes for the Official Testing of Agricultural Tractors.* Paris; Organisation for Economic Co-operation and Development, 1988. 194p.

Provides five codes for standard and restricted performance, dynamic and static testing of protective structures, and noise. Definitions, directions and sample reports are included. Also includes addresses of authorities and testing stations in Europe.

Øyjord, Egil, *International Directory of Manufacturers, Machinery, Equipment, and Instruments for Agricultural Research (The IAMFE Directory).* Ås, Norway; International Association on Mechanization of Field Experiments, 1983. 67+p.

A manufacturers/products reference for agricultural engineers, agronomists, plant breeders, and soil scientists. First part is a subject index to the over 300 manufacturers who are listed, along with their addresses, by country in the second part.

Pahwa, K. N. and I. C. Gupta. *World Literature on Reclamation and Management of Salt Affected Soils, 1950–1981.* New Delhi; Associated Pub. Co., 1982. 352p.

Over 800 references, mostly from journals, alphabetically by author. Mostly English-language references. Author, subject, source and geographical indices.

Selected ASTM Standards for Agricultural Engineering Students. Philadelphia, Pa.; American Society for Testing and Materials, 1981. 264p.

In two sections: overview of standards; four categories of twenty-three selected ASTM standards representative of those used in agricultural engineering. Table of contents, but no real index. Tables, definitions, diagrams included.

Selected Bibliography on Alcohol Fuels: 1901 through November, 1981. Sponsored by National Agricultural Library, U.S. Department of Agriculture, and the Office of Alcohol Fuels, U.S. Department of Energy. Golden, Colo.; Technical Information Office, Solar Energy Research Institute, 1982. 475p. *(SERI/SP-290-414)*

Some 1,750 references covering eight decades of alcohol fuel information. Brief entries.

Starkey, Paul. *Animal Traction Directory: Africa*. Braunschweig; Vieweg, 1988. 151p.
Gives details of several hundred organizations and publications from some fifty countries. Information was compiled from field visits, personal contacts, correspondence and literature reviews; the author warns of the changing nature of organizations and the variable reliability of information sources. A country entry gives the status of animal traction, and appropriate government ministries, projects, development organizations, and academic organizations, plus some miscellaneous information. Inclusion of references to specific publications is particularly useful.

Tools for Agriculture; A Guide to Appropriate Equipment for Smallholder Farmers. 4th ed. London; Intermediate Technology Pub., 1992. 240p.
Has chapters on seed-bed preparation, intercultivation, water lifting, post-harvest crop processing, livestock, and more. Gives sources of further information, including addresses for machine companies. Indices by manufacturer and by equipment type. Good tables and illustrations. [E387]

Wolff, Ivan A., ed. *CRC Handbook of Processing and Utilization in Agriculture*. Boca Raton, Fla.; CRC Press, 1982–83. 2 volumes in 3.
Vol. 1: Animal products, Vol. 2: Plant products. Useful tables of composition, graphs, statistics and explanatory text. Leans toward entries such as "typical formula for coffee whitener." Third World usefulness is doubtful.

World Directory of Institutions Concerned with Residues of Agriculture, Fisheries, Forestry, and Related Industries. 3d ed. Rome; Food and Agriculture Organization, 1982. 219p. (*FAO Agricultural Services Bulletin* no 21, rev. 2)
Intended to meet the "need for . . . increased appropriate utilization of the residues produced in agriculture . . ." Lists over 1,000 institutions in 102 countries. Arranged by country with subject index. Includes short polyglot keyword dictionary.

Recently Initiated Agricultural Engineering Journals

The growth of journals in agricultural engineering justifies a listing of the most important new scholarly titles. Only titles central to agricultural engineering are included, not those from closely related civil or mechanical engineering. Completely new publications are listed, excluding those with slight changes of names.

Agroforestry Systems. Vol. 1, no. 1 (1982). The Hague and Boston; M. Nijhoff. Bimonthly

Applied Engineering in Agriculture. General ed. Vol 1, no. 1 (June 1985). St. Joseph, Mich.; American Society of Agricultural Engineers. Semiannual.

Biomass. Vol. 1, no. 1 (Sept. 1981). Barking, Essex, Eng. Monthly.

Biotechnology in Agriculture and Forestry. Vol. 1, no. 1 (1986). Berlin, New York; Springer-Verlag. Irregular.

Computers and Electronics in Agriculture. Vol. 1, no. 1 (Oct. 1985). Amsterdam; Elsevier Science Publishers. Quarterly.

Drying Technology. Vol. 1, no. 1 (1983). New York; M. Dekker. Quarterly.

Irrigation and Drainage Systems. Vol. 1, no. 1 (1986). Dordrecht, Boston; M. Nijhoff. Tri-annual.

Journal of Food Engineering. Vol. 1, no. 1 (1982). London; Applied Science Publishers. Monthly.

Journal of Forest Engineering. Vol. 1, no. 1 (July 1989). Fredericton; Dept. of Forest Engineering, University of New Brunswick. Semiannual.

Pei-ching Nung yeh Kung cheng ta Hsüeh Hsüeh pao = Journal of Beijing Agricultural Engineering University. Vol. 1 (1981)+. Beijing. Quarterly.

Soil and Tillage Research. Vol. 1, no. 1 (Nov., 1980). Amsterdam; Elsevier Scientific Publ. Monthly. Published in collaboration with the International Soil Tillage Research Organization (ISTRO).

17. Primary Historical Literature, 1850–1950

SUSAN J. THOMPSON
Mann Library, Cornell University

CARL W. HALL
Engineering Information Services, Arlington, Virginia

The late 1800s witnessed the beginnings of agricultural engineering as a discipline.[1] This new field, known as rural engineering (génie rural) in French-speaking countries and agricultural engineering in English-speaking countries, incorporated the various subjects of engineering related to agricultural and rural applications.

During the early 1900s agricultural engineering departments were formed at agricultural colleges throughout the United States. The profession grew in importance during a period of transition in agriculture, when animal power was replaced by mechanical power and chemistry was adapted to agricultural production.[2] It was also the period when many of the farmstead's production and processing operations including food and fiber preparation and food processing and preservation were being moved off the farm. These changes in the nature of production accelerated after World War II.

Early research in agricultural engineering was devoted to meeting the immediate needs of the rural sector with little attention given to long-range problems. Thus, horseshoeing, harness making, hitch design, drainage, building construction, rural roads, sanitation and waste, and steam power were some of the early research concerns of agricultural engineers. This focus on problem solving resulted in a research and teaching program devoted to the practice rather than the theory of agricultural engineering. Dur-

---

1. For another discussion of historical titles and trends in the agricultural engineering literature, the reader should examine several portions of Bryan Morgan's *Keyguide to Information Sources in Agricultural Engineering* (London and New York: Mansell Pub. Ltd., 1985). This work examines several related literature activities and titles from a British viewpoint.

2. Wayne D. Rasmussen, "The Impact of Technological Change on American Agriculture, 1862–1962," *The Journal of Economic History* 22 (Dec. 1962): 578–591.

ing these early years, industry was not close to nor dependent on university and government research. Rather, inventors and entrepeneurs provided the driving force for the new tools, equipment and structures. The industries were, however, dependent on the universities and colleges for the training of future personnel, the majority of whom received a B.S. in agricultural engineering before beginning industrial employment. As the discipline grew, experiment stations, some in colleges of engineering, but most in agricultural colleges, began to participate in agricultural engineering projects.

The first English-language book to use agricultural engineering in the title predates the field as a discipline. *The Rudimental Treatise on Agricultural Engineering* was written by George H. Andrews and published in London in 1852 by J. Weale in three volumes: Volume 1, *Buildings*; Volume 2, *Motive Power and Machinery of the Steading*; Volume 3, *Field Machines and Implements*. These three volumes anticipated the major subject areas of the new discipline of agricultural engineering: buildings, power, and machinery.

The individual subjects of agricultural engineering were the topic of numerous books. Many of these early books in agricultural engineering were written by civil and mechanical engineers, by agricultural scientists with a knowledge of physics and mathematics, and by individuals skilled in the use of hand tools and equipment. These writers also lead the development of agricultural engineering as a distinct profession. One of the first books in the United States to deal with the several subjects of agricultural engineering was by Jay Brownlee Davidson of Iowa State University. In 1913, Webb Publishing Company, St. Paul, Minnesota, published Davidson's *Agricultural Engineering* (544p). The book, oriented to the general reader and secondary school students rather than the professional engineer, provided an introduction to the major subjects of agricultural engineering prior to World War I. These subjects were: surveying, drainage, irrigation, roads, farm machinery, engines and tractors, farm structures, farm sanitation and rope work. Davidson, one of the first faculty members of agricultural engineering at Iowa State, had earlier co-authored with L. W. Chase *Farm Machinery and Farm Motors*.[3]

The first monograph series devoted to agricultural engineering was published by McGraw-Hill Book Company, New York, between 1913 and 1916. The consulting editor was E. B. McCormick, a mechanical engineer and farmer who was also Dean of the Engineering Division at Kansas State

---

3. J. B. Davidson and L. W. Chase, *Farm Machinery and Farm Motors* (New York: Judd Co., 1908).

University. There were three titles in this series: *Farm Motors: Steam and Gas Engines*;[4] *Use of Water in Irrigation*;[5] and *Concrete Construction for Rural Communities*.[6]

McGraw-Hill began a second series entitled Publications in Agricultural Engineering in 1924. Daniel Scoates was the consulting editor of this series until his death in 1939. The first books in this series were:

Quincy C. Ayres and Daniel Scoates, *Land Drainage and Reclamation*, 1928. 419p.
Harris P. Smith, *Farm Machinery and Equipment*, 1929. 448p.
Fred R. Jones, *Farm Gas Engines and Tractors*, 1932. 485p.
Quincy C. Ayres, *Soil Erosion and Its Control*, 1936. 365p.
Mack M. Jones, *Farm Shop Practice*, 1939. 315p.
John C. Wooley, *Farm Buildings*, 1941. 345p.

Quincy Ayres followed Daniel Scoates as consulting editor of the McGraw-Hill Publications in Agricultural Engineering series. In addition to issuing new editions of the earlier books first published under the editorship of Scoates, the following titles were added to the series:

Harry B. Roe, *Moisture Requirements in Agriculture*, 1950. 413p.
John C. Wooley, *Planning Farm Buildings*, 1953. 306p. (Replacing the earlier *Farm Buildings*)
Harold E. Gulvin, *Farm Engines and Tractors*, 1953. 397p.
Harry B. Roe and Quincy Ayres, *Engineering for Agricultural Drainage*, 1954, 501p.
Howard F. McColly and James W. Martin, *Introduction to Agricultural Engineering*, 1955. 458p.
Harold E. Gray, *Farm Service Buildings*, 1955. 458p.
Robert H. Brown, *Farm Electrification*, 1956. 367p.

The second major agricultural engineering series of books was published by John A. Wiley and Sons, New York (Chapman and Hall, Ltd., London). This series began about five years after the first McGraw-Hill series. The Wiley series was initially co-edited by L. W. Chase and J. B. Davidson. J. B. Davidson edited the series in the twenties. The books in the series, with the dates of first editions, are:

Frederick A. Wirt, *A Laboratory Manual in Farm Machinery*, 1917. 162p.
George R. Chatburn, *Highway Engineering, Rural Roads and Pavements*, 1921. 379p.
Wilbur L. Powers and T. A. H. Teeter, *Land Drainage*, 1922. 276p.
John T. Bowen, *Dairy Engineering*, 1925, 532p.
J. Brownlee Davidson, *Agricultural Machinery*, 1931. 396p.

---

4. Andrey A. Potter, *Farm Motors: Steam and Gas Engines* (New York: McGraw-Hill Book Co. Inc., 1913).
5. Samuel Fortier, *Use of Water in Irrigation* (New York: McGraw-Hill Book Co., Inc., 1915).
6. Roy A. Seaton, *Concrete Construction for Rural Communities* New York: McGraw-Hill Book Co., Inc., 1916).

Orson W. Israelsen, *Irrigation Principles and Practices*, 1932. 424p.
William A. Foster and Deane G. Carter, *Farm Buildings*, 1932. 377p.

While universities continued to form departments of agricultural engineering between World War I and World War II, few books were written to assist in teaching at a level beyond the descriptive. Another series by John A. Wiley and Sons, the Wiley Farm Series under the editorship of A. K. Getman and C. E. Ladd, published what were to become two popular books on agricultural engineering subjects.

Archie A. Stone, *Farm Machinery*, 1928. 486p.
Archie A. Stone, *Farm Tractors*, 1932. 492p.

Both Stone's books were descriptive and served the engineer as well as the more general reader. New editions of these books continued to be used in the 1940s.

During the 1920s and 1930s, the automobile and the tractor became commonplace on the farm and electricity and telephones were being introduced in rural areas. Titles were issued that reflected these changes in the rural sector as well as the applied nature of agricultural engineering. Among the new titles were farm engineering, farm mechanics, farm enterprise mechanics, farm structure, construction and repair work, and farm or rural electrification. One of the first books to be published after World War II illustrating the farm devices of the period was *Machines for Farm, Ranch, and Plantation* by A. W. Turner and E. J. Johnson. This book was published by McGraw-Hill in its Rural Activity Series in 1948.

As the literature developed, bibliographies were written to assist those in the field to locate and comprehend the subjects. The first major bibliography was by Dorothy Williams Graf, *Agricultural Engineering, a Selected Bibliography,* published by the USDA, Washington, D.C., 1937 (373p.). Many bibliographies on subjects within agricultural engineering, such as implements, fuels, harvesting equipment, and buildings had been published previously by the USDA, Graf's was the first comprehensive bibliography of the discipline.

Comprehensive bibliographies were followed by indexing series. Two important indexing series were the *Agricultural and Horticultural Engineering Abstracts* published by the National Institute of Agricultural Engineering in England from 1951 to 1966 and the *Agricultural Engineering Indexes* published by the American Society of Agricultural Engineers. These two series incorporated the major references around the work providing a comprehensive, periodic, and accumulated record.

An effort began to coordinate international agricultural engineering research activities. In 1930, the CIGR (Commission Internationale du Génie Rurale) was formed, primarily for people in French-speaking countries and Western Europe.

A. The Historical Preservation Project

The objective of the historical component is to identify the agricultural engineering literature worthy of long-term preservation. The great need for preservation stems from the changes in the process of papermaking which occurred in the 1860s when rag was replaced by wood fiber. The result was an acidic paper that self-destructs over time by turning yellow and becoming brittle. The paper tears easily, entire pages fall away or pulverize to dust. By contrast, books from earlier periods with a high rag content are still pliable with little evidence of brittleness, yellowing, or disintegration.

The historical preservation concentrates on the agricultural engineering literature published in English between 1850 and 1950. Emphasis is placed on the literature of importance in the United States and Canada. Monographs and journals published outside the United States and Canada were not systematically evaluated. Three distinct types of publications important to agricultural engineering in the United States were identified and analyzed: (a) monographs, (b) popular and trade periodicals, and (c) scholarly and professional journals. Due to their age and rarity, all agricultural monographs published prior to 1850 have already been identified and currently have a high priority for preservation. This project also excluded land-grant publications in series and federal documents. Many of these monographs have already been preserved on microfilm and plans are currently being made for preservation of the remainder of these documents.

B. Source Documents

The historical literature was identified and analyzed using citations provided in landmark or overview works for the 1850–1949 period. These source documents were recommended by agricultural engineers each with over thirty years of professional experience in teaching and research. The subject groupings in Hummel[7] and Stewart[8] were used as guides to deter-

---

7. W. M. Hummel, "The Literature of Agricultural Engineering," *Transactions of the American Society of Agricultural Engineers* 1 (1907): 81–85. Reproduced in Chapter 13 of this book.

8. Robert E. Stewart, *Seven Decades That Changed America* (St. Joseph, Mich.: The American Society of Agriculture Engineers, 1979). pp. 30–38.

mine the scope of the literature to be included in the historical analysis (Table 17.1) and the source documents for citation analysis, listed below. These documents were chosen in consultation with select members of the Steering Committee on Agricultural Engineering for the Core Agricultural Literature Project.

Table 17.1. Major subject areas for historical agricultural engineering

Hummel[a]	ASAE Technical Divisions as of 1925[b]
Drainage	Farm Power & Equipment
Irrigation	Farm Structures
Farm Implements	Reclamation
Roads & Bridges	Rural Electric
Farm Buildings	

[a]Hummel, "The Literature of Agricultural Engineering."
[b]Stewart, *Seven Decades That Changed America.*

The process of citation analysis (explained in Chapters 14 and 15) involves a count of each journal, report series, and monograph cited in the source documents. This method provides the basic lists of titles which can then be evaluated by peer review or correlated with other studies and records. The following list includes source documents analyzed to obtain the basic lists and quantitative counts.

Source Documents for Historical Agricultural Engineering

Ayres, Quincy C. *Soil Erosion and Its Control.* New York and London; McGraw-Hill, 1936. 365p.

Ayres, Quincy C. and Daniels Scoates. *Land Drainage and Reclamation.* 2d ed. New York and London; McGraw-Hill, 1939. 496p. (1st ed., 1928)

Bainer, Roy, R. A. Kepner, and E. L. Barger. *Principles of Farm Machinery.* New York; John Wiley & Sons, 1955. 571p.

Bateman, John H. *Introduction to Highway Engineering.* 3d ed. New York; John Wiley & Sons and London; Chapman & Hall, 1939. 442p. (1st ed., 1928)

Carter, Deane G. and W. A. Foster. *Farm Buildings.* 3d ed. New York; John Wiley & Sons and London; Chapman & Hall, 1941. 404p. (1st ed., 1922)

Chatburn, George R. *Highway Engineering: Rural Roads and Pavements.* New York; John Wiley & Sons and London; Chapman & Hall, 1921. 379p.

Clarke, R. J. *Process Engineering in the Food Industries.* New York; Philosophical Library, Inc., 1957. 355p.

Culpin, Claude. *Farm Machinery.* 4th ed. London; Crosby Lockwood & Son, 1952. 622p. (1st ed., 1938)

Davidson, J. Brownlee and Leon W. Chase. *Farm Machinery and Farm Motors.* New York; Orange Judd, 1908. 513p.

Etcheverry, Bernard A. *Irrigation Practice and Engineering*. New York and London; McGraw-Hill, 1915–1916. 3 vols. (Analyze volumes 2 and 3 and selected chapters of volume 1)
Etcheverry, Bernard A. *Land Drainage and Flood Protection*. New York and London; McGraw-Hill, 1931. 327p.
Farrall, Arthur W. *Dairy Engineering*. New York; John Wiley & Sons and London; Chapman & Hall, 1942. 405p.
Frevert, Richard K., Glenn O. Schwab, Talcott W. Edminster, and Kenneth K. Barnes. *Soil and Water Conservation Engineering*. New York; John Wiley & Sons and London; Chapman and Hall, 1955. 479p.
Hall, Carl W. *Bibliography of Agricultural Engineering Books*. St. Joseph, Mich.; American Society of Agricultural Engineers, 1976. 84p. (Extracted all titles published prior to 1951.)
Heiss, Rudolf. *Lebensmitteltechnologie*. Munich; Verlag von J. F. Bergmann, 1950. 344p. (Analyzed selected chapters, English titles extracted.)
Henderson, S. M. and R. L. Perry. *Agricultural Process Engineering*. New York; John Wiley & Sons and London; Chapman and Hall, 1955. 402p.
Hienton, Truman E., Dennis E. Wiant, and Oral A. Brown. *Electricity in Agricultural Engineering*. New York; Wiley, 1958. 393p.
Israelsen, Orson W. *Irrigation Principles and Practices*. 2d ed. New York; John Wiley & Sons and London; Chapman & Hall, 1950. 405p. (1st ed., 1932) (Analyzed selected chapters)
Jones, Fred R. *Farm Gas Engines and Tractors*. 3d ed. New York and London; McGraw-Hill, 1952. 489p. (1st ed., 1932)
Jones, Mack M. *Farm Shop Practice*. New York and London; McGraw-Hill, 1939. 315p.
Matthews, R. Borlase. *Electro-Farming or The Application of Electricity to Agriculture*. London; Ernest Been Ltd., 1928. 357p.
McColly, H. F. and J. W. Martin. *Introduction to Agricultural Engineering*. New York and London; McGraw-Hill, 1955. 553p.
Powers, W. L. and T. A. H. Teeter. *Land Drainage*. New York; John Wiley & Sons, 1922. 270p.
Richey, C. B., ed. *Agricultural Engineers' Handbook*. New York; McGraw-Hill, 1961. 880p.
Smith, Harris P. *Farm Machinery and Equipment*. 4th ed. New York and London; McGraw-Hill, 1955. 514p.

C. Citation Analysis and Compilation

The historical literature in agricultural engineering is, in many instances, devoid of citations. In many cases it was necessary to use second or later editions of a source document in order to find citations. This is not unusual in early agricultural engineering publications. For instance, papers presented at the early American Society of Agricultural Engineers meetings (1907–17) that organized the literature on a particular subject "revealed a surprising depth of historical background . . . However, no papers of the period contained listed references."[9] Other distinctions also exist between

9. Ibid., p.30.

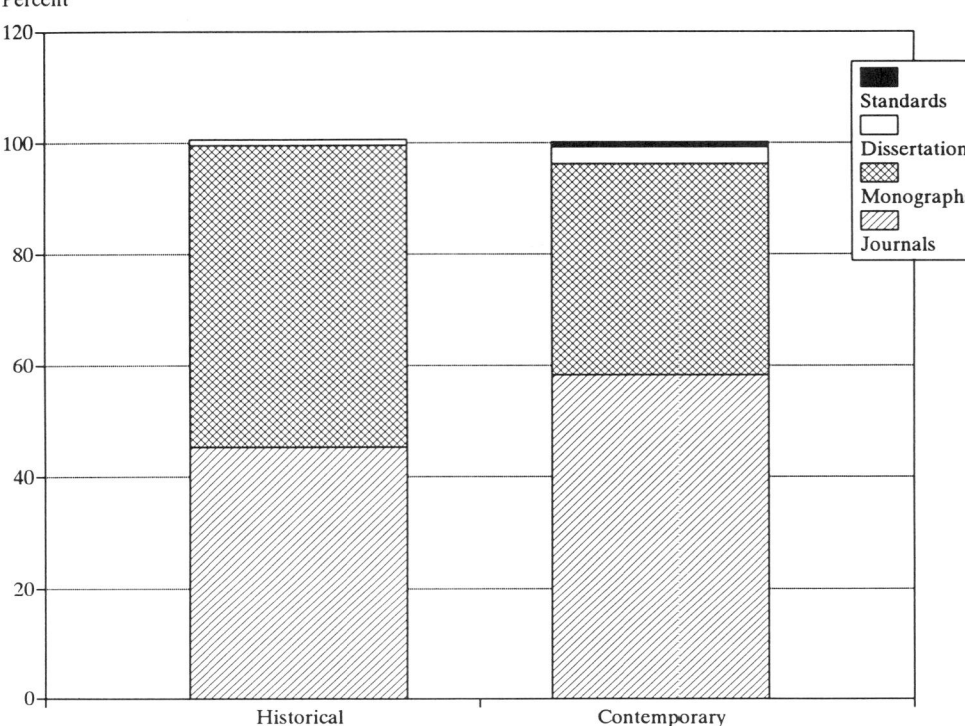

Figure 17.1. Types of literature.

the historical (1850–1950) and the contemporary (1950–90) agricultural engineering literature. Figures 17.1, 17.2, and 17.3 compare the sources of citations, the place of publication, and the types of publishers for these two periods.

Monographs were the most important source of scholarly literature in the field prior to 1950 (54.5%), while during the contemporary period journals provided the bulk of literature in the field (58.1%). Dissertations and standards, on the other hand, are not a strong source of literature in either period, although important for fundamental knowledge. Table 17.2 provides details on the types of publications most often cited during the historical period.

Because of the concentration on English language historical literature, foreign language publications were not as important as they are in the contemporary core agricultural engineering monograph list (Figure 17.2). Of the places of publication of English-language monographs, the United States dominates with 92.4% of the historical and 56.1% of the contempo-

Primary Historical Literature 347

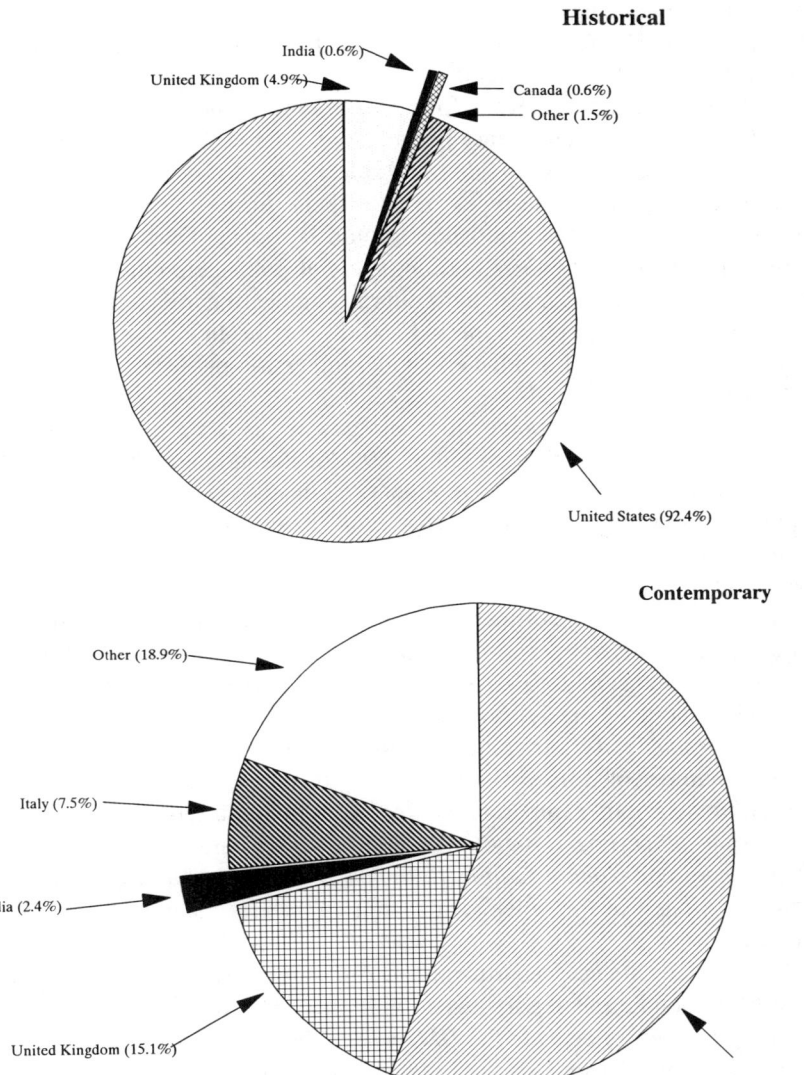

Figure 17.2. Place of publication.

rary agricultural engineering monographs published in the United States (Figure 17.2). The second ranking place of publication during both periods, the United Kingdom, increased by three-fold its contributions to the agricultural engineering literature from 4.9% during the historical period to 15.1% during the contemporary period. Actual numbers and precentages for the historical period are provided in Table 17.4.

In a comparison of the types of publishers during the two periods, the importance of university publishers should be noted. During the historical period, university publishers and commercial publishers combined to publish 61.5% of the historical agricultural engineering monographs (Figure 17.3). By the contemporary period, 50.6% of the monographs were published by commercial publishers with university publishers declining to 7.3% (Figure 17.3).

Commercial presses, universities and governments were the most important sources of publication providing 89% of the monographs in the source documents. University and government publications were usually part of a report series originating from a department, extension service, or an experiment station. A total of 204 distinct report series titles were cited from agricultural experiment stations and universities located in the United States. Agricultural experiment stations and extensions were the most important university sources. Agricultural experiment stations published 42.6% and extension 26.5%, of the 204 report series titles. The agricultural experiment stations in California, Nebraska, and Iowa were the most important sources of early agricultural engineering research. Iowa also had the greatest number of different series titles with thirteen, followed by California with ten. Another important source of monograph report series were the engineering experiment stations located at select land-grant colleges. The *Engineering Experiment Station Bulletins* from Iowa and Kansas were the most often cited in this titles series from engineering stations with 63% of the citations.

The United States Department of Agriculture was another important source of monograph report series with fifty-one report series titles. The most often cited series were the *USDA Bulletin* (9.2%), the *Farmers' Bulletin* (21.7%), the *Miscellaneous Publication* series (9.2%), the Soil Conservation Service Publication (13.6%), and the *Technical Bulletin* (11.8%). Other sources of English-language monograph report series were insignificant. Monographs cited had an average half-life of 19.28 years (with the median half-life of 15 years), while journals had a half-life of 14.63 years (with a median half-life of 14.5 years).

Commercial presses published 32% of the monographs in the core analysis. Of the 188 commercial presses cited, the most important were

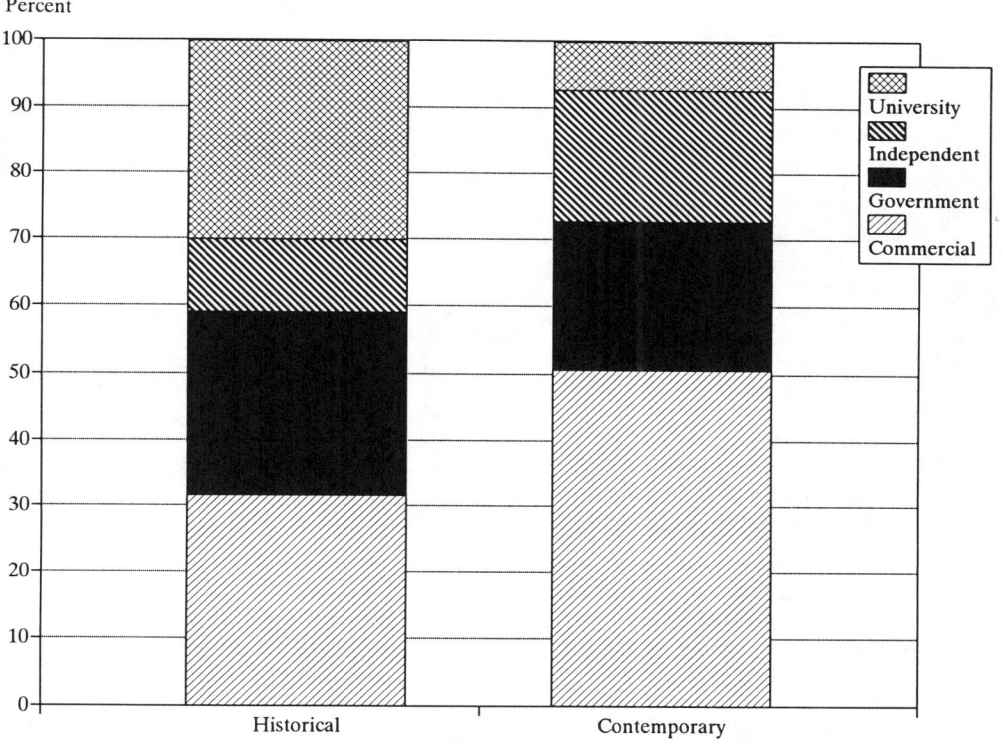

Figure 17.3. Types of publishers.

McGraw-Hill and Wiley, who together published 42.7% of the commercial monographs.

D. Identifying and Ranking the Monographs

The citations from the source documents provided the data for the citation analysis in this chapter. The criteria for identification of monographs is their importance in academic and research literature in agricultural engineering for the period 1850 to 1950. Citation analysis of these source publications provided a count of the number of times each monograph was cited. These monographs were identified using a three-step method (see Chapter 14 for a detailed description). In the first step citation analysis is used to identify key historical literature. After the key historical literature was identified it was sent to reviewers for ranking.

Table 17.2. Citations to journals, monographs, and dissertations in the twenty-three source documents

	Journals	Monographs	Dissertations	Total
Ayres (1936)	123	122	0	245
Ayres & Scoates (1939)	6	75	0	81
Bainer, et al. (1955)	262	129	5	396
Bateman (1939)	122	49	0	171
Carter & Foster (1941)	23	247	0	270
Chatburn (1921)	25	71	0	96
Clarke (1957)	25	54	0	79
Culpin (1952)	20	136	0	156
Davidson & Chase (1908)	2	44	0	46
Etcheverry (1915–1916)	399	144	2	545
Etcheverry (1931)	19	77	0	96
Farrall (1931)	30	30	0	60
Frevert, et al. (1942)	191	298	6	495
Henderson & Perry (1955)	58	105	0	163
Hienton, et al. (1958)	60	72	3	135
Israelsen, (1950)	163	149	0	312
Jones, Fred (1952)	64	180	0	244
Jones, Mack (1939)	0	121	0	121
Matthews (1928)	444	165	0	609
McColly & Martin (1955)	42	340	0	382
Powers & Teeter (1922)	0	90	0	90
Richey, ed. (1961)	528	485	6	1019
Smith (1955)	111	114	0	225
Total of all Citations	2717	3280	22	6019
Percentage of Total	45.14	54.49	0.37	100.00

Table 17.3. Place of publication of monographs cited in twenty-three source documents

	Number	Percent
United States	2946	92.4
United Kingdom	156	4.9
India	19	0.6
Canada	18	0.6
France	15	0.5
Germany	10	0.3
All Others[a]	24	0.7

[a]Australia, Austria, Belgium, Denmark, Ireland, Italy, Netherlands, New Zealand, Spain, Sweden, Switzerland.

Reviewers were asked to rank the titles identified in the analysis using the following criteria:

(1) The work has had an important influence in the subject field.
(2) The work is the first of its kind published or records major advances in the field.
(3) The work embodies an historical record of changes in the field.
(4) The title is valuable because it is a superior work of a leader in the field of agricultural engineering.

These monographs were graded by scholars using a three-category scale: category 1 = a very important historical title worthy of preservation; category 2 = worth preserving, but of secondary importance; category 3 = a title of marginal historical value. These subject specialists (listed below), all of whom have extensive research and teaching experience, are faculty members of land-grant colleges or U.S. Department of Agriculture researchers. Reviewers also made suggestions of titles for inclusion in the list. The few titles suggested were added to the list and evaluated.

Historical Agricultural Engineering Monograph Reviewers

E. Stanley Shepardson
  Ithaca, N.Y.
Harold V. Walton
  State College, Penn.
Gene C. Haugh
  Virginia Polytechnic Institute
Donnell R. Hunt
  University of Illinois
Robert A. Kepner
  Davis, Calif.
Landis L. Boyd
  Western Assoc. of Agr. Exp. Stations,
  Oakland, Calif.
Joseph K. Campbell
  Cornell University

William E. Splinter
  University of Nebraska
Howard F. McColly
  Michigan State University
Robert H. Brown
  University of Georgia
W.L. Harris
  Agricultural Research Service
  U.S. Dept. of Agriculture
  Beltsville, Md.
Glenn O. Schwab
  Powell, Ohio
Carl W. Hall
  Arlington, Virginia

In the third step, a total score was obtained for each title by adding the number of times that the title appeared in the various source documents (citation hits), plus the cumulative rankings of the reviewers. Each reviewing category (identified as C) was then weighted; category 1 (C1) rankings were weighted by a factor of 3; category 2 (C2) rankings by a factor of 2;

and category 3 (C3) rankings by a factor of 1. The final monograph ranking was calculated using the formula:

$$\text{Final ranking} = (1 \times \text{\# citation hits}) + (3 \times \text{\# C1}) + (2 \times \text{\# C2}) + (1 \times \text{\# C3})$$

Those monographs receiving the highest final rankng scores are first rank; the remainder are ranked second. The titles in the first rank should be given highest priority for preservation. Of the 684 titles on the list, 265 were in the first rank (38.7%) and 419 in the second (61.3%).

E. Historically Important Monographs in Agricultural Engineering, 1850–1950

Ranking

A

Second   Adams, Harry W. Common Sense Instruction on Gas Tractor Operation. Minneapolis, Minn.; The Jensen Printing Co., 1920. 311p.

Second   Adams, Orville. Elements of Diesel Engineering. New York; Norman W. Henley Publishing, 1936. 478p.

Second   Addison, Herbert. Hydraulic Measurements. New York; J. Wiley, 1941. 301p. (2d ed., 1946. 327p.)

Second   Addison, Herbert. Land, Water and Food. London; Chapman & Hall, 1955. 248p. (2d ed., 1961. 284p.)

Second   Addison, Herbert. A Text Book of Applied Hydraulics. London; Chapman & Hall, 1934. 409p. (4th ed., 1954. ghh724p.)

Second   Agg, T. R. American Rural Highways. New York; McGraw-Hill, 1920. 139p.

Second   Aitken, Thomas. Road Making and Maintenance: A Practical Treatise for Engineers, Surveyors, and Others. London; G. Griffin and Co. and Philadelphia; J. B. Lippincott, 1900. 440p. (2d ed., 1907. 527p.)

Second   Alden, John L. Design of Industrial Exhaust Systems. New York; Industrial Press and London; Machinery Publishing Co., 1939. 220p. (5th ed. by John L. Alden and John M. Kane, 1982. 281p.)

First   Aldrich, Darragh. The Story of John Deere: A Saga of American Industry. Minneapolis, Minn.; McGill Litho., 1942. 140p.

Second   Alford, Leon P. Bearings and Their Lubrication. New York; American Machinist, 1911. 235p.

First   Allen, Arthur H. Electricity in Agriculture. London; Pitman's Technical Printers, 1922. 127p.

First   Allen, Lewis F. Rural Architecture. New York; A. O. Moore and C. M. Saxton, 1852. 378p.

First   American Oil Company. Farm Tractors (Fuels and Lubricants). Baltimore, Md; American Oil Company, 1950. 80p.

First   American Society of Civil Engineers. Committee on Hydrology. Hydrology Handbook. New York; American Society of Civil Engineers, 1949. 184p.

First	American Society of Civil Engineers. Special Committee on Irrigation Hydraulics. Letter Symbols Glossary for Hydraulics: With Special Reference to Irrigation. New York; American Society of Civil Engineers, 1935. 39p. (American Society of Civil Engineers, Manuals of Engineering Practice No. 11) (Reprinted)
Second	American Society of Refrigerating Engineers. The Refrigerating Data Book. 2d ed. New York; American Society of Refrigerating Engineers, 1946. 1 vol.
First	Andersen, Aage J. C. Refining of Oils and Fats for Edible Purposes. New York; Academic Press, 1953. 204p. (2d ed. edited by P. N. Williams. New York; Pergamon Press, 1962. 247p.)
First	Anderson, D. H. Primer of Irrigation. Chicago; Anderson, 1905. 257p.
First	Anderson, Frederick I. Electricity for the Farm. New York; Macmillan, 1915. 265p.
Second	Anderson, Oscar. Refrigeration in America: A History of a New Technology and Its Impact. Princeton, N.J.; Princeton University Press for the University of Cincinnati, 1953. 344p.
First	Anderson, Oscar G., and Frederick C. Roth. Insecticides and Fungicides: Spraying and Dusting Equipment. A Laboratory Manual. New York; J. Wiley, 1923. 349p.
First	Andrews, George H. Rudimentary Treatise on Agricultural Engineering. London; J. Weale, 1852–1853. 3 vols.
Second	Andrews, William B., ed. Cotton Production, Marketing and Utilization. State College, Miss.; Mississippi State College Press, 1950. 476p.
First	Ardrey, Robert L. American Agricultural Implements: A Review of Invention and Development in the Agricultural Implement Industry of the United States. Part 1, General History of Invention and Improvement. Part 2, Pioneer Manufacturing Centers. Chicago; The Arthur, 1894. 236p. (Reprinted New York; Arno Press, 1972.)
Second	Armco Drainage and Metal Products, Inc. Handbook of Culvert and Drainage Practice. Middletown, Ohio; Armco Culvert Manufacturers Association, 1930. 349p. (4th ed. published as Handbook of Drainage and Construction Products, 1958. 529p.)
Second	Armco Drainage and Metal Products, Inc. Handbook of Water Control for the Solution of Problems Involving the Development and Utilization of Water. Middletown, Ohio; Armco Drainage & Metal Products, Inc., 1946. 586p.
Second	Armstrong, F. E. Farm Repair and Construction Work. Minneapolis; University of Minnesota, 1923. 38p.
Second	Arthur, William. The New Building Estimator. 11th ed. New York; David Williams Company, 1913. 705p. First Ashby, J. B., ed. British Tractors and Power Cultivators. Sussex, Eng.; Pentagon Publishers, 1949. 240p.
Second	Association for Planning and Regional Reconstruction. New Ideas for Farm Buildings. London; Farmer and Stock-Breeder, 1947. 56p.
First	Association of American Portland Cement Manufacturers. Concrete Silos. Rev. ed. Philadelphia; Association of American Portland Cement Manufacturers, 1915. 64p.
Second	Association of Official Agricultural Chemists. Official Methods of Analysis. Washington, D.C.; Association of Official Analytical Chemists, 1919. 1 vol. (15th ed., 1990. 2 vols.)
Second	Atkins, F. C. Modern Egg Farming. 2d ed. Newbury, Eng.; The Author, 1926. 120p.
First	Ayres, Quincy C. Soil Erosion and Its Control. New York and London; McGraw-Hill, 1936. 365p.

First	Ayres, Quincy C., and Daniels Scoates. Land Drainage and Reclamation. New York; McGraw-Hill, 1928. 419p. (2d ed., 1939. 496p.)

B

Second	Badger, Walter L. Heat Transfer and Evaporation. New York; The Chemical Catalog Co., 1926. 306p.
First	Bagnold, Ralph A. The Physics of Blown Sand and Desert Dunes. New York; W. Morrow & Company, 1942. 265p.
First	Bailey, Liberty J. Cyclopedia of American Agriculture. New York; Macmillan, 1907–1909. 4 vols. (4th ed., 1912. 4 vols.)
First	Bainer, Roy, R. A. Kepner, and E. L. Barger. Engineering Elements of Farm Machinery. Ann Arbor, Mich.; Edward Bros., 1953. 1 vol.
First	Bainer, Roy, R. A. Kepner, and E. L. Barger. Principles of Farm Machinery. New York; Wiley, 1955. 571p.
Second	Baker, Henry D. Manual on Thermometry. East Hartford, Conn.; United Aircraft Corp., 1950. 201p.
First	Baker, Ira O. A Treatise on Roads and Pavements. New York; Wiley & Sons, 1903. 655p. (3d ed., 1918. 666p.)
First	Ball, Robert S. Natural Sources of Power. New York; D. Van Nostrand, 1908. 348p.
Second	Banton, Ronald. Problems in Cleaning Seeds. Saginaw, Mich.; A. T. Ferrell and Company, 1946. 12p.
First	Barger, E. L., W. M. Carleton, E. G. McKibben, and Roy Bainer. Tractors and Their Power Units. New York; Wiley, 1952. 496p. (2d ed., 1963. 524p.)
First	Barkas, W. W., R. F. S. Hearmon, and H. F. Rance. Mechanical Properties of Wood and Paper. New York; Interscience and Amsterdam; North-Holland Pub. Co., 1953. 1 vol.
Second	Barnes, Ralph M. Work Methods Manual. New York; J. Wiley & Sons and London; Chapman & Hall, 1944. 136p.
Second	Barois, J. Irrigation in Egypt. Translated by A. M. Miller. Washington, D.C.; Government Printing Office, 1889. 111p.
Second	Baron, Margret. The Mechanical Properties of Cheese and Butter. London; Dairy Industries, 1952. 106p.
Second	Barr, W. M. Pumping Machinery. Philadelphia; J. B. Lippincott, 1893. 477p. (2d ed., 1908. 483p.)
First	Barre, H. T., and L. L. Sommet. Farm Structures. New York; John Wiley and Sons, 1950. 650p.
First	Barrows, H. K. Floods: Their Hydrology and Control. New York; McGraw-Hill, 1948. 432p.
Second	Barrows, Harry H., and J. V. Phillips. Agricultural Drainage in Georgia. Atlanta, Ga.; Byrd Printing, 1917. 122p.
Second	Bateman, John H. Introduction to Highway Engineering. 2d ed. New York; John Wiley & Sons and London; Chapman & Hall, 1934. 441p. (5th ed., 1948. 538p.)
Second	Baumeister, Theodore. Fans. New York and London; McGraw-Hill, 1935. 241p.
First	Baver, Leonard D. Soil Physics. New York; J. Wiley & Sons and London; Chapman & Hall, 1940. 370p. (3d ed., 1956. 489p.)
First	Beazeley, Alexander. The Reclamation of Land from Tidal Waters. London; C. Lockwood and Son, 1900. 314p.
First	Beckmann, Johan. A History of Inventions, Discoveries, and Origins.

	Translation of Beytrage zur Geschichte der Erfindungen by William Johnston. 4th ed. London; H. G. Bohn, 1946. 2 vols.
Second	Bedell, Earl L., and E. G. Gardner. Household Mechanics. Scranton, Penn.; International Textbook, 1937. 130p.
Second	Behar, Manoel F., P. R. Ewald, et al. The Handbook of Measurement and Control. Pittsburgh; Instruments Publishing Co., 1951. 1 vol. (2d ed., 1954. 216p.)
First	Bellasis, Edward S. Hydraulics. London; Rivingtons, 1903. 303p. (3d ed., London; Chapman & Hall, 1920. 348p.)
First	Bellasis, Edward S. Irrigation Works. London; E. & F. N. Spon and New York; Spon & Chamberlain, 1913. 195p.
Second	Bellasis, Edward S. River and Canal Engineering. London; E. & F. N. Spon and New York; Spon & Chamberlain, 1913. 215p.
First	Bennett, Richard, and John Elton. History of Corn Milling. London; Simpkin, Marshall & Co., 1889–1904. 4 vols.
Second	Bennion, Edmund B. Breadmaking: Its Principles and Practice. 3d ed. London and New York; Oxford University Press, 1954. 411p. (4th ed., 1967. 391p.)
First	Bennison, E. W. Ground Water. St. Paul, Minn.; Edward E. Johnson, 1947. 509p.
Second	Berry, Frederic A., E. Bollay, and Norman R. Beers. Handbook of Meteorology. New York and London; McGraw-Hill, 1945. 1068p.
Second	Beyer, Glenn H. Farm Housing in the Northeast. Ithaca; Cornell University Press, 1949. 458p.
Second	Binder, Raymond C. Fluid Mechanics. New York; Prentice-Hall, 1943. 307p. (5th ed., 1973. 401p.)
Second	Blackburn, Samuel A. Problems in Farm Woodwork. Peoria, Ill.; Manual Arts Press, 1915. 128p.
Second	Blanchard, Arthur H. Elements of Highway Engineering. New York; J. Wiley & Sons, 1915. 514p.
Second	Blanchard, Arthur H., Charles J. Bennett, Harold S. Boardman, et al., eds. American Highway Engineers' Handbook. New York; John Wiley & Sons, 1919. 1658p.
Second	Blanchard, Arthur H., and Henry B. Drowne. Text-Book on Highway Engineering. New York; J. Wiley & Sons, 1913. 762p.
Second	Bligh, W. G. The Practical Design of Irrigation Works. New York; Van Nostrand and London; Constable, 1907. 390p.
First	Bligh, William G. The Practical Design of Irrigation Works. London; Constable, 1907. 390p.
Second	Boorman, T. Hugh. Asphalts, Their Sources and Utilizations. New York; William T. Comstock Co., 1914. 191p.
First	Boss, William. The Heath Book for Threshers. Winnipeg; E. H. Heath Co., 1908. 312p.
First	Boss, William. Instructions for Traction and Stationary Engineers. Minneapolis, Minn.; Pedersen Linotype Co., 1906. 319p.
First	Bourne, John. The Steam Engine as Applied to Agricultural Purposes. London; Longman, 1856. 60p.
First	Bovey, Henry T. A Treatise on Hydraulics. New York; Wiley, 1895. 337p. (2d ed., 1901. 583p.)
First	Bowen, Eugene R. Tractor Farming. Peoria, Ill.; Avery Co., 1914. 60p.
First	Bowen, John T. Dairy Engineering. New York; Wiley, 1925. 532p.
Second	Bowie, Augustus Jesse. Practical Irrigation—Its Value and Cost. New York; McGraw-Hill and London; Spon, 1908. 232p.

Second	Brady, George S. Materials Handbook. New York; McGraw-Hill, 1929. 1 vol. (13th ed., 1991.)
First	Brate, H. R. Farm Gas Engines. London; C. Lockwood and Cincinnati, Ohio; Gas Engine Publ., 1912. 195p.
First	Breed, Charles B., and George L. Hosmer. The Principles and Practice of Surveying. New York; Wiley, 1906. 2 vols. (11th ed., 1977.)
Second	Breeder's Gazette. Farm Buildings. Chicago; Sanders Publishing Company, 1905. 185p. (4th ed., Chicago; The Breeder's Gazette, 1919. 394p.)
Second	Briggs, Howard L. Practical Bricklaying. Rev. ed. by William Carver. New York; McGraw-Hill, 1924. 223p.
First	Brooks, Frederick A. Climatic Environment: Syllabus for Agricultural Engineering 106. Davis; University of California, 1951. 153p.
Second	Brough, Charles H. Irrigation in Utah. Baltimore; Johns Hopkins University Press, 1898. 212p.
Second	Brown, Aubrey I., and Salvatore M. Marco. Introduction to Heat Transfer. New York and London; McGraw-Hill, 1942. 232p. (3d ed., 1958. 332p.)
Second	Brown, E. T. Farm Tractor: A Practical Handbook in the Selection and Management of a Tractor. London; C. Arthur Pearson Ltd., 1920. 160p.
First	Brown, George P. Drainage Channel and Waterways. Chicago; R. R. Donnelley, 1894. 480p.
First	Brown, Robert H. Farm Electrification. New York and London; McGraw-Hill, 1956. 367p.
First	Brown, Robert H. Irrigation: Its Principles and Practice as a Branch of Engineering. London; Constable and New York; Van Nostrand, 1907. 318p. (3d ed., 1920)
Second	Brown, Robert Q. Introduction to Engineering Problems. New York; Prentice-Hall, 1948. 191p.
First	Buckley, Robert B., comp. Design of Channels for Irrigation or Drainage. London; E. & F. N. Spon and New York; Spon & Chamberlain, 1911. 53p.
First	Buckley, Robert B. Facts, Figures and Formulae for Irrigation Engineers. London; Spon and Chamberlain, 1908. 239p.
First	Buckley, Robert B. The Irrigation Pocket Book. London; E. & F. N. Spon and New York; Spon and Chamberlain, 1911. 419p. (3d ed., 1920. 483p.)
Second	Buckley, Robert B. Irrigation Works in India and Egypt. London; E. & F. N. Spon and New York; Spon and Chamberlain, 1893. 336p.
Second	Burghardt, Henry D. Machine Tool Operation. New York; McGraw-Hill, 1919–1922. 2 vols. (5th ed., by Burghardt, Aaron Axelrod and James Anderson, 1959–1960. 2 vols.)
First	Burn, R. S. Practical Architecture as Applied to Farm Buildings. London; Country, 1878. 204p.
First	Burn, R. S. Practical Ventilation Applied to Public, Domestic and Agricultural Structures. Edinburgh and London; W. Blackwood and Sons, 1850. 208p.
First	Burn, R. S. Sanitary Science—As Applied to the Healthy Construction of Houses in Town and Country. Glasgow and London; W. Collins, Sons, & Co., 1872. 264p.
First	Burness, W., J. C. Morton, and G. Murray. The Equipment of the Farm. London; Bradburry, Agnew and Co., 1884. 142p.
Second	Burnett, Alexander. Tillage: A Substitute for Manure. London; Whittaker & Co., 1859. 212p.
Second	Burrell, M., ed. Power for Cultivation and Haulage on the Farm. London; Benn, 1928. 64p.

Second	Butler, Walter P., comp. Irrigation Manual. Huron, S.D.; Huronite Printing House, 1892. 252p.
Second	Butterworth, Benjamin, comp. The Growth of Industrial Art. Washington, D.C.; U.S. Government Printing Office, 1892. 200p. (Reprinted New York; Knopf, 1972.)
Second	Byers, Horace R. Synoptic and Aeronautical Meteorology. New York and London; McGraw-Hill, 1937. 279p. (Subsequent eds. entitled General Meteorology. 4th ed., 1974. 461p.)
Second	Byrne, Austin T. A Treatise on Highway Construction. New York; J. Wiley and Sons, 1892. 686p. (5th ed., 1908. 1040p.)

C

Second	Campbell, Henry C. How to Use Cement for Concrete Construction for Town and Farm. Chicago; Stanton and Van Vliet Co., 1920. 380p.
Second	Carpenter, Rolla C. Experimental Engineering and Manual for Testing. 2d ed. New York; J. Wiley, 1892. 709p. (7th ed., 1913. 1132p.)
Second	Carrier, Else H. The Thirsty Earth: A Study in Irrigation. London; Christophers, 1928. 222p.
Second	Carslaw, Horatio S., and J. C. Jaeger. Conduction of Heat to Solids. Oxford; Clarendon Press, 1947. 386p. (2d ed., 1959. 510p. 2d ed. reprinted 1986.)
First	Casson, Herbert N. Cyrus Hall McCormick (1809–1884): His Life and Work. London; A. C. McClurg, 1909. 264p.
First	Casson, Herbert N. Horse, Truck and Tractor. Chicago; F. G. Browne & Co., 1913. 200p.
First	Casson, Herbert N. The Romance of the Reaper. London and New York; Page & Co., 1908. 184p.
First	Chamberlain, W. I. Tile Drainage. Medina, Ohio; A. I. Root, 1891. 148p.
First	Chandler, Albert E. Elements of Western Water Law. San Francisco; Technical Publishing Co., 1913. 150p.
Second	Chatburn, George R. Highway Engineering: Rural Roads and Pavements. New York; John Wiley & Sons and London; Chapman & Hall, 1921. 379p.
First	Church, Austin H. Centrifugal Pumps and Blowers. New York; J. Wiley and London; Chapman & Hall, 1944. 306p.
Second	Church, Irving P. Mechanics of Engineering. New York; J. Wiley, 1890. 834p. (2d ed., 1915. 854p.)
Second	Churchill, Ruel V. Fourier Series and Boundary Value Problems. New York and London; McGraw-Hill, 1941. 206p. (4th ed. by Ruel V. Churchill and James W. Brown, 1987. 292p.)
Second	Clapp, William H., and Donald S. Clark. Engineering Materials and Processes: Metals and Plastics. Scranton, Pa.; International Textbook Co., 1938. 543p. (2d ed., 1949. 526p.)
First	Clark, Neil M. John Deere—He Gave the World the Steel Plow. Moline, Ill.; Deere and Co., 1937. 61p.
Second	Clarkson, Ralph P. Practical Talks on Farm Engineering. Garden City, NY; Page and Co., 1915. 223p.
Second	Clayton, C. H. J. Land Drainage: From Field to Sea. London; Country Life, 1919. 192p.
Second	Cleghorne, William S. H. Farm Buildings and Building Construction in South Africa. London and New York; Longmans, Green and Co., 1916. 325p.
First	Clerk, Dugald. The Gas and Oil Engine. London and New York; Longmans, Green and Co., 1902. 588p.

Second	Cobleigh, Rolfe. Handy Farm Devices and How to Make Them. New York; O. Judd and London; Kegan Paul, Trench, Trubner & Co., 1910. 288p.
First	Cochrane, Robert H. Farm Machinery and Tractors. Christchurch, New Zealand; Whitcombe and Tombs, 1951. 175p.
First	Coleman, T. E. Stable Sanitation and Construction. London; E. & F. N. Spon, 1897. 226p.
Second	Collins, Archie F. Farm and Garden Tractors: How to Buy, Run, Repair and Take Care of Them. New York; F. A. Stokes Co., 1920. 279p.
Second	Collins, Hubert E. Boilers, Pipes and Piping, Pumps. New York; McGraw, 1908. 435p.
First	Committee on the Relation of Electricity to Agriculture. CREA Handbook. Chicago; Committee on the Relation of Electricity to Agriculture, 1930. 28p.
Second	Cook, Glen C., comp. 380 Things to Make for Farm and Home, Vol. I; 500 More Things to Make for Farm and Home, Vol. II; and 600 Things to Make for the Farm and Home, Vol. III. Danville, Ill.; The Interstate, 1941–1952. 3 vols.
Second	Cook, Glen C. Practical Methods in Teaching Farm Mechanics. Danville, Ill.; The Interstate, 1936. 201p. (2d ed. with Clyde Walker, 1952. 658p.)
First	Cook, Glen C., L. L. Scranton, and H. F. McColly. Farm Mechanics Text and Handbook. Danville, Ill.; The Interstate, 1935. 450p. (1954 and subsequent eds. by Lloyd J. Phipps, H. F. McColly, L. L. Scranton, and G. C. Cook. 1959 ed. by Lloyd J. Phipps, et al. 752p.)
Second	Corn Industries Research Foundation. Corn in Industry. New York; Corn Industries Research Foundation, 1937. 63p. (3d ed., 1952. 60p.)
Second	Cotton, A. T. Irrigation in India. Manchester; Guardian Letterpress, 1875. 43p.
First	Cotton, Richard T. Insect Pests of Stored Grain and Grain Products: Identification, Habits and Methods of Control. Minneapolis, Minn.; Burgess Publishing Co., 1941. 242p. (2d ed. entitled Pests of Stored Grain and Grain Products, 1963. 318p.)
Second	Council on Fertilizer Application. Methods of Applying Fertilizer. Washington, D.C.; Council on Fertilizer Application, 1948. 27p. (Pamphlet No. 149)
First	Courtney, Wilshire S. The Farmers' and Mechanics' Manual. New York; E. B. Treat, 1868. 506p. (Revised and enlarged by George E. Waring, 1880. 562p.)
First	Cox, W. Gibbons. Irrigation with Surface and Subsurface Water and Land Drainage. Sydney; Angus & Robertson, 1906. 297p.
First	Craig, Dudley P. Steam Power and Internal Combustion Engines. New York and London; McGraw-Hill, 1931. 482p. (2d ed. Dudley P. Craig and Herbert J. Anderson, 1937. 571p.)
Second	Crane, Robert L., comp. Farm Labor Savers. New York; Greenberg, 1948. 116p.
Second	Cranshaw, Fred D., and E. W. Lehmann. Farm Mechanics. Peoria, Ill.; The Manual Arts Press, 1922. 423p.
First	Creager, William P., Joel D. Justin, and Julian Hinds. Engineering for Dams. New York; J. Wiley & Sons and London; Chapman & Hall, 1945. 3 vols.
Second	Croft, Terrell. Wiring for Light and Power. New York; McGraw-Hill, 1917. 426p. (4th ed., 1924. 551p.)
First	Croft, Terrell W., ed. Steam-Engine Principles and Practice. New York; McGraw-Hill, 1922. 513p. (2d ed., 1939.)
Second	Cruess, William V. Commercial Fruit and Vegetable Products. New York; McGraw-Hill, 1924. 530p. (4th ed., 1958. 884p.)

First   Culpin, Claude. Farm Machinery. London; Crosby Lockwood & Son, 1938. 405p. (11th ed. Oxford and Boston; Professional Books, 1986. 450p.)
Second   Culpin, Claude. Farm Mechanization: Costs and Methods. London; C. Lockwood, 1951. 147p.
First   Cumings, Richard C. The American Ice Harvests: A Historical Study in Technology, 1800–1918. Berkeley; University of California Press, 1949. 184p.
First   Currie, Barton W. The Tractor and Its Influence Upon the Agricultural Implement Industry. Philadelphia, Penn.; Curtis Publishing, 1916. 228p.
Second   Curtis, C. E. Farm Buildings for Landowners, Agents and Tenants. London; Vinton & Co., 1912. 152p.

D

Second   Dalla Valle, Joseph M. Micromeritics: The Technology of Fine Particles. New York; Pitman, 1943. 428p. (2d ed., 1948. 555p.)
Second   Dalzell, James R. Air Conditioning—Insulation. Chicago; American Technical Society, 1937. 301p.
Second   Dalzell, James R., and Gilbert Townsend. Concrete Block Construction for Home and Farm. Chicago; American Technical Society, 1951. 216p. (2d ed., 1957. 216p.)
First   Daniels, Farrington, and John A. Duffie. Solar Energy Research. Madison; University of Wisconsin Press, 1955. 290p.
Second   Darby, Henry Clifford. The Draining of the Fens. Cambridge, Eng.; The University Press, 1940. 312p. (2d ed., 1956. 314p.)
First   Davidson, Jay Brownlee. Agricultural Engineering. St. Paul, Minn.; Webb Publishing Co., 1913. 554p. (Rev. ed., 1918.)
First   Davidson, Jay Brownlee. Agricultural Machinery. New York; Wiley & Sons and London; Chapman & Hall, 1931. 396p.
First   Davidson, Jay Brownlee, G. W. McCuen, and R. U. Blasingame. Changes in Quality Values of Farm Machines. St. Joseph, Mich.; American Society of Agricultural Engineers, 1933. 165p.
First   Davidson, Jay Brownlee, and Leon W. Chase. Farm Machinery and Farm Motors. New York; Orange Judd Company, 1908. 513p.
First   Davidson, Jay Brownlee, Edwin L. Hansen, Howard F. McColly, and Archie A. Stone. A Report of Agriculture and Agricultural Engineering in China. Chicago; International Harvester, 1949. 259p.
First   Davidson, Jay Brownlee, G. W. McCuen, and R. U. Blasingame. Report of an Inquiry Into Changes in Quality Values of Farm Machines Between 1910–14 and 1932. St. Joseph, Mich.; American Society of Agricultural Engineers, 1933. 165p.
Second   Davie, W. Galesworthy. Old Cottages and Farmhouses in Surry. London; B. T. Batsford, 1908. 69p.
Second   Davies, Cornelius. Mechanized Agriculture. London; Temple Press, 1950. 122p.
Second   Davies, John G. The Principles of Cane Sugar Manufacture. London; N. Rodger, 1938. 144p.
First   Davis, Arthur P. Irrigation Works Constructed by the United States Government. New York and London; Chapman and Hall, 1917. 413p.
Second   Davis, Calvin V. Handbook of Applied Hydraulics. New York and London; McGraw-Hill, 1942. 1084p. (3d ed., 1969. 1 vol.)
Second   Davis, Kary C., ed. Farm Enterprise Mechanics. Chicago; J. B. Lippincott, 1935. 408p.

First	Davis, Raymond E. Elementary Plane Surveying. New York and London; McGraw-Hill, 1936. 424p. (4th ed., 1969. 574p.)
First	Davis, Raymond E., and Francis S. Foote. Surveying Theory and Practice. New York and London; McGraw-Hill, 1928. 1016p. (6th ed., 1981. 992p.)
Second	Davis, Raymond E., and Joe W. Kelly. Short Course in Surveying. New York; McGraw-Hill, 1942. 330p.
Second	Dawber, Edward Guy, and W. Galesworthy Davie. Old Cottages and Farmhouses in Kent and Sussex. London; B. T. Batsford, 1900. 28p.
Second	Day, William. Mechanical Science and the Prize System Related to Agriculture. London; Harrison, 1857. 54p.
Second	Deakin, Alfred. Irrigated India and Ceylon: An Australian View. London; W. Thacker & Co., 1893. 322p.
Second	Dean, George A. Essays on Construction of Farm Buildings and Labourers' Cottages. London; Simpkin, Marshall & Co., 1949. 65p.
First	Deere and Company. The Operation, Care, and Repair of Farm Machinery. 3d ed. Moline, Ill.; J. Deere, n.d. 192p. (28th ed., 1959. 279p.)
First	Deering Harvester Company. Official Retrospective Exhibition of the Development of Harvesting Machinery for the Paris Exposition of 1900. Chicago; Deering Harvester Company, 1900. 125p.
Second	Degler, Howard E. Internal-Combustion Engines. New York; J. Wiley & Sons, 1938. 411p.
Second	Dempsey, G. Drysdale. On the Drainage of Lands, Towns, and Buildings. Rev. ed. London; Crosby Lockwood, 1887. 351p. (Revised by D. Kennear Clark)
Second	Demsey, G. Drysdale. Rudimentary Treatise on the Drainage of Towns and Buildings. London; John Weale, 1849. 176p. (6th ed. London; Lockwood & Co., 1875. 245p.)
First	Denison, Merrill. Harvest Triumphant: The Story of Massey-Harris. New York; Dodd Mead, 1949. 340p.
First	Denton, John B. Agricultural Drainage: A Retrospect of Forty Years Experience. London; E. & F. N. Spon, 1883. 94p.
First	Dickey, George D., and Charles L. Bryden. Theory and Practice of Filtration. New York; Reinhold Publishing, 1946. 346p.
Second	Dickinson, Sherman. Job Operations in Farm Mechanics. 3d ed. Dansville, Ill.; Interstate, 1936. 164p.
Second	Dike, Paul H. Thermoelectric Thermometry. Philadelphia; Leeds & Northrup Co., 1954. 90p. (Technical Publication No. EN-33A)
Second	Dobson, Edward. A Rudimentary Treatise on the Manufacture of Bricks and Tiles. London; John Weale, 1850. 2 vols. (13th ed. revised and amplified by Alfred B. Searle. London; C. Lockwood and Son, 1921. 273p.)
Second	Dodge, Russell A., and Milton J. Thompson. Fluid Mechanics. New York and London; McGraw-Hill, 1937. 495p.
Second	Doyle, Kinsley D. Agriculture and Irrigation in Continental and Tropical Climates. London; Constable and Co., 1921. 268p.
First	Drachmann, Aage G. Ancient Old Mills and Presses. Copenhagen; Levin and Munksgaard, 1932. 181p.
First	Drew, James M. Farm Blacksmithing. St. Paul, Minn.; Webb Publishing Co., 1901. 91p. (2d ed., 1910.)
Second	Drew, James M. Ropework, Knots, Hitches, Splices and Halters. St. Paul, Minn.; Webb Publishing Co., 1936. 69p.
First	Dudgeon, E. C. Growing Crops and Plants by Electricity. London; S. Rentell and Co., 1912. 36p.

Second Duncan, John. Applied Mechanics for Beginners. London; Macmillan, 1902. 324p.
Second Dusinberre, George M. Numerical Analysis of Heat Flow. New York; McGraw-Hill, 1949. 227p.

E

Second E. I. Du Pont de Nemours & Company. Blasters' Handbook. Compiled under the direction of Arthur La Motte. Wilmington, Del.; E. I. Du Pont de Nemours & Co., 1922. 187p. (3d ed., 1930. 238p.)
Second E. I. Du Pont de Nemours & Company. Blasting Ditches. Wilmington, Del.; Du Pont de Nemours & Co., 1939. 48p.
First Earp, Unus F. Rural Electrification Engineering. New York; McGraw-Hill, 1950. 313p.
Second Easterbrook, L. F., ed. Farming and Mechanized Agriculture. London; Todd Publication Co., 1945. 415p.
Second Eberlein, H. D., and R. W. Ramsdell. Small Manor Houses and Farmhouses in France. New York; Lippincott, 1926. 303p.
Second Eckel, Edwin C. Cements, Limes, and Plasters. New York; J. Wiley & Sons, 1905. 712p. (3d ed., 1928. 699p.)
First Edminster, Frank C. Fish Ponds for the Farm. New York; C. Scribner's Sons, 1947. 114p.
First Eihinger, S. R., and M. S. Hutton. Steam Traction Engineering: A Book for Operating Engineers. London and New York; D. Appleton, 1916. 317p.
First Ekblaw, Karl J. T. Farm Concrete. New York; Macmillan, 1917. 295p.
First Ekblaw, Karl J. T. Farm Structures. New York; Macmillan, 1914. 347p.
Second Ellington, Karl J. Modern Pise-Building. Lindsborg, Kan.; Bethany Printing Company, 1924. 116p.
First Elliott, Charles G. Engineering for Land Drainage. New York; J. Wiley, 1903. 232p. (3d ed. Rev., 1919. 363p.)
First Elliott, Charles G., and J. J. W. Billingsley. Practical Farm Drainage. Indianapolis, Ind.; J. J. W. Billingsley, 1882. 2 vols. (Subsequent eds. New York; J. Wiley & Sons and London; Chapman & Hall. 2d ed., 1908. 188p.)
First Ellis, Lynn W., and Edward A. Rumely. Power and the Plow. Garden City, NY; Doubleday, Page & Co., 1911. 318p.
First Etcheverry, Bernard A. Irrigation Practice and Engineering. New York and London; McGraw-Hill, 1915–1916. 3 vols. (2d ed. with S. T. Harding, 1933. 1 vol.)
First Etcheverry, Bernard A. Land Drainage and Flood Protection. New York and London; McGraw-Hill, 1931. 327p. (Republished by Stanford University Press, 1940.)
First Ewart, John. A Treatise on the Arrangement and Construction of Agricultural Buildings. London; Brown, Green and Longmans, 1851. 28p.
Second Ewing, Scott. Soil Corrosion and Pipe Line Protection. New York; American Gas Association, 1938. 277p.

F

Second Fairbanks, Morse and Company. Catechism of Electrical Machinery. new ed. Chicago; Fairbanks, Morse and Company, 1957. 47p.
Second Faires, Virgil M. Design of Machine Elements. New York; Macmillan, 1934. 468p. (4th ed., 1965. 624p.)
First Farm Equipment Institute. Land of Plenty. Chicago; Farm Equipment Institute, 1950. 64p.

First	Farm Journal and Farmer's Wife. Electricity on Farms: An Analysis of the United States by Counties. Philadelphia; Farm Journal and Farmer's Wife, 1939. 95p.
Second	Farmer, F. Rhodes. Thirsty Earth. London; Longmans, 1934. 300p.
First	Farrall, Arthur W. Dairy Engineering. New York; John Wiley & Sons and London; Chapman & Hall, 1942. 405p. (2d ed., 1953. 477p.)
Second	Farrington, George H. Fundamentals of Automatic Control. London; Chapman & Hall, 1951. 285p.
First	Faulkner, Edward H. Plowman's Folly. Norman; University of Oklahoma Press and New York; Grosset & Dunlap, 1943. 155p.
Second	Field, Albert M., R. W. Olson, and V. E. Nylin. Farm Mechanics. London and New York; The Century Co., 1928. 385p.
Second	Fiske, G. B., comp. Poultry Architecture. New York; Judd, 1902. 130p.
Second	Fleming, Burton P. Practical Irrigation and Pumping. New York; Wiley and London; Chapman & Hall, 1915. 226p.
First	Flosdorf, Earl W. Freeze-Drying. New York; Reinhold, 1949. 280p.
First	Flynn, Patrick J. Irrigation Canals and Other Irrigation Works. San Francisco; G. Spaulding & Co., 1892. 398p.
First	Fortier, Samuel. Use of Water in Irrigation. New York; McGraw-Hill, 1915. 265p. (3d ed., 1926. 420p.)
First	Foster, Edgar E. Rainfall and Runoff. New York; Macmillan, 1948. 467p.
First	Foster, William A., and Deane G. Carter. Farm Buildings. New York; Wiley, 1922. 377p. (3d ed., 1941. 404p.)
First	French, Henry F. Farm Drainage. New York; C. M. Saxton, Barker & Co., 1860. 384p. (Republished New York; O. Judd Co., 1884 and 1919.)
First	French, Thomas E., and F. W. Ives. Agricultural Drawing and Design of Farm Structures. New York; McGraw-Hill, 1915. 130p.
First	Frevert, Richard K., et al. Engineering in Soil and Water Conservation. Ann Arbor, Mich.; Edwards Bros., 1953. 1 vol. (Subsequent editions entitled Soil and Water Conservation Engineering, New York; Wiley. 3d ed. by Glenn R. Schwab, et al., 1981. 525p.)
Second	Friese, John F. Farm Blacksmithing. Peoria, Ill.; The Manual Arts Press, 1921. 92p.
Second	Frost, Harwood. The Art of Roadmaking. New York; The Author, 1910. 544p.

G

Second	Gale, Stanley. Drainage of Land, Estates and Buildings. London; Chapman and Hall, 1949. 248p.
First	Ganguillet, E., and W. R. Kutter. A General Formula for the Uniform Flow of Water in Rivers and Other Channels. Translated from German by Rudolph Hering and John C. Trautwine, Jr. New York; J. Wiley and Sons and London; E. & F. N. Spon, 1889. 240p.
Second	Gay, Charles M., and Harry Parker. Materials and Methods of Architectural Construction. New York; J. Wiley & Sons, 1932. 639p. (3d ed. with John W. MacGuire, 1958. 724p.)
First	Geiger, Rudolf. The Climate Near the Ground. Translation by Milroy N. Steward and others of Das Klima der Bodennahen Luftschicft. Cambridge; Harvard University Press, 1950. 482p. (2d ed. from 4th German ed. Translated by Scripta Technics, 1965. 611p.)

Second	General Electric Company. Farm Wiring Guide. Bridgeport, Conn.; General Electric Company, 1945. 52p.
Second	General Motors Corporation. A Power Primer. Detroit, Mich.; General Motors Corporation, Department of Public Relations, 1955. 101p.
Second	Gerard, Geoffrey. Electricity for Farmers. London; Technical Press, 1949. 111p.
Second	Gerhard, William Paul. Disposal of Household Wastes (Farm Houses and Country Homes). New York; Van Nostrand, 1890. 195p.
Second	Gibson, Arnold H. Hydraulics and Its Applications. New York; D. Van Nostrand Company, 1908. 757p. (5th ed., London; Constable, 1952. 813p.)
First	Giese, Henry. Farm Fence Handbook. Chicago; Republic Steel Corporation, 1938. 63p.
Second	Gilbreth, F. B. Field Systems. New York; E. & F. N. Spon, 1908. 194p.
Second	Gillette, Halbert P., and Charles S. Hill. Concrete Construction: Methods and Cost. New York and Chicago; M. C. Clark Publishing Co., 1908. 690p.
Second	Glese, Henry. A Practical Course in Concrete. Chicago; Portland Cement Association, 1948. 67p.
Second	Glock, Haman. Concrete Construction on the Farm. Cleveland, Ohio; The Ohio Farmer, 1907–1908. 32p.
Second	Golding, Edward W. The Electrification of Agriculture and Rural Districts. London; English Universities Press, 1939. 244p.
First	Golding, Edward W. The Generation of Electricity by Wind Power. London; E. & F. N. Spon, 1955. 318p. (2d ed., 1976. 332p.)
Second	Goleze, Alfred R. Reclamation in the United States. New York; McGraw-Hill, 1952. 451p. (2d ed., Caldwell, Idaho; Caxton, 1961. 486p.)
First	Gray, Alfred S. Sprinkler Irrigation Handbook. 5th ed. Glendora, Cal.; Rain Bird Sprinkler Mfg. Corp., 1952. 40p. (10th ed. compiled by A. W. Fry and Alfred S. Gray, 1971. 43p.)
First	Gray, Harold E. Farm Service Buildings. New York and London; McGraw-Hill, 1955. 458p.
Second	Green, John L. English Country Cottages. London; Rural World Publishing Co., 1899. 235p.
Second	Green, John L. The Rural Industries of England. London; E. Marlborough & Co., 1895. 205p.
Second	Green, John L. Village Industries: A National Obligation. London; Rural World Publishing Co., 1915. 180p.
Second	Greene, Arthur M. The Elements of Refrigeration. New York; Wiley & Sons, 1916. 472p.
Second	Greenhill, Michael, and Evelyn Dunbar. Book of Farmcraft. London; Green and Co., 1942. 96p.
First	Greeno, Follett L., ed. Obed Hussey, Who, of All Inventors, and Bread Cheap (Reaper). Rochester, N.Y.; Rochester Herald Publishing, 1912. 228p.
Second	Griffith, Ira S. Carpentry. 3d ed. Peoria, Ill.; The Manual Arts Press, 1919. 188p.
First	Griffith, Ira S. Essentials of Woodworking. Peoria, Ill.; The Manual Arts Press, 1915. 190p.
Second	Grigson, G. English Farmhouse and Its Neighborhood. London; Max Parrish & Co., 1948. 128p.
First	Grimaldi, Filippo. Instructions for the Use and Preservation of Portable Engines and Threshing Machines. London; R. Garrett & Sons, 1881. 96p.

First Grover, Nathan C., and Arthur W. Harrington. Stream Flow: Measurements, Records and Their Uses. New York; Wiley, 1943. 363p.
First Gulvin, Harold E. Farm Engines and Tractors. New York and London; McGraw-Hill, 1953. 397p.
Second Gunn, Edwin. Farm Buildings: New and Adapted. Surbiton, Eng.; H. C. Long, 1935. 86p. (3d ed., London; C. Lockwood, 1946. 138p.)

H

First Hall, William H. Irrigation Development. Sacramento, Cal.; James J. Ayers, Supt. State Printing, 1886. 622p.
Second Hallock, Edward F. Tractor Engines. Cincinnati, Ohio; American Automobile Digest, 1920. 233p.
First Halsted, Byron D. Barn Plans and Outbuildings. New York; O. Judd Co., 1886. 235p. (2d ed., 1903. 388p.)
Second Halsted, Byron D., comp. Farm Conveniences. New York; Orange Judd Co., 1884. 240p. (3d ed., 1918. 240p.)
Second Hammond, Rolt. Water, Drainage, and Community. London; J. M. Dent & Sons, 1945. 92p.
Second Hanna, Frank W., and Robert C. Kennedy. The Design of Dams. New York and London; McGraw-Hill, 1931. 456p. (2d ed., 1938. 478p.)
First Hansen, E. L., H. F. McColly, A. A. Stone, and J. B. Davidson. Introducing Agricultural Engineering to China. Chicago; International Harvester Co., 1949. 259p.
Second Harcourt, Robert H. Elementary Forge Practice. 2d ed. Peoria, Ill.; Manual Arts Press, 1920. 158p. (3d ed., 1938. 182p.)
First Hardenberg, W. A., and Samuel Baker. Design of Dams, Irrigation. Scranton, Penn.; International Textbook Co., 1933. 115p.
Second Harding, Louis A. Mechanical Equipment of Buildings. New York; J. Wiley & Sons, 1916–1917. 2 vols. (2d ed. entitled Heating, Ventilating and Air Conditioning. New York; J. Wiley & Sons and London; Chapman & Hall, 1932. 963p. Previously published as vol. I of Mechanical Equipment of Buildings.)
Second Harding, Sidney T. Operation and Maintenance of Irrigation Systems. New York; McGraw-Hill, 1917. 271p.
Second Harding, Sidney T. Water Rights for Irrigation. Palo Alto, Cal.; Stanford University Press, 1936. 176p.
Second Harger, Wilson G. Handbook for Highway Engineers. 2d ed. New York; McGraw-Hill, 1916. 609p. (4th ed., 1927. 1721p.)
Second Harger, Wilson G. Rural Highway Pavements, Maintenance and Reconstruction. New York; McGraw-Hill, 1917. 271p.
Second Harris, D. G. Irrigation in India. London; H. Milford, Oxford University Press, 1923. 102p.
Second Hausbrand, Eugen. Evaporating, Condensing and Cooling Apparatus (Verdampfen, Kondensieren und Kouhlen). Translated by A. C. Wright. London; Scott, Greenwood & Co., 1908. 400p. (4th and 5th eds., revised by Basil Heastie. 5th ed., London; E. Benn Ltd.,1933. 503p.)
Second Hawkins, J. C. The Mechanical Equipment of Farms. London; E. & F. N. Spon, 1949. 418p.
First Hays, Willet M. Farm Development. New York; Orange Judd Company, 1910. 391p.

Primary Historical Literature 365

First   Hayward, Charles B. Gasoline Tractors. Chicago; American Technical Society, 1919. 169p.

Second   Heldt, Peter M. The Gasoline Automobile. 2d ed. Nyack, N.Y.; P. M. Heldt, 1916–1918. 3 vols. (Vol. 1 entitled The Gasoline Motor. 14th ed. of vol. 1 published separately as High-Speed Combustion Engines, 1948. 759p. Vol. 2 published separately as Motor Vehicles and Tractors, 1929. 678p.)

Second   Heldt, Peter M. Motor Vehicles and Tractors. Nyack, N.Y.; By the Author, 1929. 678p.

Second   Heldt, Peter M. Torque Converters or Transmissions. Nyack, N.Y.; P. M. Heldt, 1942. 406p. (5th ed., Philadelphia; Chilton Co., 1955. 496p.)

Second   Henderson, Melvin. Farm Mechanics in the Program of Vocational Agriculture. Springfield, Ill.; Board for Vocational Agriculture, 1949. 79p.

First   Henderson, Silas M., and R. L. Perry. Engineering Elements of Agricultural Processing. Ann Arbor, Mich.; Edwards, 1952. 204p.

Second   Hercules Powder Company. Hercules Dynamite on the Farm. Wilmington, Del.; Hercules Powder Company, 1933. 60p.

First   Hetzel, Frederic V. Belt Conveyors and Belt Elevators. New York; J. Wiley & Sons, 1922. 333p. (3d ed. by Frederic V. Hetzel and Russell K. Albright, 1941. 439p.)

Second   Higgins, J. F. Shop Plans of Useful Woodworking Equipment for Our Farms. Tallahassee; Florida Department of Agriculture, 1940. 136p.

Second   Hildebrand, Arthur E. Hilde's Manual for Demonstrating and Selling Tractors and Special Equipment for Agricultural and Industrial Use. Chicago; Ford Dealers' Service Bureau, 1926. 512p.

Second   Hill, Ernestine. Water into Gold. Melbourne; Robertson and Mullens, 1937. 328p. (8th ed., 1949. New and revised ed., 1958. 291p.)

Second   Hine, Howard J. Farm Mechanics Notebook. London; Farmer & Stock-breeder, 1948. 117p.

Second   Hine, Howard J. The Farm Workshop. London; C. Lockwood, 1953. 245p.

Second   Hine, Howard J. Good Farming by Machine. London; Hodder and Stoughton for the English University Press, 1948. 176p.

First   Hine, Howard J. Tractors on the Farm: Use and Maintenance. 3d ed. London; Farmer & Stock-Breeder, 1947. 128p. (5th ed., 1955. 247p.)

Second   Hinton, Richard J. Irrigation in the United States. 2d ed. Washington, D.C.; Government Printing Office, 1890. 386p. (1st ed., 49th Congress Miscellaneous Document No. 15.)

Second   Hirschfield, Clarence F., and Tomlinson C. Ulbricht. Gas Engines on the Farm. New York; Wiley and London; Chapman & Hall, 1913. 239p.

Second   Hiscox, Gardner D. Modern Steam Engineering in Theory and Practice. New York; N. W. Henley Publishing Co., 1907. 487p. (3d ed., 1913.)

First   Hjorth, Herman. Basic Woodworking Processes. New York and Milwaukee, Wis.; The Bruce Publishing Company, 1933. 221p. (3d ed. revised by Ewell W. Fowler, 1961. 224p.)

First   Hobbs, George W. The Gasoline Automobile. New York; McGraw-Hill, 1915. 259p. (2d ed. completely revised by Benn G. Elliott and Earl L. Consoliver, 1919. 483p. 5th ed. by Benjamin G. Elliott, 1939. 754p.)

First   Hogentogler, Chester A., ed. Engineering Properties of Soils. New York and London; McGraw-Hill, 1937. 434p.

Second   Holmstrom, John G. Modern Blacksmithing and Horseshoeing. Chicago; Alhambra Book Co., 1900. 202p. (2d ed., Chicago; F. J. Drake & Co., 1913. Republished New York; Drake Publishers, 1971.)

Second	Holmstrom, John G. Scientific Horse, Mule and Ox Shoeing. Chicago; F. J. Drake & Co., 1902. 119p.
Second	Holmstrom, John G. Standard Blacksmithing, Horseshoeing, and Wagon Making. St. Paul, Minn.; Webb Publishing Co., 1907. 211p.
Second	Holmstrom, John G., et al. American Blacksmithing, Toolsmiths and Steelworks Manual. Chicago; F. J. Drake for Sears, Roebuck & Co., 1911. 240p.
Second	Holtman, Dudley F. Wood Construction. New York; McGraw-Hill, 1929. 711p.
Second	Hool, George A. Reinforced Concrete Construction. New York; McGraw-Hill, 1912-1913. 2 vols. (3d ed., 1927-1928.)
Second	Hool, George A., and Harry E. Pulver. Concrete Practice. New York; McGraw-Hill, 1926. 369p.
Second	Hopkins, Alfred. Modern Farm Buildings. New York; McBride & Co., 1913. 206p. (2d ed., 1920. 237p.)
Second	Hopkins, George M. Farm Mechanics for Amateurs. London; Low, 1903. 376p.
First	Houghton, Albert A. Practical Silo Construction. New York; Norman W. Henley Publishing Co., 1911. 69p.
First	Houk, Ivan E. Irrigation Engineering. New York; Wiley, 1951-1956. 2 vols.
Second	Housden, Charles E. Practical Hydraulic, Water Supply and Drainage. London and New York; Longmans, Green, 1907. 105p.
Second	Howe, C. B. Agricultural Drafting. New York; Wiley, 1913. 63p.
Second	Hoyt, John C., and Nathan C. Grover. River Discharge. New York; J. Wiley, 1907. 137p. (4th ed., 1916. 210p.)
Second	Hubbard, Prevost. Dust Preventives and Road Binders. New York; J. Wiley & Sons, 1910. 416p.
Second	Hubbard, Prevost. Highway Inspectors' Handbook. New York; J. Wiley & Sons, 1919. 372p.
Second	Hubbard, Prevost. Laboratory Manual of Bituminous Materials. New York; J. Wiley & Sons, 1916. 153p.
Second	Hughes, Hector J., and Arthur T. Safford. A Treatise on Hydraulics. New York; Macmillan, 1911. 505p. (2d ed., 1926. 356p.)
Second	Hughes, Thomas P. Principles of Forging and Heat Treatment of Steel. Minneapolis; Burgess Rosberry, 1928. 87p. (2d ed., 1935. 115p.)
Second	Hughes, William J. Traction Engines Worth Modelling. London; P. Marshall, 1950. 160p.
Second	Huntington, Whitney C. Building Construction. New York; J. Wiley & Sons and London; Chapman & Hall, 1929. 596p. (6th ed. by Donald C. Ellison, W. C. Huntington, and Robert E. Mickadeit, 1987. 434p.)
Second	Hutchinson, Thomas. Machinery on the Farm. London; Blackie, 1949. 198p.
First	Hutchinson, William T. Cyrus Hall McCormick. New York and London; The Century Co., 1930. 2 vols.
First	Hutton, Frederick R. The Gas Engine. New York; J. Wiley & Sons and London; Chapman & Hall, 1903. 483p. (3d ed., 1908. 103p.)

I

First	Illuminating Engineering Society. IES Lighting Handbook. 2d ed. New York; Illuminating Engineering Society, 1952. 1 vol. (6th ed., 1981. 2 vols.)
First	Institute of Makers of Explosives. Explosives in Agriculture. New York; Institute of Makers of Explosives, 1931. 87p.
Second	Insulation Board Institute. Farm Building Insulation. Chicago, Ill.; Insulation Board Institute, 1948. 48p.

Second Investment Bankers Association of America. Reclamation Securities Commission. Synopses of Drainage Laws; A Handbook. Columbia, Miss.; Stephens Publishing Co., 1918. 365p.
First   Israelsen, Orson W. Irrigation Principles and Practices. New York; John Wiley & Sons and London; Chapman and Hall, 1932. 422p. (3d ed. with Vaughn E. Hansen, 1962. 447p.)

J
Second Jakob, Max. Heat Transfer. New York; J. Wiley, 1949–1957. 2 vols.
Second James, George W. Reclaiming the Arid West. New York; Dodd, Mead & Co., 1917. 441p.
Second Jasny, Naum. Research Methods on Farm Use of Tractors. New York; Columbia University Press and Oxford; Oxford University Press, 1938. 273p.
First   Jeffery, Joseph A. Text-Book of Land Drainage. New York; Macmillan Co., 1916. 256p.
First   Jennings, Burgess H., and Edward F. Obert. Internal Combustion Engines. Scranton, Pa.; International Textbook Co., 1944. 471p. (2d, 3d and 4th eds. by Obert. 4th ed. entitled Internal Combustion Engines and Air Pollution. New York; Intext Education Pub., 1973. 740p.)
Second Jensen, Gerard J. G. Cast-Iron House Drainage. London; Sanitary Publications, 1908. 218p.
First   Johnson, Elmer J., and Alvin H. Hollenberg. Servicing and Maintaining Farm Tractors. New York; McGraw-Hill, 1950. 230p.
Second Johnson, John B. The Materials of Construction. New York; Wiley and London; Chapman & Hall, 1897. 787p. (8th ed. revised by M. O. Withey and James Aston and entitled Johnson's Materials of Construction, 1947. 867p.)
First   Johnstone, Don, and William P. Cross. Elements of Applied Hydrology. Ronald Press; 1949, 276p.
Second Jones, Edward R. Notes on Drainage. Madison, Wis.; The Author, 1908. 164p.
Second Jones, Forrest R. Gas Engines. New York; Wiley, 1909. 447p.
First   Jones, Fred R. Farm Gas Engines and Tractors. New York and London; McGraw-Hill, 1932. 485p. (4th ed., 1963. 518p.)
First   Jones, Mack M. Farm Shop Practice. New York and London; McGraw-Hill, 1939. 315p.
First   Jones, Mack M. Shopwork on the Farm. New York and London; McGraw-Hill, 1945. 486p. (2d ed., 1955. 626p.)
Second Jones, Sydney R. English Village Homes and Country Buildings. London; B. T. Batsford, Ltd., 1936. 120p. (2d ed., New York; C. Scribner's Sons and London; B. T. Batsford, Ltd., 1937.)
Second Judge, Arthur W. The Testing of High Speed Internal Combustion Engines. New York; Van Nostrand, 1925. 392p. (4th ed., London; Chapman & Hall, 1955. 494p.)
First   Justin, Joel D. Earth Dam Projects. New York; J. Wiley & Sons and London; Chapman & Hall, 1932. 345p.

K
Second Kanowitz, S. B. Crushing, Grinding and Pulverising. New York; McGraw-Hill, 1950.
Second Kanthack, Francis E. The Principles of Irrigation Engineering. London; Longmans, Green & Co., 1924. 299p.

Second	Keene, Edward S. Mechanics of the Household. New York; McGraw-Hill, 1918. 391p.
Second	Kendall, Reginald G. Land Drainage. London; Faber and Faber, 1950. 133p.
Second	Kent, William. The Mechanical Engineer's Pocket-Book. New York; J. Wiley, 1895. 1087p. (12th ed. entitled Mechanical Engineers' Handbook, 1950. 2 vols.)
Second	Kenton, J. Bailey. The Farm Homesteads of England. London; Chapman and Hall, 1864. 178p. (2d ed., 1865. 188p.)
Second	Kershaw, Joseph W. Elementary Internal Combustion Engines. London; Longmans, Green & Co., 1912. 174p.
First	Ketchum, Milo S. The Design of Walls, Bins, and Grain Elevators. New York; The Engineering News Publishing Co., 1907. 393p. (3d ed., New York; McGraw-Hill, 1919. 556p.)
First	Kidder, Frank E. The Architect's and Builder's Pocket-Book. New York; J. Wiley & Sons, 1887. 649p. (18th ed., 1936. 2315p.)
Second	Kidner, Roger W. A Short History of Mechanical Traction and Travel. Chislehurst, Eng.; Oakwood Press, 1946–1947. 2 vols.
Second	King, David W. Homes for Home Builders. New York; O. Judd, 1886. 251p.
Second	King, Franklin H. Elementary Lessons in the Physics of Agriculture. Madison, Wis.; The Author, 1894. 184p.
First	King, Franklin H. Irrigation and Drainage: Principles and Practice of Their Cultural Phases. New York and London; Macmillan & Co., 1899. 502p. (4th ed., 1906.)
First	King, Franklin H. A Text Book of the Physics of Agriculture. 2d ed. Madison, Wis.; The Author, 1901. 604p. (6th ed. published by Mrs. F. H. King, 1914.)
First	King, Franklin H. Ventilation for Dwellings, Rural Schools and Stables. Madison, Wis.; The Author, 1908. 1128p.
First	King, Horace W. Handbook of Hydraulics. New York; McGraw-Hill, 1918. 424p. (6h ed. by Ernest F. Brater and Horace W. King, 1976. 604p.)
Second	King, Horace W., and Chester O. Wisler. Hydraulics. New York; Wiley, 1922. 237p. (5th ed., 1948. 351p.)
First	King, James A. Tile Drainage. Minneapolis; Chesher Printing Co., 1918. 63p. (4th ed., Mason City, Iowa; Mason City Brick & Tile Co., 1946. 129p.)
First	King, Matthew L. Silos: Construction and Service. St. Paul, Minn.; Webb Publishing Co., 1913. 100p.
Second	Kirkham, John E. Highway Bridges: Design and Cost. New York and London; McGraw-Hill, 1932. 395p.
First	Klippart, John H. The Principles and Practice of Land Drainage. Cincinnati, Ohio; R. Clarke & Co., 1861. 454p. (3d ed., 1888.)
Second	Koester, Frank. Electricity for the Farm and Home. New York; Sturgis and Walton Co. and McMillan, 1913. 279p.
First	Kranich, Frank N. G. Farm Equipment for Mechanical Power. New York; Macmillan, 1923. 405p.
First	Krynine, Dimitri P. Soil Mechanics. New York and London; McGraw-Hill, 1941. 451p. (2d ed., 1947. 511p.)

L

Second	Lacey, Joseph M. Hydrology and Ground Water. London; C. Lockwood & Son, 1926. 159p.

First	Laurens, Henry. The Physiological Effects of Radiant Energy. New York; The Chemical Catalog Company, 1933. 610p.
First	Lawrence, Charles P. Economic Farm Buildings. London; The Library Press Ltd., 1919. 180p.
Second	Lea, Frederick C. Hydraulics for Engineers and Engineering Students. London; E. Arnold, 1909. 536p. (6th ed., 1938. 757p.)
Second	Lebowitz, Samuel H. Pre-Service Course in Machine Science. New York; J. Wiley & Sons, 1943. 440p.
First	Leffel, James & Co. The Construction of Mill Dams. Springfield, Ohio; J. Leffel, 1874. 336p. (2d ed. entitled Leffel's Construction of Mill Dams and Bookwalter's Millwright and Mechanic. Springfield, Ohio; J. Leffel & Co., 1881. 283p.)
Second	Leland, E. H. Farm Homes In-Doors and Out-Doors. New York; Orange Judd Co., 1881. 204p.
Second	Lemstrom, Selim. Electricity in Agriculture and Horticulture. New York; D. Van Nostrand, 1904. 72p.
Second	Lewis, Alfred Dale. Irrigation and Settlement in America. Pretoria; Government Printing Office, 1915. 238p.
Second	Libbey-Owens-Ford Glass Company. How to Use the Sun Angle Calculator. Toledo, Ohio; Libbey-Owens-Ford Glass Company, 1951. 31p.
Second	Ligutti, L. G., and J. C. Rawe. Rural Roads to Security. Milwaukee, Wis.; Bruce Publishing Co., 1940. 387p.
Second	Lincoln Electric Company. Procedure Handbook of Arc Welding, Design and Practice. Cleveland, Ohio; The Lincoln Electric Company, 1933. 434p. (12th ed., 1973. 600p.)
Second	Link-Belt Company. Link-Belt Silent Chain for the Efficient Transmission of Power. Philadelphia, Penn.; Link-Belt Company, 1914. 111p.
Second	Linsley, Ray K., Max A. Kohler, and Joseph L. H. Paulhus. Applied Hydrology. New York; McGraw-Hill, 1949. 689p.
First	Lockwood, Joseph F. Flour Milling. Liverpool and New York; Northern Publishing, 1945. 511p. (4th ed., Stockport, Eng.; H. Simon, 1960. 526p.)
Second	Longamecker, E. W. The Practical Gas Engineer. 3d ed. Chicago; The Author, 1903. 152p. (8th ed., 1910. 165p.)
Second	Loving, Morris W. Concrete Pipe for Irrigation and Drainage. Chicago; American Concrete Pipe Association, 1939. 99p.
Second	Lowe, L. T. Students Handbook to Farming Implements and Machinery. Worcester, Eng.; Littlebury, 1950. 140p.
Second	Luckiesh, Matthew. Applications of Germicidal, Erythemal and Infrared Energy. New York; D. Van Nostrand, 1946. 463p.
Second	Lynde, Carlton J. Home Waterworks: A Manual of Water Supply in Country Homes. New York; Sturgis & Walton, 1911. 270p.

M

First	Ma, Fengchow C., T. Takasaka, and Chin-wen Yang. A Preliminary Study of Farm Implements Used in Taiwan Province. Taipei; Chinese-American Joint Commission on Rural Reconstruction, 1955. 331p. (2d ed., 1958. 333p.)
First	MacGregor, Wallace F. Science of Successful Threshing. 5th ed. Racine, Wis.; J. I. Case Threshing Machine Company, 1907. 222p. (7th ed., 1915. 263p.)
Second	Macintire, Horace J. The Principles of Mechanical Refrigeration. New York; McGraw-Hill, 1922. 252p. (2d ed., 1928. 317p.)

Second	Mackenzie, Nicol F. Notes on Irrigation Works. New York; D. Van Nostrand Co., 1910. 111p.
First	Madison, Richard D., ed. Fan Engineering. Buffalo, NY; Buffalo Forge Co., 1925. 610p. (8th ed., edited by Robert Jorgensen, 1983. 1000p.)
Second	Maggard, James H. Rough and Tumble Engineering. Iowa City, Iowa; Republican Printing Company, 1901. 143p.
First	Maggard, James H. The Traction Engine. Philadelphia; D. McKay, 1898. 128p. (3d ed., 1908. 293p.)
Second	Malden, Walter J. Farm Buildings and Economical Agricultural Appliances. London; K. Paul, Trench, Troubner, 1896. 192p.
Second	Maleev, Vladimir L. Machine Design. Scranton, Penn.; International Textbook Co., 1946. 581p. (4th ed. published as Mechanical Design of Machines by Martin J. Siegel, Vladimir L. Maleev and James B. Hartman, 1965. 576p.)
First	Marks, L. S., ed. Mechanical Engineers' Handbook. New York; McGraw-Hill, 1916–1967. 7 vols. (Volumes 6 and 7 edited by T. Baumeister.)
Second	Martin, George A. Farm Appliances; A Practical Manual. New York; O. Judd, 1892. 198p. (2 ed., 1913. 192p.)
Second	Massey, George B. The Engineering of Excavation. New York; Wiley, 1923. 376p.
Second	Massingham, Harold J. Country Relics: Old Tools, Craftsman. Toronto and New York; Macmillan, 1939. 240p.
Second	Mathews, John L. The Conservation of Water. Boston; Small, Maynard & Co., 1910. 289p.
First	Matthews, R. Borlase. Electro-Farming or The Application of Electricity to Agriculture. London; Ernest Benn Ltd., 1928. 357p.
Second	Mauldin, Hurst, and William A. Cochran Jr. Electricity for the Farm. Rev. ed. Birmingham; Alabama Power Co., Rural and Towns Division, 1952. 144p.
Second	Mawson, E. O. Pioneer Irrigation. London; C. Lockwood & Son, 1904. 206p.
Second	Maxwell, William H. Drainage Work and Sanitary Fittings. London; St. Bride's Press, 1911. 142p.
Second	McAdams, William H. Heat Transmission. New York and London; McGraw-Hill, 1933. 383p. (3d ed., 1954. 532p.)
First	McColly, Howard F., and James W. Martin. Introduction to Agricultural Engineering. New York and London; McGraw-Hill, 1955. 458p.
Second	McConnell, Primrose. Farm Equipment, Building and Machinery. London; Cassell, 1910. 124p.
First	McCormick, Cyrus. The Century of the Reaper. Boston and New York; Houghton-Mifflin Co., 1931. 307p.
First	McHardy, Douglas N. Farm and Industrial Tractors. London; C. Lockwood and Son, 1930. 244p.
First	McHardy, Douglas N. Modern Farm Buildings. London; C. Lockwood and Son, 1932. 227p.
Second	McHardy, Douglas N. Modern Farm Machinery. London; Methuen & Co., 1924. 255p.
First	McHardy, Douglas N. Power Farming for Crops and Stock. London; Philip Palmer Press, 1938. 208p.
Second	McMillan, Franklin R. Basic Principles of Concrete Making. New York; McGraw-Hill, 1929. 99p.
First	Mead, Daniel W. Hydrology. New York; McGraw-Hill, 1919. 647p. (2d ed., 1950. 728p.)

Second Mead, Elwood. Irrigation Institutions. New York and London; Macmillan, 1903. 392p. (Reprinted New York; Arno Press, 1972.)
Second Mead, Elwood. Irrigation in Northern Italy (Part 2). London; Wm. Wesley & Son, 1907. 86p.
First Meinzer, Oscar E., ed. Hydrology. New York; Dover, 1942. 712p.
First Mercer, Henry C. Ancient Carpenters' Tools. Doylestown, Penn.; The Bucks County Historical Society, 1929. 328p. (5th ed., New York; Horizon Press for the Bucks County Historical Society, 1975. 339p.)
Second Mercier, Charles A. A Manual of the Electro-Chemical Treatment of Seeds. London; University of London Press, 1919. 134p.
Second Merriman, Mansfield. A Treatise on Hydraulics. New York; J. Wiley and Sons, 1889. 381p. (10th ed. by Mansfield Merriman and Thaddeus Merriman, 1916. 565p.)
Second Merriman, Mansfield, ed. Mechanics of Materials. 10th ed. New York; Wiley, 1905. 507p. (11th ed., 1914. 524p.)
Second Middleton, George A. T. Building Materials. London; B. T. Batsford, 1905. 420p.
Second Middleton, George A. T. Drainage of Town and Country Houses. London; Batsford, 1908. 176p.
Second Middleton, William E. K. Meteorological Instruments. Toronto; University of Toronto Press, 1941. 213p. (3d ed., 1953. 286p.)
First Miles, Manly. Land Draining: A Handbook for Farmers on the Principles and Practice of Farm Draining. New York; O. Judd, 1892. 199p.
Second Mills, J. Warner. Mills' Irrigation Manual for Lawyers, Irrigation Officers, Engineers and Water Users. Denver, Col.; Mills Publishing Company, 1907. 635p.
Second Miyawaki, Atsushi. Condensed Milk. New York; J. Wiley & Sons and London; Chapman & Hall, 1928. 380p.
Second Molitor, David A. Hydraulics of Rivers, Weirs and Sluices. New York; J. Wiley and Sons, 1908. 135p.
Second Moon, Parry H. The Scientific Basis of Illuminating Engineering. New York and London; McGraw-Hill, 1936. 608p.
Second Moore, Henry I. Silos and Silage. London; Farmer & Stock-Breeder Publications, 1941. 112p. (2d ed., 1950. 116p.)
Second Moritz, Ernest A. Working Data for Irrigation Engineers. New York; J. Wiley and London; Chapman & Hall, 1915. 395p.
Second Morrison, Ivan G. Farm Tractor Maintenance. Danville, Ill.; Interstate, 1946. 202p.
Second Morrison, Ivan G. Repairing Farm Machinery. Danville, Ill.; Interstate, 1940. 181p.
Second Mortensen, Martin. Management of Dairy Plants. New York; Macmillan, 1938. 407p.
Second Mosier, Jeremiah G., and A. F. Gustafson. Soil Physics and Management. Philadelphia and London; J. B. Lippincott, 1917. 442p.
Second Mullins, J. Irrigation Manual. London and New York; E. & F. N. Spon for the Madras Government, 1890. 223p.
First Munn, B. The Practical Land Drainer. New York; C. M. Saxton & Co., 1856. 190p.
First Murphy, Glenn. Similitude in Engineering. New York; Ronald Press, 1950. 302p.

First   Muskat, Morris. The Flow of Homogeneous Fluids through Porous Media. New York; McGraw-Hill, 1937. 763p.

N

Second   Nadler, Maurice. Modern Agricultural Mathematics. New York; Orange Judd Publishing, 1940. 315p.

Second   Nardo, A. Farm Houses, Small Chateaux, and Country Churches in France. Cleveland, Ohio; J. H. Hansen Pub., 1924. 173p.

First   National Institute of Agricultural Engineering (Great Britain). Tractor Ploughing. Rev. ed. London; H. M. Stationery Office, 1948. 40p.

First   National Research Council (United States). International Critical Tables of Numerical Data, Physics, Chemistry and Technology. New York; McGraw-Hill for the National Research Council, 1926–1930. 7 vols. (Reprinted 1986.)

Second   Newell, Frederick H. Irrigation in the United States. New York; T. Y. Crowell, 1902. 417p. (2d ed., 1906. 433p.)

First   Newell, Frederick H. Irrigation Management. New York; D. Appleton, 1916. 306p.

First   Newell, Frederick H., comp. Proceedings of the 1st Conference of Engineers of the Reclamation Service. Washington, D.C.; Government Printing Office, 1904. 361p. (U.S. Geological Survey, Water-Supply and Irrigation Paper No. 93)

First   Newell, Frederick H., comp. Proceedings of the 2d Conference of Engineers of the Reclamation Service. Washington, D.C.; Government Printing Office, 1905. 267p. (U.S. Geological Survey, Water-Supply and Irrigation Paper No. 146)

First   Newell, Frederick H. Water Resources: Present and Future Uses. New Haven, Conn.; Yale University Press, 1920. 310p.

First   Newell, Frederick H. Water Supply for Irrigation. Washington, D.C.; Government Printing Office, 1894. 99p.

First   Newell, Frederick H., and Daniel W. Murphy. Principles of Irrigation Engineering. New York; McGraw-Hill, 1913. 293p.

Second   Newhouse, F., M. Ionides, and G. Lacey. Irrigation. London; Longmans, 1950. 67p.

First   Nicholson, Harry H. The Principles of Field Drainage. Cambridge, Eng.; Cambridge University Press, 1942. 165p. (2d ed., 1953. 163p.)

Second   Northend, Mary H. Remodeled Farmhouses. Boston; Little, Brown, and Co., 1915. 264p.

O

Second   O'Brien, Morrough P., and George H. Hickox. Applied Fluid Mechanics. New York and London; McGraw-Hill, 1937. 360p.

Second   Ogden, Henry N. Rural Hygiene. New York; MacMillan Co., 1911. 434p.

Second   Ogden, Henry N. Sewer Construction. New York; J. Wiley & Sons, 1908. 335p.

Second   Olin, Walter H. American Irrigation Farming. Chicago; A. C. McClurg Co., 1913. 364p.

Second   Ower, Ernest. The Measurement of Air Flow. London; Chapman & Hall, 1927. 199p. (5th ed. by Ernest Ower and R. C. Pankhurst. Oxford and New York; Pergamon Press 1977. 362p.)

First   Oxley, T. A. Scientific Principles of Grain Storage. London; The Northern Publishing Co., 1948. 103p.

P

Second	Page, Victor W. The Model T and A Ford Car, Truck and Tractor Conversion Sets and Fordson Farm Tractor. London; Hodder & Stoughton, 1928. 574p.
First	Page, Victor W. The Model T Ford Car Including the Fordson Farm Tractor. New York; Norman W. Henley Publishing Co., 1926. 495p.
First	Page, Victor W. The Modern Gas Tractor. New York; Norman W. Henley, 1914. 475p. (4th ed., 1922. 573p.)
Second	Parker, Marvin M. Farm Welding. Hayti, Mo.; 1950. 200p. (3d ed., New York; McGraw-Hill, 1958. 262p.)
Second	Parker, Philip M. The Control of Water as Applied to Irrigation, Power and Town Water Supply Purposes. London; G. Routledge & Sons, 1913. 1055p. (2d ed., 1925.)
Second	Parkes, Louis C. House-Drainage, Sewerage and Sewage Disposal in Relation to Health. London; H. K. Lewis, 1909. 150p.
First	Parsons, John L. Land Drainage. Chicago; Myron C. Clark Publishing Co. and London; E. & F. N. Spon, 1915. 165p.
Second	Passmore, J. B. The English Plough. London; Oxford University Press, 1930. 88p.
Second	Patch, A. J., ed. Welding Helps for Farmers. Cleveland, Ohio; James F. Lincoln, 1947. 431p.
Second	Paustian, Paul W. Canal Irrigation in Punjab. New York; AMS Press and Columbia University Press, 1930. 179p.
Second	Peet, Louise J., and Lenore E. Sater. Household Equipment. New York; J. Wiley & Sons and London; Chapman & Hall, 1934. 315p. (8th ed. by Louise J. Peet, Mary S. Pickett, and Mildred G. Arnold, 1979. 583p.)
Second	Pence, William D., and Milo S. Ketchem. Surveying Manual. New York; McGraw-Hill, 1915. 388p. (5th ed., 1932. 363p.)
Second	Pennington, A. M. Farm Silos, Granaries, and Tanks. London; Concrete Pub., 1942. 88p.
First	Perry, John J. Chemical Engineers' Handbook. New York and London; McGraw-Hill, 1934. 2609p. (6th ed., 1984. 1 vol.)
Second	Peter, Dimitur T. Practical Repairing of Land Machines. Sophia; Zemsnav, 1948. 280p.
First	Pickels, George W. Drainage and Flood-Control Engineering. New York and London; McGraw-Hill, 1925. 450p. (2d ed., 1941. 476p.)
First	Pinches, Harold E. Introduction to Agricultural Engineering. Ann Arbor, Mich.; Edwards Brothers, 1941. 79p.
Second	Plummer, Harry C., and Edwin F. Wanner. Principles of Tile Engineering: Handbook of Design. Washington, D.C.; Structural Clay Products Institute, 1947. 453p.
Second	Polson, Joseph A. Internal Combustion Engines. New York; J. Wiley & Sons and London; Chapman & Hall, 1931. 475p. (1st ed., corrected 1933. 2d ed., 1942. 554p.)
Second	Poole, Cecil P. The Gas Engine. New York; Hill Publishing Co., 1909. 97p.
Second	Portland Cement Association. Irrigation with Concrete Pipe. Chicago; Portland Cement Association, 1948. 55p. (2d ed., 1952.)
First	Potter, Andrey A. Farm Motors: Steam and Gas Engines. New York; McGraw-Hill, 1913. 261p. (3d ed., 1925. 299p.)
Second	Potter, Andrey A., and James P. Calderwood. Elements of Steam and Gas Power Engineering. New York; McGraw-Hill, 1920. 304p. (4th ed., 1938.)

First	Powers, Wilbur L., and T. A. H. Teeter. Land Drainage. New York; J. Wiley & Sons, 1922. 270p. (2d ed., 1932. 353p.)
Second	Preston, Thomas. The Theory of Heat. London and New York; Macmillan, 1894. 719p. (4th ed., edited by J. Rogerson Cotter, 1929. 836p.)
Second	Purvis, G. H. Agricultural Implements. London; E. Benn, Ltd., 1923. 110p.
Second	Putnam, Xenophon W. The Gasoline Engine on the Farm. New York; The Norman W. Henley Publishing Co., 1913. 527p. (2d ed., 1919.)

R

First	Radebaugh, Gustav H. Standard Mechanical Practices in Repairing Farm Machinery and Equipment. Milwaukee, Wis.; The Bruce Publishing Company, 1923. 260p.
Second	Radford, William A., ed. Book of Farm Improvements and City Buildings. Chicago; Excelsior Printing Co., 1917. 160p.
Second	Radford, William A., ed. Farm and Building Book. Chicago; J. Thomas & Co. and Norfolk, Vir.; North Carolina Pine Association, 1915. 160p.
Second	Radford, William A. Materials Lists of Designs Appearing in Farm and Building Book. Chicago; J. Thomas & Co., 1915. 219p.
Second	Radford, William A. The New Handy Book of Up-To-Date Barn Plans. Chicago; Farm Press Publication Co., 1907. 160p.
Second	Radford, William A., ed. Our Farm and Building Book. Chicago; Excelsior Printing Co., 1914. 160p.
Second	Radford, William A. Practical Country Building. Wausau, Wis.; Northern Hemlock & Hardwook Manufacturing Association, 1912. 192p.
Second	Radford, William A., ed. Radford's Combined House and Barn Plan Book. Chicago; The Radford Architecture Co., 1908. 287p.
Second	Radford, William A. Radford's Handy Book of Practical Barn Plans and All Kinds of Out-Buildings. Chicago; The Radford Architecture Co., 1911. 160p.
Second	Radford, William A., ed. Radford's Practical Barn Plans. Chicago and New York; The Radford Architecture Co., 1909. 287p.
Second	Radford, William A., A. S. Johnson, and B. L. Johnson. Framing: House, Barn, Roof. Chicago; The Radford Architecture Co., 1919. 338p.
Second	Ramsower, Harry C. Equipment for the Farm and Farmstead. Boston; Ginn and Co., 1917. 523p.
Second	Ransome, James S. Modern Wood-Working Machinery. London; William Rider & Son, 1896. 236p.
Second	Rathbun, John B. Gas Engine Troubles and Installation. Chicago; Charles C. Thompson Co., 1911. 440p. (2d ed. Chicago; Stanton and Van Vliet Co., 1917. 448p.)
Second	Rathbun, John B. Gas, Gasoline and Oil Engines. Chicago; Stanton and Van Vliet Co., 1919. 341p.
Second	Rathbun, John B. Practical Hand Book of Gas, Oil and Steam Engines. Chicago; Charles C. Thompson Co., 1913. 370p.
Second	Republic Steel Corp. How to Lay Steel Roofing. Chicago; Republic Steel Corp., 1938. 94p.
First	Reynolds, Percy A. Farm Mechanization Handbook. London; English University Press for Temple Press, 1948. 184p. (2d and 3d eds. by Thomas H. Cradock, 1952 and 1957. 339p. and 470p.)
Second	Richardson, Clifford. Asphalt Construction for Pavements and Highways. New York; McGraw-Hill, 1913. 155p.

Second	Richardson, Clifford. The Modern Asphalt Pavement. 2d ed. New York; J. Wiley & Sons, 1908. 629p.
First	Richter, Herbert P. Practical Electrical Wiring. New York and London; McGraw-Hill, 1939. 503p. (13th ed. by Herbert P. Richter and W. Creighton Schwan, 1984. 685p.)
Second	Richter, Herbert P. Practical Electricity and House Wiring. Chicago; F. J. Drake and Co., 1934. 183p.
Second	Richter, Herbert P. Wiring Simplified. 15th ed. Chicago; Park Publishing Co., 1943. 122p. (32d ed., St. Paul, Minn.; Park Publishing, 1977. 159p.)
Second	Ripper, William. Ripper's Steam Engine Theory and Practice. 8th ed. London and New York; Longmans, Green and Company, 1932. 841p.
Second	Roadhouse, Chester L., and James L. Henderson. The Market-Milk Industry. New York and London; McGraw-Hill, 1941. 624p. (3d ed., Westport, Conn.; AVI Publishing Co., 1971. 677p.)
First	Robb, B. B., and F. G. Behrends. Farm Engineering. Vol. 1—Farm Mechanics. New York; Wiley and London; Chapman and Hall, 1924. 454p.
Second	Roberts, Howard A. The Farmer: His Own Builder. Philadelphia; David McKay, 1918. 302p.
Second	Roberts, I. P. The Farmstead: The Making of the Rural Home and the Layout of the Farm. New York; MacMillan, 1902. 350p.
First	Roe, Harry B. Moisture Requirements in Agriculture: Farm Irrigation. New York; McGraw-Hill, 1950. 413p.
First	Roe, Harry B., and Quincy C. Ayres. Engineering for Agricultural Drainage. New York; McGraw-Hill, 1954. 501p.
Second	Roehl, Louis M. Agricultural Woodworking. Milwaukee, Wis.; Bruce Publishing Co., 1916. 137p.
Second	Roehl, Louis M. The Farmer's Shop Book. Milwaukee, Wis.; The Bruce Publishing Company, 1924. 429p. (10th ed., 1953. 452p.)
Second	Roehl, Louis M. Farm Woodwork. Milwaukee, Wis.; Bruce Publishing Co., 1919. 136p.
First	Roehl, Louis M. Fitting Farm Tools. Milwaukee and New York; The Bruce Publishing Company, 1930. 102p.
Second	Roehl, Louis M. Harness Repairing. Milwaukee, Wis.; Bruce Publishing Co., 1921. 53p.
Second	Roehl, Louis M. Household Carpentry. New York; Macmillan, 1927. 196p.
Second	Roehl, Louis M. Problems for School and Home Workshop. Milwaukee, Wis.; Bruce Publishing Co., 1935. 88p.
Second	Roehl, Louis M. Problems in Carpentry. St. Paul, Minn.; Webb Publishing Co., 1913. 111p. (2d ed., 1918.)
First	Roehl, Louis M. Rope Work. Milwaukee, Wis.; Bruce Publishing Co., 1921. 47p.
Second	Roehl, Louis M. Shop Management in Rural Schools. Milwaukee, Wis.; Bruce Publishing Co., 1934. 96p.
First	Rogin, Leo. The Introduction of Farm Machinery in its Relation to the Productivity of Labor in the Agriculture of the United States during the Nineteenth Century. Berkeley; University of California Press, 1931. 260p. (Reprinted)
Second	Rommel, George M. Farm Products in Industry. New York; Rae D. Henkle, 1928. 318p.
First	Rouse, Hunter. Elementary Mechanics of Fluids. New York; J. Wiley, 1946. 376p.

Second	Rouse, Hunter. Fluid Mechanics for Hydraulic Engineers. New York; McGraw-Hill, 1938. 422p.
Second	Russell, George E. Text Book on Hydraulics. New York; H. Holt and Company, 1910. 183p. (5th ed. 1942. 468p.)
First	Russell, William. Scientific Horseshoeing for Leveling and Balancing the Action and Gait of Horses. Cincinnati, Ohio; Robert Clarke Co., 1899. 340p.

S

Second	Sanders, J. H. Practical Hints About Barn Building. Chicago; Sanders, 1893. 284p.
Second	Savage, William G. Rural Housing. London; T. F. Unwin, 1915. 297p.
First	Schaenzer, Joseph P. Rural Electrification. Milwaukee, Wis.; Bruce Publishing Co., 1935. 266p. (5th ed., 1955. 378p.)
Second	Schmidt, G. A., W. A. Ross, and M. A. Sharp. Teaching Farm Shop Work and Farm Mechanics. London and New York; The Century Co., 1927. 288p.
Second	Schneider, Norman H. Electric Light for the Farm. New York; Spon & Chamberlain, 1911. 1 vol.
Second	Schuyler, James D. Reservoirs for Irrigation, Water Power, and Domestic Use. New York; J. Wiley, 1901. 414p.
First	Schwarzkopf, Ernst. Plain and Ornamental Forging. New York; J. Wiley & Sons, 1916. 267p. (2d ed., 1930. 281p.)
First	Scoates, Daniel. Farm Buildings. College Station, Texas; The Author, 1937. 2 vols.
Second	Scotland. Committee on Farm Buildings for Scotland. Farm Buildings for Scotland. London; H. M. Stationary Office, 1946. 90p. (Great Britain Ministry of Works Post-War Building Studies No. 22)
First	Scott, James H. Flour Milling Processes. New York; D. Van Nostrand Co., 1936. 416p. (2d ed., London; Chapman & Hall, 1951. 670p.)
First	Scott, John G. S. The Complete Text-Book of Farm Engineering. Volume 1, Drainage and Embanking; Volume 2, Irrigation and Water Supply; Volume 3, Farm Roads, Fences and Gates; Volume 4, Farm Buildings; Volume 5, Barn Implements and Machines; Volume 6, Field Implements and Machines; Volume 7, Agricultural Surveying. London; C. Lockwood and Co., 1885. 7 vols.
Second	Scott-Moncrieff, Colin C. Irrigation in Southern Europe. London; Spon, 1868. 371p.
Second	Seaton, Roy. Concrete Construction for Rural Communities. New York; McGraw-Hill, 1916. 223p. (2d ed., 1918.)
First	Segler, G. Pneumatic Grain Conveying. Braunschweig; G. Segler, 1951. 174p.
Second	Selvidge, Robert W., and J. M. Allton. Blacksmithing. Peoria, Ill.; Manual Arts Press, 1925. 156p.
Second	Severns, William H. Heating, Ventilating, and Air Conditioning Fundamentals. New York; J. Wiley & Sons and London; Chapman & Hall, 1937. 467p. (2d ed., 1949. 666p.)
Second	Seymour, Hartland. Crushing and Grinding Machinery. London; E. Benn Ltd., 1924. 143p.
Second	Sharma, K. R. Irrigation Engineering. 2d ed. Jullundur; India Printers, 1948–1949. 3 vols.
Second	Sharp, M. A., and W. A. Sharp. Principles of Farm Mechanics. New York; Wiley and London; Chapman & Hall, 1930. 269p.

Second   Shearer, Herbert A. Farm Buildings with Plans and Descriptions. Chicago; F. J. Drake & Co., 1917. 256p.
Second   Shearer, Herbert A. Farm Mechanics. London; Geoffrey Parker & Gregg and Chicago; F. J. Drake & Co., 1919. 250p.
Second   Sheldon, Samuel. Dynamo Electric Machinery. New York; D. Van Nostrand, 1900. 281p. (8th ed. by Samuel Sheldon and Erich Hausmann, 1901. 328p.)
Second   Sherwood, George. Farm Tractor Handbook. London; Iliffe & Sons, Ltd., 1918. 202p.
Second   Sherwood, Thomas K. Absorption and Extraction. New York and London; McGraw-Hill, 1937. 278p.
Second   Skilton, C. P. British Windmills and Watermills. London; Collins, 1948. 47p.
First   Slight, James, and Scott R. Burn. The Book of Farm Implements and Machines. Edinburgh and London; W. Blackwood and Sons, 1858. 648p.
Second   Smith, Edward. The Peasant's Home, 1760–1875. London; E. Stanford, 1876. 139p.
Second   Smith, G. Geoffrey, ed. The Modern Diesel. 11th ed. revised and rewritten by Donald H. Smith. London; Iliffe, 1949. 277p. (14th ed. edited by D. S. D. Williams, R. J. B. Keig, and John M. Dickson-Simpson. London; Newnes-Butterworths, 1972. 248p.)
First   Smith, Harris P. Farm Machinery and Equipment. New York and London; McGraw-Hill, 1929. 448p. (6th ed., 1976. 488p.)
Second   Smith, Robert H. Agricultural Mechanics. Philadelphia and Chicago; J. B. Lippincott Co., 1925. 357p.
Second   Smythe, William E. The Conquest of Arid America. New York; Harper, 1900. 325p. (2d ed., New York and London; Macmillan, 1907. 360p.)
Second   Spalding, Frederick P. A Text-Book on Roads and Pavements. New York; Wiley & Sons, 1894. 213p. (4th ed., 1912. 408p.)
First   Spangler, Merlin G. Soil Engineering. Scranton, Pa.; International Textbook Co., 1951. 458p. (3d ed. by Merlin G. Spangler and Richard L. Handy. New York; Intext Educational Publishers, 1973. 748p.)
Second   Spinks, William. House Drainage Manual. London; Biggs and Co., 1897. 306p.
First   Spon, Ernest. The Present Practice of Sinking and Boring Wells. New York; E. & F. N. Spon, 1875. 217p.
Second   Steinmetz, Charles P. Theory and Calculation of Transient Electric Phenomena and Oscillations. New York; McGraw Publishing Co., 1909. 572p. (3d ed., 1920. 696p.)
First   Stepanoff, Alexey J. Centrifugal and Axial Flow Pumps. New York; J. Wiley, 1948. 428p. (2d ed., 1957. 462p.)
Second   Stephens, Henry and R. S. Burn. The Book of Farm Buildings. London and Edinburgh; W. Blackwood & Sons, 1861. 562p.
Second   Stephenson, James H. Farm Engines and How to Run Them. Chicago; F. J. Drake, 1910. 243p. (2d ed., 1918. 252p.)
First   Stephenson, James H. Traction Farming and Traction Engineering. Chicago; F. J. Drake & Co., 1917. 330p.
Second   Stewart, Henry. Irrigation for the Farm, Garden, and Orchard. New York; Orange Judd Co., 1877. 264p. (2d ed., 1886. 276p.)
Second   Stewart, John T. Engineering on the Farm. Chicago and New York; Rand, McNally and Co., 1923. 538p.
First   Stone, Archie A. Farm Machinery. New York; J. Wiley & Sons and London; Chapman & Hall, 1928. 466p. (2d ed., 1942. 524p.)

First	Stone, Archie A. Farm Tractors. New York; Wiley and London; Chapman & Hall, 1932. 492p.
Second	Stone, Archie A., Kendrick S. Hart, and Robert H. Smith. Suggested Unit Course in Farm Tractor Maintenance. Albany; The University of the State of New York, the State Education Department, Agricultural Bureau and the Bureau of Industrial and Technical Education, 1942. 170p.
Second	Strange, William Lumisden. Notes on Irrigation, Roads and Buildings and on the Water Supply of Towns. London; G. Routledge and Sons, 1920. 849p.
Second	Streeter, Robert L. Internal Combustion Engines. New York; McGraw-Hill, 1915. 418p. (4th ed., 1933. 539p.)
Second	Struck, Ferdinand T. Construction and Repair Work on the Farm. Boston and New York; Houghton Mifflin Co., 1923. 382p.
Second	Sutermeister, Edwin, ed. Casein and Its Industrial Applications. New York; Chemical Catalog Company, 1927. 296p. (2d ed. edited by Edwin Sutermeister and Frederick L. Browne. New York; Reinhold, 1939. 433p.)
Second	Swoope, Coates W. Lessons in Practical Electricity. 10th ed. New York; D. Van Nostrand, 1908. 494p. (18th ed. revised by Erich Hausmann, 1948. 769p.)

T

First	Taylor, Donald W. Fundamentals of Soil Mechanics. New York; J. Wiley, 1948. 700p.
Second	Taylor, Francis M. G. Du Plat. The Reclamation of the Land from the Sea. London; Constable & Co., 1931. 153p.
Second	Taylor, Frederick W., Sanford E. Thompson, and Edward Smulski. Concrete, Plain and Reinforced. 4th ed. New York; J. Wiley & Sons, 1925–1928. 2 vols.
Second	Teele, Ray P. Irrigation in the United States. New York and London; D. Appleton, 1915. 252p.
First	Tennessee Valley Authority. Barn Haydrier. Prepared by John A. Schaller, Nolan Mitchell and W. H. Dickerson, Jr. Knoxville, Tenn.; Agricultural Engineering Development Division, Tennessee Valley Authority, 1945. 129p.
Second	Teodoro, Anastasio L. Primary Power for the Philippines. Manilla, Philippines; The Author, 1931. 327p.
First	Terzaghi, Karl. Theoretical Soil Mechanics. New York; Wiley and London; Chapman and Hall, 1943. 510p.
First	Terzaghi, Karl, and Ralph B. Peck. Soil Mechanics in Engineering Practice. New York; J. Wiley, 1948. 566p. (2d ed., 1967. 729p.)
First	Thomas, George. Early Irrigation in the Western States. Salt Lake City; University of Utah, 1948. 63p.
Second	Thomas, Harold E. The Conservation of Ground Water. New York; McGraw-Hill, 1951. 327p.
First	Thomas, John J. Farm Implements and Farm Machinery. New York; O. Judd, 1879. 312p. (2d ed., 1883.)
First	Thomas, John J. Farm Implements and the Principles of Their Construction and Use. New York; Harper, 1854. 267p. (2d ed., New York; O. Judd Co., 1879. 312p.)
Second	Thompson, Silvanus P. Elementary Lessons in Electricity & Magnetism. London; Macmillan, 1881. 446p. (7th ed., 1915. 706p.)
Second	Thomsen, Thomas C. The Practice of Lubrication. New York; McGraw-Hill, 1920. 607p. (4th ed., 1951. 617p.)

Second   Thorne, David W., and H. B. Peterson. Irrigated soils: Their Fertility and Management. Philadelphia; Blakiston Co., 1949. 288p. (2d ed., 1954. 392p.)
First   Thurston, Robert H. The Animal as a Machine and a Prime Motor. New York; J. Wiley & Sons, 1894. 97p.
Second   Tillson, George W. Street Pavements and Paving Materials. New York; J. Wiley and London; Chapman & Hall, 1900. 532p.
First   Tolman, Cyrus F. Ground Water. New York and London; McGraw-Hill, 1937. 593p.
First   Townsend, Charles R. Mechanized Logging. Montreal; Canadian Pulp and Paper Association, 1938. 90p.
Second   Tripp, Guy E. Electric Development as an Aid to Agriculture. New York; The Knickerbocker Press, G. P. Putnam's Sons, 1926. 78p.
Second   Trist, Philip J. O. Land Reclamation. London; Faber and Faber, 1948. 178p.
First   Tromp, L. A. Machinery and Equipment of the Cane Sugar Factory. London; N. Rogers, 1936. 644p.
First   Tschebotarioff, Gregory P. Soil Mechanics, Foundations, and Earth Structures. New York; McGraw-Hill, 1951. 655p.
Second   Turneaure, Frederick E., and E. R. Maurer. Principles of Reinforced Concrete Construction. New York; J. Wiley, 1907. 317p. (4th ed., 1932. 461p.)
First   Turner, Arthur W., and Elmer J. Johnson. Machines for the Farm, Ranch, and Plantation. New York; McGraw-Hill, 1948. 793p.
First   Turner, C. N. Farm Electrical Equipment Handbook. New York; Edison Electric Institute, 1950. 224p.
First   Tustison, Francis E., and Arthur G. Brown. Instructional Units in Hand Woodwork. Milwaukee, Wis. and New York; The Bruce Publishing Company, 1930. 222p. (Rev. ed. by Tustison, Brown and Louis Barocci, 1965. 185p.)
Second   Tustison, Francis E., and Ray F. Kranzusch. Metalwork Essentials. New York and Chicago; The Bruce Publishing Company, 1936. 176p.

U
Second   Union Pacific Railroad Company. Department of Traffic. Agricultural Development. Irrigation Guide. Omaha, Neb.; Union Pacific Railroad Company, 1947. 31p.
First   United States. Department of Agriculture. Interbureau Committee on Technology. Technology on the Farm. Washington, D.C.; U.S. Government Printing Office, 1940. 224p.
Second   Unwin, William C. A Treatise on Hydraulics. London; A. and C. Black, 1907. 327p.

V
Second   Van Leuven, Edwin P. Cold Metal Working. New York and London; McGraw-Hill, 1931. 275p.
First   Vaughan, Lawrence M., and Lowell S. Hardin. Farm Work Simplification. New York; J. Wiley, 1949. 145p.
Second   Veen, Johan van. Dredge, Drain, Reclaim: The Art of a Nation. The Hague; M. Nijhoff, 1948. 165p. (4th ed., 1955. 200p.)
Second   Vernon, A. Estate Fences: Their Choice, Construction and Cost. London; E. & F. N. Spon, 1909. 420p.
First   Von Loesecke, Harry W. Drying and Dehydration of Foods. New York; Reinhold, 1943. 302p. (2d ed., 1955. 300p.)

W

Second	Wade, C. P. G. Mechanical Cultivation in India. Delhi; Burma-Shell Co., 1935. 124p.
Second	Wakeling, Arthur. Fix it Yourself. New York; Popular Science Publishing Co., 1929. 256p.
First	Walker, Reginald D. The Principles of Underdrainage. London; Chapman & Hall, 1929. 225p.
Second	Walker, William H. Principles of Chemical Engineering. New York; McGraw-Hill, 1923. 637p. (3d ed. with Warren K. Lewis, William H. McAdams and Edwin R. Gilliland, 1937. 749p.)
First	Wangaard, Frederick F. Mechanical Properties of Wood. New York; Wiley, 1950. 377p.
Second	Waring, George E. Draining for Profit and Draining for Health. New York; O. Judd, 1867. 244p. (2d ed., 1902. 252p.)
Second	Waring, George E. The Sanitary Condition of City and Country Dwelling Houses. New York; D. Van Nostrand, 1877. 145p. (3d ed., 1910. 71p.)
Second	Waring, George E. The Sanitary Drainage of Houses and Towns. New York; Hurd and Houghton, 1876. 336p. (11th ed., Boston; Houghton Mifflin, 1904. 366p.)
Second	Waring, George E. Sewerage and Land-Drainage. New York; D. Van Nostrand Co., 1889. 406p. (3d ed., 1891. 406p.)
Second	Waring, George E. Village Improvements and Farm Villages. Boston; J. R. Osgood & Co., 1877. 200p.
Second	Warington, Robert. Lectures on Some of the Physical Properties of Soil. Oxford; Clarendon, 1900. 231p.
Second	Warren, C. H. English Cottages and Farm Houses. New York and Toronto; Wm. Collins Sons & Co., 1948. 48p.
Second	Webre, Alfred L., and Clark S. Robinson. Evaporation. New York; The Chemical Catalog Co., 1926. 506p.
First	Weisbach, Julius, and Gustav Herrmann. The Mechanics of Pumping Machinery. Translated from by the German by Karl P. Dahlstrom. New York; Macmillan, 1897. 300p.
First	Weiss, Howard F. The Preservation of Structural Timber. Madison, Wis.; 1914. 312p. (2d ed. New York; McGraw-Hill, 1916. 361p.)
Second	Western Electric Company. The Farmer's Electrical Handbook. New York; Western Electric Co., 1917. 160p.
Second	Westinghouse Electric Corp. Farmstead Wiring. Pittsburgh, Penn.; Westinghouse Electric Corp., 1947. 43p.
Second	Westinghouse Electric and Manufacturing Company. Rural Electrification Department. How to Apply Motors and Controls to Farm Jobs. Pittsburgh, Penn.; Westinghouse Electric and Manufacturing Company, 1951. 15p.
Second	Wheeler, William H. The Drainage of Fens and Low Lands by Gravitation and Steam Power. London and New York; E. & F. N. Spon, 1888. 175p.
Second	Whinery, Samuel. Specifications for Street Roadway Pavements. New York; Engineering News Publishing Co., 1907. 56p. (2d ed., New York; McGraw-Hill, 1913. 116p.)
Second	Whitman, Roger B. Tractor Principles: The Action, Mechanism, Handling, Care, Maintenance, and Repair of the Gas Engine Tractor. New York; D. Appleton & Co., 1920. 281p.
Second	Widstoe, John A. The Principles of Irrigation Practice. New York; Macmillan, 1914. 496p.

Second	Widstoe, John A. Success on Irrigation Projects. London; Chapman & Hall and New York; Wiley, 1928. 153p.
First	Wiel, Samuel C. Water Rights in the Western States. San Francisco; Bancroft-Whitney Company, 1905. 619p. (3d ed., 1911. 2 vols.)
First	Wilcox, Lucius M. Irrigation Farming. New York; Orange Judd Co., 1895. 311p. (2d ed., 1902. 494p.)
Second	Willcocks, William. The Ancient System of Irrigation in Bengal and Its Application to Modern Problems. London; E. & F. N. Spon, 1930. 136p.
Second	Willcocks, William. Egyptian Irrigation. London; E. & F. N. Spon and New York; Spon & Chamberlain, 1899. 467p. (3d ed. with J. I. Craig, 1913. 2 vols.)
Second	Willcocks, William. The Irrigation of Mesopotamia. London; E. & F. N. Spon and New York; Spon & Chamberlain, 1911. 136p. (2d ed., 1917.)
Second	Williams, Gardner S., and Allen Hazen. Hydraulic Tables. New York; Wiley, 1905. 75p. (3d ed., 1933. 115p.)
First	Willoughby, George A. Practical Electricity for Beginners. Peoria, Il.; The Manual Arts Press, 1921. 104p.
Second	Willows, Richard S., and E. Hatschek. Surface Tension and Surface Energy. London; J. & A. Churchill, 1915. 80p. (3d ed., 1923. 134p.)
First	Wilson, Herbert M. Manual of Irrigation Engineering. New York; J. Wiley & Sons, 1893. 351p. (6th ed. entitled Irrigation Engineering, 1910–1912. 573p. 7th ed. by Arthur P. Davis and Herbert M. Wilson, 1919. 640p.)
Second	Wilson, John D., and S. O. Werner. Simplified Roof Framing. New York; McGraw-Hill, 1927. 122p. (2d ed., 1948. 160p.)
Second	Winder, Thomas. Handbook of Farm Buildings, Ponds, etc. & Their Appurtenances. London; Country Gentlemen's Association, 1908. 157p.
Second	Wirt, Frederick A. A Laboratory Manual in Farm Machinery. New York; J. Wiley & Sons, 1917. 162p.
First	Wolff, Alfred R. The Windmill as a Prime Mover. New York; J. Wiley, 1885. 159p. (2d ed., 1900. 161p.)
Second	Wolff, Henry W. Rural Reconstruction. London; Selwyn & Blount, Ltd., 1921. 363p.
First	Woodcroft, Bennet. Reaping Machine. London; George E. Eyre and William Spottiswoode, 1853. 108p.
Second	Woods, Katherine S. Rural Crafts of England. London; Harrap, 1949. 267p.
Second	Woodward, G. E., and F. W. Woodward. Woodward's Graperies and Horticultural Buildings. New York; Woodward, 1867. 139p.
First	Wooley, John C. Farm Buildings. Columbia, Mo.; The University Cooperative Store, 1936. 266p. (2d ed., New York and London; McGraw-Hill, 1946. 354p. 3d ed. entitled Planning Farm Buildings, 1953. 303p.)
First	Wooley, John C. Repairing and Constructing Farm Buildings. New York; McGraw-Hill, 1952. 261p.
First	Wooley, John C., and R. P. Beasley. Farm Water Management. Columbia, Mo.; Lucas Brothers, 1950. 170p.
Second	Woolrich, Willis R., and E. L. Carpenter. Manual of Mechanical Processing of Cottonseed. Knoxville, Tenn.; University of Tennessee, Engineering Experiment Station, 1935. 149p.
Second	Woolridge, Sidney W., and David L. Linton. Structure, Surface and Drainage in South-East England. London; G. Philip, 1939. 124p. (Institute of British Geographers Publications No. 9 and 10) (2d ed., 1955. 176p.)
First	Wright, Forrest B. Electricity in the Home and on the Farm. New York; J.

	Wiley & Sons and London; Chapman & Hall, 1935. 320p. (3d ed., 1950. 380p.)
First	Wright, Forrest B. Rural Water Supply and Sanitation. New York; J. Wiley & Sons and London; Chapman & Hall, 1939. 288p. (3d ed., Huntington, N.Y.; R. E. Krieger Publishing Co., 1977. 305p.)
First	Wright, Frederic B. A Practical Handbook on the Distillation of Alcohol from Farm Products. 2d ed. New York; Spon & Chamberlain and London; E. & F. N. Spon, 1918. 271p.
First	Wright, William J. Greenhouses: Their Construction and Equipment. New York; Orange Judd Co., 1917. 269p. (2d ed., 1946.)

Z

Second	Zworykin, Vladimir K., and E. G. Ramberg. Photoelectricity and Its Applications. New York; J. Wiley, 1949. 494p.

F. Scholarly and Professional Journals

By means of the citation analysis method previously described, a basic list of 373 scholarly and professional journals was compiled. Of the 373 scholarly journals identified, 58.4% were single citations and 12.3% were cited no more than two times in a single source document. Another 17.4% of the scholarly journals on the basic list were foreign language journals or journals outside the major subject areas of agricultural engineering, such as *American Fruit Grower*.

A list of 45 scholarly journals were identified as primary journals which were then evaluated by subject specialists in select fields, usually faculty members of agricultural colleges or U.S. Department of Agriculture scholars with long research and teaching experience. The reviewers were:

Robert H. Brown
   University of Georgia
Joseph K. Campbell
   Cornell University
William Chancellor
   University of California-Davis
S. S. DeForest
   Coraopolis, Pennsylvania

William Fox
   Mississippi State University
Carl W. Hall
   Arlington, Virginia
Lester Larsen
   Lincoln, Nebraska
Robert Tweedy
   Batavia, Illinois

Reviewers were asked to rank the journal titles in terms of the importance of preserving them for historical research by assigning each journal to one of these ranks: 1 = top-ranked journal; 2 = journal of lesser importance, although worth preserving; 3 = journal of local interest only and of less value than 1 and 2.

A total score was obtained for each scholarly journal title using the reviewers' rankings. The journal titles receiving the highest scores were rated first rank (A) and all others received a ranking of second (B). This resulted in a statistically valid core list of forty-two scholarly and professional journals. These are identified and ranked in Column 1 of the scholarly and professional journals list. The reviewers' first rank (Column 1) has twenty-five titles or 60% of the titles.

Core Scholarly and Professional Journals

	1	2	3	4
Agricultural engineering	A	A	*	M
American Society of Agricultural Engineers, Transactions	A	A	*	Pm
Implement & tractor	A	A	*	
American Society of Civil Engineers, Transactions	A	A	*	M
Society of Automotive Engineers, Quarterly Transactions	A	B	*	
Society of Automotive Engineers, Journal	A	A	*	M
American Society of Mechanical Engineers, Transactions	A	B	*	M
Farm implement news	A	A	*	
Farm mechanization	A	B	*	
American Society of Mechanical Engineers, Papers	A	B	*	
American Geophysical Union, Transactions	A	A	*	M
Science	A	B	*	M
Food engineering	A	B	*	M
American Society for Testing Materials, Proceedings	A	B	*	M
American Institute of Chemical Engineers, Transactions	A	A	*	M
American Society of Civil Engineers, Proceedings	A	A	*	M
Engineering	A	B	*	M
Journal of agricultural science	A	A	*	M
Institution of Civil Engineers (G.B.), Proceedings	A	A	*	Pm
Civil engineering	A	A	*	M
National land and irrigation journal	A	B	*	Pm
Thresherman's review	A	B	*	
Heating & ventilating	A	B	*	M
Reclamation era	A	B	*	M
Agricultural chemicals	A	A	*	M
Engineering news-record	B	A	*	M
Highway Research Board, Proceedings	B	A	*	Pm
California agriculture	B	A	*	
Journal of the Franklin Institute	B	B	*	M
New Zealand journal of science and technology	B	B	*	M
Engineering news	B	B	*	M
Journal of the Royal Society of Arts	B	B		M
Implement record	B	B	*	
Progressive agriculture in Arizona	B	B	*	
Physical Society (London), Proceedings	B	B		M
Journal of the New England Water Works Association	B	B		M
Reclamation record	B	B	*	M
Public roads	B	B		M
Electro-farming	B	B		

General Electric review	B B	M
Engineering record, building record and sanitary engineer	B A	M
Industrial chemist and chemical manufactures	B A	

Column 1 = Ranking by reviewers' scores.
Column 2 = Ranking by citation analysis.
Column 3 = * is a recommendation for preservation.
Column 4 = M means already microfilmed; Pm means partially microfilmed.

Of the thirty-four scholarly journals recommended for preservation, 85% have already been microfilmed.

Agricultural Engineering, the journal of the American Society of Agricultural Engineers, was both the highest ranked by the reviewers and the most often cited journal in the twenty-five source documents.

G. Trade and Popular Periodicals

Popular and trade periodicals are those magazines used by practitioners and commercial firms. The citation analysis of seminal works does not offer help in identifying and determining the historical value of popular literature. Citation analysis deals with scholarly and scientific publications and the references to the popular periodical literature are few. Other methods had to be used to identify popular periodicals in agricultural engineering and establish their relative historical merits. Journal lists, library catalogs, and historical writings concerned with agricultural periodicals and agricultural history were used. This literature served as the source for the compilation of U.S. and Canadian popular agricultural engineering and related industry periodicals.

Sources Used to Compile a List of Trade and Popular Periodicals in Historical Agricultural Engineering, Pre-1950

N. W. Ayer & Sons American Newspaper Annual and Directory. (Philadelphia: N. W. Ayer & Son, 1910–1929). Sections on "Farm implements, tractors, etc., agricultural implements, and blacksmiths and horse shoers." Selected years.

N. W. Ayer & Sons American Directory of Newspapers and Periodicals. (Philadelphia: N. W. Ayer & Son, 1930–1950). Sections on "Farm implements and tractors, and power farming." Selected years.

Bailey, Liberty Hyde, ed. Cyclopedia of American Agriculture. (New York: Macmillan, 1907–1909). "List of current agricultural journals," Vol. 4, pp. 78–87.

Bailey, Liberty Hyde and Bailey, Ethel Zoe, compilers. R U S: A Biographical Register of Rural Leadership in the United States and Canada. (Ithaca, N.Y.: [n.p.], 1925). "Directory of journals devoted to agriculture and rural life," pp. 745–758.

Bailey, Liberty Hyde, compiler. *R U S: A Register of the Rural Leadership in the United States and Canada.* (Ithaca, N.Y.: [n.p.], 1920). "Directory of journals devoted to agriculture and rural life," pp. 514–526.

Batten, George. *Batten's Agricultural Directory.* (New York: George Batten Company, 1908).

Dictionary Catalog of the National Agricultural Library, 1862–1965. (New York: Rowman and Littlefield, 1967). These subject headings were examined: Electricity in agriculture, Vol. 20; Gas and oil engines, Vol. 26; Traction-engines, Vol. 61; Drainage, Vol. 19; Farm buildings, Vol. 21; Farm engines, Vol. 21; Farm equipment, Vol. 21; Agricultural engineering, Vol. 1; Agriculture-Implements and machinery, Vol. 2; Poultry houses and equipment, Vol. 49; Irrigation, Vol. 33.

Duke, Dorothy M. *Agricultural Periodicals Published in Canada, 1836–1860.* (Ottawa: Canada Department of Agriculture, Information Division, 1961). Chapter VI, Physical Sciences, pp. 46–48.

Fusonie, Alan M., compiler. *Heritage of American Agriculture: A Bibliography of Pre-1860 Imprints,* U.S. Department of Agriculture. (Washington, D.C.: National Agricultural Library, 1975). Library List No. 98.

List of Periodicals Currently Received in the Library of the U.S. Department of Agriculture. (Washington, D.C.: U.S. Department of Agriculture, 1909). USDA Library Bulletin No. 75.

List of Serials Currently Received in the Library of the U.S. Department of Agriculture. (Washington, D.C.: U.S. Department of Agriculture, 1922). Department Circular 187.

Periodicals and Society Publications Currently Received by the Department Library. (Washington, D.C.: U.S. Department of Agriculture, 1894). 8p. USDA Library Bulletin No. 2.

Stuntz, Stephen Conrad. *List of the Agricultural Periodicals of the United States and Canada Published During the Century July 1810 to July 1910.* (Washington, D.C.: U.S. Department of Agriculture, 1941). 190p. USDA Miscellaneous Publication No. 398.

U.S. Department of Agriculture. *Report.* (Washington, D.C.: U.S. Department of Agriculture, 1868–1971). "Agricultural and Horticultural Periodicals" in reports for 1867, pp. 404–409; 1870, pp. 544–548.

U.S. Patent Office. Report. *Agriculture.* (Washington, D.C.: The Office, 1846). "List of Agricultural Journals," p. 1165.

Wilson, Allen D. *Agricultural Periodicals in the United States.* (Master's thesis, University of Illinois, 1930).

A list of popular periodicals was prepared from these sources and the titles ranked. The criteria used to rank these titles were: 1 = top-ranked periodical based on the number of years of publication. Periodicals published for twenty years or more were considered of greater importance than those periodicals that were published for five years or less. Titles older than 1850 were given this top ranking; 2 = those of lesser importance, but with national or international importance; 3 = titles of local interest only; lesser value than 1 and 2. Dr. Robert Brown, University of Georgia, provided comments and valued judgments on some of the titles.

United States and Canadian Trade and Popular Periodicals in Agricultural Engineering, 1850–1950

// at end of citation = ceased publication on this date.

2 Agricultural equipment dealer. Chicago.
3 Agricultural news letter. Wilmington, Del. E.I. DuPont deNemours & Co. m. v. 1–6. 1932–Dec. 1938.
2 Agricultural wheel. Lebanon, Mo. w. v. 1–4? 1888–1892.
2 Agrimotor magazine. Chicago. m. v. 1–5, no. 11. Aug. 1917–Sept. 15, 1922//.
3 Alabama rural electric news. Montgomery. m? v. 1- Jan. 1948–
3 American blacksmith and motor shop. Buffalo, N.Y. m. 1901–
2 American farm equipment. Omaha, Neb., Waverly, Iowa. m. v. 1–35, no. 9. 1898–1932//.
2 American farm implement. Indianapolis, Ind. m. v. 1–2? 1883–1885//?
3 American farmer and mechanic. Brooklyn, N.Y., A.E. Carter. m? v. 1 Oct. 1862–Jan. 1863.
1 American farming and farm implements. St. Petersbourgh, Fla. s-m. v. 1- Sept. 1, 1909–
2 American implement herald. Indianapolis, Ind. 1894–//?
1 American thresherman and farm power. Madison, Wis. m. v. 1–35. 1898–1932//.
  Began as
    (a) American thresherman.
  Continued as
    (b) American thresherman and farm power.
  Last issues entitled
    (c) American farm equipment.
1 Arid America. Denver, Colo. m. v. 1–9. 1887–1897//.
3 Arkansas agricultural and mechanical journal. Little Rock.
3 Arkansas REA news. Little Rock. m? v. 1- Nov. 1946–
1 Better farm equipment and methods. St. Louis. m. v. 1, no. 1- Sept. 1928–1944.
  Continued as
    (a) Better Farming Methods.
2 Better farming for better living. Chicago. Oliver Corporation.
3 Blacksmith and wheelwright. New York. m. v. 1–106, no. 3. 1880–1932//.
1 Business in farming. Chicago. m. no. 1–75, 1912–15; Series 2, v. 1–9, no. 5, 1915–May 1923//.
  Also known as
    (a) Business in Farming Bulletin. 1915–1920 as National Gas Engine Assoc. Bull.; 1920–22 as Gas Engine and Farm Power Assoc. Bull.
3 California advocate. San Francisco. m. v. 1–3, no. 129. Oct. 1896–July 30, 1898//?
  Combined with
    (a) National irrigation.
3 California irrigationist. San Francisco. s-m. v. 1- 1891–1892?
3 Canadian blacksmith and woodworker. Winnipeg, Man. m. 1910–
3 Canadian farm implements. Winnipeg, Man. m. v. 1- July, 1904–July, 1910.
2 The Canadian implement and vehicle trade.
  Established as
    (a) Farm machinery. Sarnia, Ont. m. v. 1- July 1900–
  Continued as

(b) Canadian implement trade. Sarnia, Ont. (1901), Toronto (1901–1902). m. v. 1901–1903.
   Continued as
   (c) The Canadian implement and vehicle trade. Toronto. m. v. -21, no. 3. 1903–July 1910.
2 The Canadian implement trade journal. Toronto. m. v. 1–49, no. 8. 1900–1935//.
   Absorbed
   (a) Power farming of Canada. v. 29, no. 8. Dec. 1916.
3 Canadian power farmer. Winnipeg, Man.
   Established as
   (a) Canadian thresherman, v. 1–3. 1903–1905.
   Continued as
   (b) Canadian thresherman and farmer, v. 3–24. 1906–1919.
   Continued as
   (c) Canadian power farmer. v. 25–28. 1920–23//.
3 Canadian tractor farming. London, Ont. m. v. 1– Jan. 1919–
3 Capital co-op. Bismark, N.D.
2 Carriage and implement journal. Toronto and Winnipeg, Man. (1900–1901), Toronto (1902–1905). m. v. 1– 1900–1905//.
3 C.E.C. news. Fulton, Mo. Callaway Electric Cooperative. v. 1- Feb. 1948–
3 Cement and engineering news. Chicago. m. v. 1–36, no. 11; July 1896–Nov. 1924//.
1 Chilton tractor journal. Philadelphia, Pa. m. v. 1–20 July 1912–1927.
   Also known as
   (a) Chilton Tractor and Equipment Journal
   Superseded by
   (b) Tractor and Equipment Journal
3 Chugach current. Anchorage. Chugach Electric Association.
3 Clay-Union sparks. Vermillion, S.D. v. 1-
3 Concrete review. Philadelphia. m. v. 1–4, no. 9. Feb. 1907–May 1911//.
1 C.R.E.A. newsletter (Committee on the relation of electricity to agriculture or National Committee . . .). Chicago. w? no. 1- Apr. 16, 1928–?
3 Colorado rural electric news. Denver.
2 Confessor. Washington, D.C. v. 1- Jan. 1923– (confidential news sheet circulated among members of the ASAE)
3 The Connector. Henderson, Ky. Henderson-Union Rural Electric Cooperative Corporation.
3 Co-op REA news. Somerset, Ky. 1950–
3 Co-op spotlight. McKee, Ky.
3 Country living magazine. Columbus, O. Ohio Rural Electric Cooperatives.
3 Current flashes. Glasgow, Ky.
3 Current hi-lites. Garrison, N.D. (removed from North Dakota rural electric magazine)
3 Currently speaking. Melrose, Minn. Stearns Co-Operative Electric Association. v. 2, no. 2- Feb. 1947–?
1 Drainage journal. Indianapolis. v. 1–24. 1879–1902//.
   Established as
   (a) Drainage and farm journal. v. 1–11. 1879–1889.
   Merged with
   (b) Irrigation age. 1903.
2 The Eastern dealer in implements and vehicles. Philadelphia. s-m. v. 1– 1907–July 1910.
3 The electric cooperator. Paintsville, Ken.

3 Electric farmer. Lincoln. v. 1- Jan. 1947–?
   Established as
      (a) Nebraska electric farmer. v. 1–5.
2 Electric farmer. Tell City, Ind.
3 Electrical ruralist. Cleveland. v. 1 no. 1–7. May–Nov. 1937//.
3 Electrik. Thompson, Ia. Winnebago Rural Electric Cooperative Association.
1 Export implement age. Philadelphia. m. v. 1–15, no. 5. Oct. 1899–Mar. 1907//. (in English, French, German, and Spanish)
2 Farm and shop. Indianapolis, Ind. v. 1–2, no. 13. Mar. 15, 1853–Sept. 15, 1854//?
1 The farm builder. Ardsley, N.Y. v. 1- Jan./Feb. 1949–
   Established as
      (a) Farm and Home Builder. v. 1–6, no. 6. 1949–1954.
1 Farm building news and views. Ft. Atkinson, Wis. v. 1, no. 3- 194–?
2 Farm cement news. Chicago; Pittsburgh, Pa. v. 1–9, no. 1. 1909–1911//.
2 Farm chemurgic journal. Dearborn, Mich. v. 1, n. 1–3. Sept 1937–Dec. 1938//.
1 Farm electrification. New York. 1- 1947–
   Supersedes
      (a) Edison Electric Institute's Rural electrification bulletin.
1 Farm engineering. Chicago. v. 1–7. 1913–1918//.
   Continued in part as
      (a) Farm mechanics.
3 Farm equipment dealer. Toronto. m. v. 1- Oct. 1945–
1 Farm equipment dealer monthly. St. Joseph, Mich. m. v. 1–25. June 1903–Dec. 1927//.
2 Farm equipment merchandiser. Middletown, O. m. v. 1- Feb. 1938–
1 Farm equipment news. Kansas City, Mo. v. 1, no. 1- Jan. 1946–
   Supersedes
      (a) Farm machinery and equipment, and NRFEA monthly digest.
1 Farm equipment retailing. St. Louis. m. v. 1- Jan. 1946–
3 Farm, furnace and factory. Roanoke, Va. m. v. 1–6? Aug.? 1895–1900. (June 1899 omitted)
2 Farm ideas. Minneapolis. m. v. 1–2. June 1937–Dec. 1938//.
1 Farm implement news. v. 1–79, no. 9. 1882–Apr. 25, 1958//.
   Established as
      (a) Farm implement and country hardware trade. Chicago. m. v. 1- 1883–1884.
   Continued as
      (b) Farm implement. Chicago. m. v. (-1885).
   Continued as
      (c) Farm implement news. Chicago. m. v. 1–31, no. 27. 1886–July 7, 1910. (Published Farm implement news daily during the Cincinnati convention of the National Association of agricultural implement and vehicle manufacturers,1899 or 1900.)
   Absorbed by
      (d) Implement and Tractor.
1 Farm implements.
   Established as
      (a) Farm implement herald. Minneapolis and St. Paul. m. v. 1–2? 1887–1888.
   Continued as
      (b) Farm implements and hardware. Minneapolis and St. Paul. m. v. 3–5? 1889–1891.
   Continued as

  (c) Farm implements. Minneapolis and St. Paul. m. v. 6–24, no. 7. 1892–July 1910. (Issued daily edition during Minneapolis convention of National association of agricultural implement and vehicle manufacturers, Oct. 15–17, 1902.) DA 16–24.
1 Farm machinery. St. Louis and Kansas City. m. (1887–94). w. (1895–1910). v. 1-[24] no. 978. 1887–July 5, 1910. (1887 is old ser. v. 15) Beginning 1895? there is no volume numbering. (Issued daily edition at Kansas City and Minneapolis conventions of National association of vehicle manufacturers, Oct. 30– Nov. 1, 1901, and Oct 15–17, 1902; issued also in monthly ed., 1901–1904?)
1 Farm machinery and equipment. St. Louis. m. no. 1–1,944 1886–1945//.
  Partially supersedes
   (a) Farm machinery.
  Continued as
   (b) Farm Equipment News.
1 Farm machinery and hardware. St. Louis. m. 1878–
2 Farm mechanics. Chicago. Farm Mechanics Co. m. v. 1–27 1919–1932//.
2 Farm power. Iowa ed. Des Moines. v. 1- Jan. 1947–
3 Farm power. Ithaca, N.Y. m. v. 1– Aug. 1946–Sept. 1947.
3 Farm power. Madison, Wis. 1936–
2 Farm tools. Chicago. w. v. 1- 1892–1893.
2 Farm tractor. Kansas City, Mo. m. v. 1–3, no. 2. Dec. 1915–July, 1918//.
2 Farmer and mechanic. Leesburg, Va. v. 1– Feb. 1876–.
2 Farmer and mechanic. Raleigh, N.C. w. v. 1–35. 1877–1910.
3 The Farmer and mechanic. Bangor, Maine. v. 1- 1836–Jan. 1837.
3 The Farmer and mechanic. Cincinnati, Ohio. s-m. v. 1–11? Sept. 1832–1836//.
3 The Farmer and mechanic. Lewistown Falls, Maine. m. v. 1, no. 1–8. Mar.–May 1853//.
3 The Farmer and mechanic. Paris, Ky. m. v. 1– Apr. 1855–1856.
3 The Farmer and mechanic. Toronto, Ont. m. v. 1–3? Sept.? 1848–1850.
1 The Farmers' and mechanics' advocate. St. Louis. w. v. 1. Dec. 1842–1843?
1 Farmers' and mechanics' journal. Alexander, N.Y. w. v. 1–3? Nov. 1837–June 1840.
  Continued as
   (a) Batavia times and Farmers' and mechanics journal. Batavia, N.Y. Merged with
   (b) Batavia spirit of the times. 1843.
3 The Farmers' and mechanics' journal. Chagrin Falls, O. v. 1–2 1842– //.
3 The Farmers' and mechanics' journal. Harrisburg, Pa. v. 1. Aug. 12–Dec. 15, 1827.
  Merged with The intelligencer and continued as
   (a) Pennsylvania intelligencer and Farmers' and mechanics' journal.
3 Farmers' and mechanics' journal. Springfield, Ill. bi-m. v. 1– 1882–1883.
3 Farmers' and mechanics' journal and Newbern price current. Newbern, N.C. m. v. 1- Jan. 1868–
1 Farmers' and mechanics' repository. June 1843.
1 Farmers' and planters' friend. Philadelphia. no. 1–7. 1821.
2 The Farmers' and planters' guide.
  Established as
   (a) Farmer and mechanic. Baltimore. m. v. 1–2. Jan. 1864–Dec. 1865.
  Continued as
   (b) The Maryland farmer. Baltimore. m. v. 3–24, no. 8. Jan. 1866–Aug. 1887.
  Combined with New farm, Sept. 1887, and continued as
   (c) Maryland farmer and New farm. Baltimore. m. (1887–Oct. 1889, Mar.?

1891–May 1892). w. (Nov. 6, 1889–Feb.? 1891). v. 24, no. 9–v. 29, no. 5. Sept. 1887–May 1892.
Continued as
    (d) Maryland farmer. Baltimore. m. v. 29, no. 6 v. 33. June 1892–1899.
Continued as
    (e) The Farmers' and planters' guide. Baltimore. m. v. 34- 1899–July 1910.
3 Ford tractor equipment news. New York. q. 1940–
2 The gas engine. Cincinnati, O. m. v. 1–23 1898–1921.
Superseded by
    (a) Oil Field Engineering.
2 Good roads. Boston. m. v. 1–7, no. 3. Jan. 1892–Mar. 1895//.
3 Good roads for Wisconsin. Madison. m.
1 Hardware and farm equipment. Kansas City, Mo. v. 1– 1895– .
Supersedes
    (a) Implement hardware bulletin.
2 Hardware and implement journal. Dallas. s-m. 1896–
3 Hi-lights. Milnor, N.D. R.S.R. Electric Cooperative. (removed from North Dakota rural electric magazine)
1 Home friend and illustrated mechanic. Kansas City. v. 1–34, no. 1. 1916–March 1938//?
Supersedes
    (a) Illustrated mechanics. v. -34, no. 6. -Mar. 1937. Supersedes
    (b) Illustrated rural mechanics. -v. 10, no. 1. -Apr. 1925. Supersedes
    (c) Rural mechanics. v. 6, no. 3–v.9, no.- June 1921– Supersedes
    (d) Farm and home mechanics. v. 1–6, no. 2. 1916–1921.
2 Horse shoers' journal. Detroit. m. 1875–
3 Illinois REA news. Association of Illinois Electric Cooperatives. Madison, Wis. v. 1- 1943–
3 Illinois rural electrification bulletin. Springfield. v. 1, no. 2– July 1936– .
2 Illustrated mechanics. Kansas City, Mo. m. v. 1–11 1916–Jan. 1927//.
Established as
    (a) Farm and home mechanics. v. 1–6 1916–1921.
Continued as
    (b) Rural mechanics. v. 6–9 1921–1925.
1 The Implement age. Philadelphia and Baltimore (1892–1902). m. (1892–1897). s-m. v. 1–19, no. 2. 1892–1902. (Published Implement age daily during the convention of the National association of agricultural implement and vehicle manufacturers, Oct. 19–21, 1898.)
Continued as
    (a) Implement and tractor age.
2 Implement and farm journal. Kansas City, Mo. m. v. 1- 1883–1892.
2 Implement and hardware news. Sioux Falls, S.D. m. v. l. Mar. 1905–
1 Implement and tractor. Kansas City, Mo. w. v. 16–53. 1902–1938//.
Supersedes
    (a) Implement and tractor trade journal.
1 Implement and tractor age. Philadelphia and Springfield, O. s-m. v. 19–55, June 1, 1902–1918//.
1 Implement and tractor trade journal. Kansas City, Mo. w. 1886–1901//.
Continued as
    (a) Implement and tractor.
2 Implement and vehicle journal. Dallas, Tex. m. (1904). s-m. (1905–1910). v. 1- Nov. 1904–July 1910.

Supersedes
    (a) Implement and vehicle edition of the Texas trade review. Issued the first of each month during 1904.
1 The Implement and vehicle news. Dubuque, Ia. and St. Louis (1901), St. Louis and Cincinnati, O. (1902–1903), Cincinnati and New York (1904), Cincinnati, O. (1905–1910). m. v. 1–10, no. 4. Apr. 1901–July 1910.
2 Implement and vehicle record.
    Established as
    (a) Pacific coast implement and vehicle record. San Francisco. m. v. 1- April 1904–1905.
    Continued as
    (b) Implement and vehicle record. San Francisco. m. 1906–July 1910.
3 Implement dealer. Council Bluffs, Ia. m. v. 1- 1898–1908.
3 The Implement dealers' bulletin.
    Established as
    (a) Dealers' bulletin. Abilene, Kan. v. 1– Oct.? 1904– .
    Continued as
    (b) The Implement dealers' bulletin. Abilene, Kan. m. Mar. 1905– .
2 Implement record, tractors and farm equipment. San Francisco. m. v.22, no. 10–33. Oct. 1925–1936.
1 Implement trade journal. Kansas City, Mo. 1893–1902.
2 Industrial plow. Mexico, Mo. m. v. 1. 1871.
3 Inter county REA co-op link. Danville, Ky. 1950– .
3 Iowa rural electric news. Des Moines.
2 Iron age farm and garden news. Grenloch, N.J. q. v. 1–4, no 3. 1907?-Summer 1910. (Issued Spring, Summer, Autumn, Winter.)
2 Iron, hardware and implement trade.
    Established as
    (a) Hardware and implement trade. Chicago. w. v. 1. 1877.
    Continued as
    (b) Iron, hardware and implement trade. Chicago. w. v. 2–3. 1877–1878//?
1 Irrigated west. v. 1- Dec. 1896– .
1 Irrigation. San Francisco. m. v. 1–2, no. 1. Nov. 1909–May 1910//.
    Merged with
    (a) Orchard and fruit.
    Continued as
    (b) Orchard and farm.
    Combined with
    (c) Irrigation.
1 The Irrigation age. Salt Lake City, Ut., San Francisco, and Denver, Colo. (1891–1892), Chicago (1893–1910). s-m. (1891–1892), m. (1893?–1910). v. 1–25, no. 9. Apr. 15, 1891–July 1910.
    Absorbed
    (a) Drainage journal. Jan. 1903.
    Absorbed
    (b) Modern irrigation. April 1904.
3 Irrigation aid. Pearsall, Tex. m. v. 1. 1905.
1 Irrigation and modern farming.
    Established as
    (a) Irrigation. Denver, Colo. m. v. 1–3, no. 1. Apr. 1903–Apr. 1905.
    Continued as
    (b) Irrigation and its applications to agriculture. Denver, Colo. m. v. 3, no. 2–

v. 6, no. 2. May 1905–Nov. 1906.
Continued as
(c) Irrigation and modern farming. Denver, Colo. w. v. 6, no. 3. Dec. 29, 1906//.
3 Irrigation and the new west sugar beet journal. San Francisco. Oswald Wilson, ed., 1917–
3 Irrigation champion. Garden City, Kan. m. v. 1–3? 1894–1897//.
3 Irrigation engineering and maintenance. Port Lavaca, Tex. v. 1– .
1 Irrigation era. Denver, Colo. m. v. 1–8 1887–1903.
Continued as
(a) Modern irrigation.
3 Irrigation farmer. Salina, Kan. (1894–1895), Ottawa, Kan. (1895–1898). m. v. 1–4? 1894–1898//.
2 The Irrigation fruit grower and agriculturist. v. 1–6, no. 4. 1905–Ar. 1911.
Established as
(a) The Western slope fruit grower. Paonia, Colo. m. v. 1- Nov. 1905–1906.
Continued as
(b) Colorado fruit grower. Paonia, Colo. (1907–1908), Grand Junction, Colo. (1909–June 1910). m. v. 2?-5, no. 6. Jan. 1907–June 1910.
Continued as
(c) The Irrigation fruit grower and agriculturist. Denver, Colo. m.
2 Irrigation market. New York. m. v. 1. 1893–1894.
3 The irrigation review. Calgary, Alberta m. v. 1–7. Apr. 1920–1926//.
3 Irrigation review. Denver, Colo. m. v. 1, no. 1–3. Sept. 1897–Feb. 1898.
3 Irrigation world. Greensburg, Kan. m. v. 1. 1894–1895.
3 The irrigator. DeLand, Fla. (1893–1894), Winter Park, Fla. (1894), Orlando, Fla. (1895). m. v. 1–2? 1893–1895.
3 The Irrigator. North Yakima, Wash. m. v. 1. 1910.
2 Irrigator and colonist. Chicago. m. v. 1– Aug. 13, 1904–Feb. 1905.
Merged with
(a) Field and farm. Denver, Colo. Feb. 1905.
3 Irrigon irrigator. Irrigon, Oreg. w. 1904– .
3 Journal of irrigation. Las Animas, Colo. v. 1– 1888–1889//?
1 Journal of agriculture, containing the best current production in promotion of agricultural improvement, including the choicest prize essays issued in Europe and America. New York. m. v. 1–3. July 1845–June 1848//.
Superseded by
(a) Plough, the loom, and the anvil.
3 Kansas electric farmer. Topeka.
3 Kansas highways. Topeka. q. v. 1–5, no. 5. Jan. 1918–May 1926//.
Suspended 1921–24.
3 KEM kilowatt kapers. Linton, N.D. (removed from North Dakota rural electric magazine)
3 The Kilowatt. Finlayson, Minn. North Pine Electric Cooperative.
Supersedes
(a) REA news bulletin.
2 Land and water. Los Angeles. m. 1892–1897//.
3 Licking valley. West Liberty, Ky.
3 Light post. Columbia City, Ind.
3 Live sparks. Lawrenceburg, Ky.
2 Machinery. New York. m. 1894, 2d series 1902–1914.

2 Maxwell's talisman. Chicago. (irrigation) v. 1–13, no. 3. 1902–Mar. 1913//.
3 McKenzie electric co-op. Watford City, N.D. (removed from North Dakota rural electric magazine)
3 Mechanic and farmer. Pictou, Nova Scotia, Feb. 1842.
1 The mechanic and farmer's journal. Bangor, Me. Dec. 1842–1843.
3 Meeker REA pioneer. Litchfield, Minn. v. 1– May 1950–
2 The mile post. Live Oak, Fla. Power Farming Association of America. v. 1– 1920– .
2 Modern highway. St. Joseph, Mo. m. v. 1–7. 1916–Dec. 1922. 1916–19 as Jefferson highway declaration.
3 Mississippi planter and mechanic. Jackson. m. v.1–2, no. 5. Jan. 1857– May 1858//?
1 Modern irrigation.
   Established as
      (a) Arid America. Denver. m. v. 1– 1887 (1896)-Nov. 1897.
   Continued as
(b) Irrigation era. Denver. m. v. 8, no. 11–v. 14, no. 5? Nov. 1897–1903.
   Continued as
      (c) Modern irrigation. Denver. m. v. 14, no. 6?- 1903–Apr. 1904//.
   Absorbed by
      (d) Irrigation age, Apr. 1904.
   Successor to
      (e) Arid west and farm herald.
1 Modern power farming. Milwaukee, Wis. no. 1- 1938–
3 Monona county REA news. Onawa, Wis. Monona County Rural Electric Cooperative. v. 5, no. 15– Nov. 1948– .
3 Montana rural electric news. Denver, Colo.
3 Mor-lite. Flasher, N.D. Mor-Gran-Sou Electric Cooperative. v. 1, no. 1– Jan. 1949– .
1 National irrigation. Washington, D.C. m. v. 1–8. 1897–1903//.
   v. 1–5 as National advocate.
2 National land and irrigation journal. Chicago. v. 1–7. 1909–1913//.
   Established as
      (a) National irrigation journal. v. 1–2. 1909–1910.
   Merged with
      (b) Irrigation age. v. 3–7. 1910–1913//.
3 Nebraska electric farmer. Lincoln. v. 1–5 Jan. 1947–51.
   Continued as
      (a) Electric farmer.
3 New agriculture. San Francisco. m. v. 1- Nov. 1, 1917– (v. 18, no. 11- Aug. 1936– sub-title: Western irrigation and sugar beet journal)
3 New Mexico electric news. Artesia.
3 The Nodak neighbor. Grand Forks, N.D. Nodak Rural Electric Co-operative. v. 7, no. 2- May 1946– .
3 North Dakota rural electric magazine. Bismarck.
3 North shore volts and jolts. Two Harbors, Minn. July 1984– .
3 Northern lights. Phillips, Wis. Price Electric Co-op. v. 1, no. 1- Nov. 1949– .
1 Northwest farm equipment journal. Minneapolis. v. 40– 1926– .
   Established as
      (a) Farm implements. v. 1–32. 1887–1918.
   Continued as
      (b) Farm implements and tractors. v. 32–40. 1918–1926.
3 Northwest implement trade. Fargo, N. D. v. 1–4? 1904–1906//?

3 Northwest rural electric news. Seattle, Wash. v. 1-.
3 Northwest ruralite. Portland, Oreg.
3 The Outlet. Columbus, N.D. Burke-Divide Electric Cooperative. v. 1- .
3 Our REA news. Williston, N.D. v. 1- Jan. 1949- .
2 Pacific farm power. San Francisco. v. 1, no. 1- Jan. 1937- .
1 Pennsylvania cultivator and mechanic and iron and coal register. Harrisburg. m. v. 1- Aug.-Dec. 1848//?
1 Plough, the loom and the anvil. Philadelphia; New York. m. v. 1-10. July 1848- December 1857//.
2 Plow. New York. v. 1, no. 1-12. Jan.-Dec. 1852//.
2 Plow and tractor. Moline, Ill. v. 1-6, no. 2. Oct. 1916-Apr. 1921//.
1 Plowboy. Cincinnati. v. 1, no. 1-6. Apr.-Dec. 1844//.
3 Plowboy and country farmer. Atlanta, Ga. v. 1-14, no. 4. 1889-June 1902//?
3 Plowman. Racine, Wis. v. 1-7. 1912-1917//?
1 Power. New York. v. 1- 1880- .
   Other titles
      (a) Steam. 1880-1884.
      (b) Power-steam. 1885-1891.
   Absorbed
      (c) Engineer April 14, 1908.
   Continued as
      (d) Power and the engineer. April 14, 1908-Dec. 1991.
   Continued as
      (e) Power.
1 Power farming. Detroit; St. Joseph, Mich., Power Farming Press. 1913?- 1928//.
   Established as
      (a) Threshermen's review. 1892-1913.
   Continued as
      (b) Threshermen's review and power farming. 1913-?
1 Processing equipment news. Ottumwa, Ia. v. 1- Mar. 1947- .
2 Profitable retailing. Chicago. m. 1922- (hardware, implements)
2 Pryor's farm equipment journal. Minneapolis. m. 1918- .
1 Quarterly journal of agriculture, mechanics, and manufacturer. New York. q. v. 1-2, no. 2. 1834-July 1835//?
2 Quonset farm building news. Detroit. Stran-Steel Division, Great Lakes Steel Corporation.
13 REA co-op synchronizer. Burlington, Kan. Coffey County Rural Electric Cooperative Association.
3 REA flashes. Coeur d'Alene, Ida.
3 REA hi-line. Albert Lea, Minn. North Pine Electric Cooperative. Sept. 1940- .
3 The REA news. Alexandria, Minn. Douglas County. v. 1, no. 1- Mar. 1946- .
3 The REA news. Alexandria, Minn. Millo Lace Region. v. 1, no. 1- Jan. 1950- .
3 The REA news. Alexandria, Minn. Tri-County. v. 1, no. 1- Dec. 1950- .
3 The REA news. Danube, Minn. v. 1- Aug. 1949- .
   Supersedes
      (a) The static.
3 The REA news. Litchfield, Minn. v. 1. May 1949-Apr. 1950.
   Superseded by
      (a) Meeker REA pioneer.
3 The REA news. St. James, Minn. v. 1, no. 1- Sept. 1949- .
   Supersedes
      (a) The live wire.

3 The REA news. Alexandria, Minn. v. 1, no. 1- Sept. 1951– (general ed.)
3 REA rural lite. Fleming-Mason RECC.
3 The REA voice. Tonganoxie, Kans. v. 1, no. 1- Apr. 1948–
1 Reclamation and farm engineering. Chicago. National Reclamation Publishing Company. m. v. 1–4, 7–10. Dec. 1921–Sept. 1926//.
2 Right of way. Harrisburg, Penn. bi-m. 1915– (farm implements, tractors)
2 Road maker. Moline, Ill. m.
2 Rural electric dealer. New York. v. 1–8, no. 6. Sept. 1920–Feb. 1928//.
  Began as
    (a) Farm light and power.
1 Rural electric minuteman. Washington, D.C. National Rural Electric Cooperative Association.
3 Rural electric Missourian. St. Louis. Missouri State Rural Electrification Association. v. 1- Jan. 1948– .
3 Rural electric news. Jackson, Miss.
3 Rural electric times. Lewiston, Mo. Lewis County Rural Electric Co- operative Association. v. 1, no. 1– May 1949– .
2 Rural electrification bulletin. New York. v. 1, no. 1–v. 3, no. 5. Mar. 1943–Sept. 1946.
  Superseded by
    (a) Farm electrification.
1 Rural electrification exchange. New York. q. v. 1- Jan. 1938– .
1 Rural electrification magazine. Washington, D.C.; National Rural Electrification Assoc. v. 1– Oct. 1942–44 as Assoc. Bulletin. Also titled Rural Electrification.
3 Rural Georgia. Monticello. Georgia Electric Membership Corporation. 194–?
3 Rural Kentuckian. Louisville. Kentucky Rural Electric Coop Corporation. v. 1- Aug. 1948– .
  Began as
    (a) Kentucky electric co-op news. v. 1–5, no. 5. Aug. 1948–May 1952.
3 Rural light. Cumming, Ga. Mar. 1947– .
  Merged with
    (a) Rural Georgia.
3 Rural line news. Owenton?
1 Rural mechanic. Indianapolis, Ind. m. v. 1–2? 1894–1897//?
  Merged with
    (a) Threshermen's review. May, 1987.
1 Rural mechanics. Kansas City. m. v. 6–9. 1916–21.
  Superseded by
    (a) Farm and home mechanics.
3 Rural power magazine. Lafayette, La. Southwest Louisiana Electric Membership Corp. v. 1, no. 1– May 1950.
3 Rural power pictorial. Baltimore, Md. Consolidated Gas Electric Light and Power Co. of Baltimore. no. 12– June 1935– .
3 Rural Minnesota news. Alexandria.
3 Rural Virginia. Richmond. Virginia REA. v. 1, no. 2– Dec. 1946– .
2 Scientific farmer. Mossmain, Mont. v. 1–17 1906–1941.
  Supersedes
    (a) Dry farming magazine. v. 1, no. 1 & 2.
    (b) Campbell's scientific farmer. 1908–22.
3 Sho-me live wire. Marshfield, Mo. Sho-Me Power Corporation. v. 1, no. 1– Aug. 9, 1949– .
3 Sign post. New England, N.D. (removed from North Dakota rural electric magazine)

2 Sorgo journal and farm machinist. Cincinatti, O. q.
3 South Carolina electric co-op news. Cayce.
3 South Dakota high-liner. Madison.
3 Southern good roads. Lexington, N.C. v. 1–22 1910–20//.
1 Southern hardware. Chattanooga, Tenn; Atlanta, Ga. W. R. C. Smith Publishing. 1927– .
   Began as
      (a) Tradesman. v. 1–72. 1879–1915.
   Continued as
      (b) Iron tradesman. 1915–1918.
   Continued as
      (c) Southern hardware and implement journal. 1918–1926.
1 Southwest hardware and implement journal. Dallas, Tex. m. 1896– .
3 Southwest rural electrification news. Palisade, Neb. no. 1- Sept. 1948–
3 Southwestern farm and orchard. Las Cruces, N. Mex. m. 1894– .
3 Speed up the farm. Rockford, Ill. Emerson-Brantingham Implement Co., Agricultural Extension Dept. v. 1, no. 1–6. June 1917–July 1918//.
3 The Tennessee magazine. Nashville. Tennessee Rural Electric Cooperative Association.
   Supersedes
      (a) The Tennessee rural electric bulletin.
3 Texas cooperative electric power. Austin. v. 1– July 1944–Jan. 1945.
3 Texas stock farm and irrigation. Brownwood. w. 1895– .
1 Thresher world and farmers' magazine. Milan, O.; Chicago.
   Began as
      (a) Thresher world. v. 1–7. 1897–1903.
   Merged with
      (b) American thresherman. v. 8–10. 1903–1905.
   Continued as
      (c) American farm equipment.
2 Tractor. Cincinnati, O. Tractor Publishing Co. 1918–//?
1 Tractor and equipment journal. New York. v. 21–22. 1928–1929//.
   Began as
      (a) Chilton tractor and equipment news. v. 1–20. 1918–1928.
1 Tractor and gas engine review. Madison, Wis. m. v. 1–18. 1908–1925//.
   Merged with
      (a) American thresherman.
   Continued as
      (b) American farm equipment.
2 Tractor and implement topics. m. v. 1–4, no. 5. 1917–June 1920//.
   Supersedes
      (a) Tractor and trailer. New. York. Tractor Publishing Co. v. 1–3.
1 Tractor builder. Cleveland, O. m. 1920– .
1 Tractor farming. Chicago. International Harvester Company of America. m. v. 1– 1916– .
   Suspended May–Dec. 1932.
3 Trail-lights. Stanley, N.D. v. 1– (removed from North Dakota rural electric magazine)
3 Verendrye electric co-op. Velva, N.D.
3 Warren co-op news. Bowling Green, Ky. Warren Rural Electric Cooperative Corporation. v. 1, no. 1– Sept. 1948– .

3 Watts new. Dickinson, N.D. v. 1– (removed from North Dakota rural electric magazine)
1 The weekly implement trade journal.
   Established as
      (a) Implement trade journal. Kansas City, Mo. m. (1886–1895), w. (1896), s-m. (1897–1901).
   Continued as
      (b) The weekly implement trade journal. Kansas City, Mo. w. v.—24, no. 27. 1902–July 2, 1910. (Issued daily edition during Kansas City convention of the National Association of Agricultural Implement and Vehicle Manufacturers, Oct. 30–Nov. 1, 1901.)
1 Western farm equipment. San Francisco. 1953– .
   Established as
      (a) Implement record. San Francisco. 1904–1953.
1 Western farmer. Spokane, Wash. and Portland, Ore. v. 1–27. 1899–Aug. 1926//.
   Began as
      (a) Washington, Idaho and Oregon farmer.
3 Western water news. San Francisco. v. 1– 1949– .
   Established as
      (a) California water news. v. 1– Jan. 1949–June 1951.
1 Wind and water. Chicago. v. 1– May 1896– (windmills and farm implements)
3 Wisconsin REA news. Madison.
3 Wyoming rural electric news. Denver, Colo.
3 Wyoming stockman-farmer. v. 1–13. 1899–Sept. 1912//.
   Established as
      (a) Wyoming industrial journal.
   Continued as
      (b) Mid-west industrial journal.

H. Foreign Language Monographs

Although the focus of the historical preservation project was English-language books and journals published in the United States and Canada, a number of monographs were cited that had been written in languages other than English. The largest number of non-English monographs were published in French (40.8% of the titles), followed by German (19.0%), Russian (17.9%), and Spanish (8.4%). This list is provided for information only.

Non-English Monographs, Historical Agricultural Engineering, 1850–1950

A

Albertini, Renzo. La Vita Pastorale Sul Gruppo Ortles-Cevdale. Trento; Arti Grafiche Saturnia Presentazione, 1955. 119p.

Aleksandrov, Grigorii Ia. Book for Tractor Brigade. Moscow; Moskovskiy Rabochiy, 1954. 510p. (In Russian. 2d ed., 1956.)

Aleksandrov, V. I. Fertilizer and Seeding Machinery and Manure Spreader. Moscow; Gos. Nay.-Tekh. Izd-Vo Mach. Lit., Sverdlovsk, 83p. (In Russian)

Alfred, Emil. Technik im Bauernhof. Munich; H. Neureuter, 1952. 94p.

Anakin, I. A. Repair and Maintenance of Agricultural Machines. Sverdlovsk; Gos. Nay.-Tekh. Izd-Vo Mach. Lit., 1955. 332p. (In Russian)

Aniferov, F. E. Machinery and Instruments on Socialistic Peasant Lands. Moscow; Gos. Izd-Vo Sel. Lit-Pub Kiev, 1954. 134p. (In Russian)

Aniferov, F. E. Machinery and Instruments for Cultivating Vegetables for Food. Moscow; Gos. Izd-Vo Sel. Lit-Pub Kiev, 1955. 166p.

Antipov-Karataev, Ivan N. Influence of Prolonged Irrigation on Soil Fertility. Moscow; Izd-Vo Akad. Nauk CCCP, 1955. 204p. (In Russian)

Aranda Heredia, E. Posibilidades del las Maquinas Agricolas. Madrid; Editorial Escuela Especial de Ingenieros Agronomos, 1950. 104p.

Arkhangel'skii, B. E. Tractor KD-35 and KDTS-35. Moscow; Gos. Izd-Vo Sel. Lit-Pub Kiev, 1954. 552p. (In Russian.)

Arkhipov, P. P. Agricultural Production Buildings and Their Erection. Moscow; Gos. Izd-Vo Lit-Pub Kiev, 1955. 325p. (In Russian.)

Artobolevskii, Ivan I., ed. Agricultural Machinery. Moscow; Foreign Literature Press, 1954. 259p. (In Russian)

Astrua, Giuseppe. Il Risanamento Della Edilizia Rurale. Milano; U. Hoepli, 1955. 248p.

B

Balle, Niels. Forbraendingsmotorer og Traktor. Kopenhavn; Teknologisk Instituts Forlag, 1949. 133p.

Ballu, Tony. Machines Agricoles. Paris; Bailliere, 1948. 592p.

Ballu, Tony. Le Mechanisme Agricole. Paris; Presses Universitaires de France, 1951. 128p.

Ballu, Tony. Traction Mecanique en Agriculture. Paris; Librarie Agricole de la Maison Rustique, 1948. 372p.

Barral, Jean A. Drainage des Terres Arables. 3d ed. Paris; Librarie Agricole de la Maison Rustique, 1862. 2 vols.

Barral, Jean A. Irrigations, Engrais Liquides et Ameliorations foncieres Permanentes. Paris; Librarie Agricole de la Maison Rustique, 1862. 782p. (1)

Barral, Jean A. Legislation du Drainage, des Irrigations et Autres Ameliorations Foncieres Permanentes. Paris; Librairie Agricole de la maison Rustique, 1862. 663p.

Barral, Jean A. Manuel du Drainage des Terres Arables. Paris; Impr. National, 1854. 4 vols. (3d ed., 1884.) (1)

Barzykin, V. M. Basic Mechanization of Private Rural Enterprises. Moscow; Gos. Izd-Vo Sel. Lit-Pub Kiev, 1955. 318p. (In Russian.)

Bauzil, V. Traité d'Irrigation. Paris; Eyrolles, 1952. 413p.

Behr, Ricardo. Hidraulica Agricola. Barcelona and Madrid; Salvat Editores, S. A., 1954. 283p.

Bellincioi, Giovanni. La Technica Dell-Irrigazione. Torino; G. Lavagnolo, 1949. 224p.

Bergmann, Hellmuth. Kleinbauerngehofte; Eine Bauliche und Betriebswirtschaftliche Analyse. Berlin; Verlag Technik, 1952. 121p.

Bergos Masso, Juan. Construcciones Urbanas Y Rurales. Barcelona; Bosch, 1945. 562p.

Bernal Martinez, Enrique. Hidrologia de Tierra; El Agua y sus Applications. Madrid; Dossat, 1955. 383p.
Bertin, Jean Louis Henri. Manuel de l'Irrigateur. Paris; Dusacq, 1850. 380p. Bock, O., and A. Nawrath. Die Ziegelei. Berlin; P. Parey, 1955. 176p.
Boitard, Pierre. Les Instrumens Aratoires, Collection Complete de Tous les Instrumens d'Agriculture et de Jardinage, Francais et Etrangers, Anciens et Nouvellement Inventes ou Perfectionnes. Paris; A. Ledoux, 1833. 196p.
Boitard, Pierre. Nouveau Manuel Complet des Instruments d'Agriculture. Paris; Roret, 1844. 274p.
Botter, Francesco L. Nuovo Sistema di Trazione Degli Strumenti Aratori dei Fratelli Selmi. Bologna; 1867. 117p.
Bouma, G. J. A., and J. Vlieger. Pleatseboek. Amsterdam; Wed. W. J. Ahrend, 1949. 239p.
Bruent, R. Cavas y Bodegas. Barcelona; Salvat, 1939. 340p.
Bryas, Charles R. A. De. Expose des Travaux de Drainage e Dessechement. Paris; Imp. de Mallet-Bachelier, 1855. 261p.
Buchner, F. W. Verhandeling Over den Invloed der Noord-Hollansche Droogmakerijen na 1608 op de Gezondheid der Ingezetenen. Utrecht; 1826. 303p.
Bulakh, Vladimir L., N. A. Solomentsev, and V. A. Chekmarev. Basic Hydrology and Agricultural Irrigation. Leningrad; Gidrometeorolo-gicheskoe Izd-vo, 1955. 311p. (Hall cites 2d ed., 1963. 366p. In Russian.)
Burlev, M. S. Worker-Repair of MTC. Sverdlovsk; Gos. Nay.-Tekh. Izd-Vo Mach. Lit., 1955. 511p. (In Russian.)

C

Candura, Giovanni. Lezion di Meccanica Agraria. Citta di Castello; Soc. Top. Leonardo da Vinci, 1949. 80p.
Casanova, A. M. Manuel de la Charrue. Paris; Librarie Agricole de la Maison Rustique, 1861. 176p.
Castelli, Mario. Construcciones Rurales. Barcelona, Spain; Editorial Gustavo Gili, 1944. 364p.
Cazeaux, Pierre-Euryale. D'Influence de Irrigations dans le Midi de la France. Paris; Bouchard-Huzard, 1841. 74p.
Chassiron, F. C. M. de. Essais sur la Legislation les Reglemens Necessaires aux Dessechemenas Faire ou ` Conserver en France. Paris; Impr. de Mme Huzard, 51p.
Cohendy, Michel. Notice sur les Enterprises de Dessechements de lacs et de Marais dans la Generalite D'Auvergne. Clermont; Imprint. de F. Thibaud, 1870. 43p.
Coupan, Gaston. Machines de Culture, Preparation des Terres, Epandage des Engrais et des Semences, Entretien des Cultures. Paris; J. B. Bailliere et Fils, 1907. 420p. (Hall cites 2d ed., 1919. 510p.)
Coupan, Gaston. Machines de Recolte. Paris; J. B. Bailliere et Fils, 1912. 464p. (Hall cites 2d ed., 1919. 510p.)
Coupan, Gaston. Maquinas de Labranza. Barcelona; Salvat, 1926. 520p.
Coupan, Gaston. Les Moteurs Agricoles. Paris; J. B. Bailliere, 1904. 484p. (1)
Crette de Paullel, Francois. Memoire sur le Dessechement des Marais et L'Utilite Qu'on Peut Tirer des Marais Desseches. Paris; So. D'Agr. de Laon, 1789. 103p.

D

Danguy, Jacques. Constructions Rurales. Paris; J. B. Bailliere, 1904. 442p. (3d ed., 1923. 462p.)

Davidovich, S. M. Tracktory i Automobili (Tractors and Automobiles). Moscow; Sel'khozgiz, 1946. 727p. (10th ed., Moscow; Gos. Izd-Vo Sel. Lit-Pub Kiev, 1957. 671p. In Russian.)

Davydov, Lev D. National Tractors. Moscow; Profizdat, 1950. 371p. (In Russian.)

Deby, Julien-Marc. Manuel Pratique Irrigation. Bruxelles; G. Stapleaux, 1850. 175p.

Delacroix, S. C. Faits des Drainage: Debit des Terres Drainees. Paris; Librarie Agricole de la Maison Rustique, 1859. 84p.

Delonca, Emile. Le Canal d'Ille. Perpignan, France; Impr. du Midi, 1949. 170p.

Dencker, Carl H. Landwirtschaftliche Stoff- und Maschinenkunde: Allgemeinverstandlicker Leitfaden der Physikalischen Grundlagen und der Landmaschinenkunde fur den Unterricht und den Bauernhof. Berlin; P. Parey, 1936. 262p. (2d ed., 1948. 256p.)

Dobler, August. Dorfgenossenschaft und Dorfgenossenschaftshaus. Stuttgart; W. Kohlhammer Verlag, 1941. 118p.

Drechsler, Arthur, and Anton Walter. Steirische Landbaufible. Salzburg; O. Buller, 1948. 111p.

Dubakh, Aleksandr D. Hydrology of Peat Bogs. Leningrad; Len. Otdedenie, 1936. 110p. (In Russian.)

Dupuit, Jules. Etudes Theoriques et Pratiques sur le Mouvement des Eaux Courantes. Paris; Carillah-Gooeury et V. Dalmont, 1848. 275p. (2d ed. Paris, Dunod, 1863. 304p.)

Dupuit, Jules. Traite Theorique et Pratique de la Conduite et de la Distribution des Eaux par J. Dupuit . . . suivi d'un Extrait de l'Essai sur les Moyens de Conduire, d'Elever et de Distrubuer les Eaux, par Genieys . . . et de la Description des Filtres Naturels de Toulouse par d'Aubuisson. Paris; Carilian-Goeury et V. Dalmont, 1854. (2d ed., Etudes Theorique et Pratique sur le Movement des Eaux dans les Canary Decouvets de Travers des Terranes Permeables, 1863.)

E

Engels, Hebert. Handbuch des Wasserbaues fur das Studium und die Praxis. Leipzig; W. Engelmann, 1923. 2 vols.

Engles, Otto, and Hermann Schnitt. Mineraldunger und Landmaschinen als Hauptstutzen der Grossdeutschen Landwirtschaft. Berlin; Allgemeiner Industrie Verlag Knorre & Co., 1943. 299p.

Evreinov, M. G. Applications of Electricity for Rural Economy. Moscow; Sel'Khozgiz, 1948. 455p. (In Russian.)

Eylands, Arni G. Buvelar og Raektun. Reykjavik; Bokautgafa Menningarsjoos, 1950. 475p.

F

Fabre, Jean-Antoine. Essai sur la Maniere la Plus Advantageuse de Construire les Machines Hydrauliques. Paris; A. Jombert Jeune, 1783. 402p.

Faure, L. Drainage et Assainissement Agricole des Terres. Paris; C. Baeranger, 1903. 279p.

Febre, Jean-Antoine. Traité Complet sur la Theorie et la Pratique du Nivellement. Draguignan, France; Fabre, 1790. 368p.

Fontenay, Cher Royer de. Manuel Pratique des Constructions Rustiques (Dans les Constructions Rurales). Paris; Roret, 1836. 271p.

G

Gabay, Adil. L'Application du Tracteux aux Travaus de Terrassement et d'Excavation. Lausanne; R. Rouge, 1946. 187p.

Gausebeck, Aenne. Landfrau und Kamerad Maschine. Essen; W. Girardet, 1950. 264p.
Gerasimov, S. A. Beet-Harvesting Machinery Spg-1. Sverdlovsk; Gos. Nay.-Tekh. Izd-Vo Mach. Lit, 1950. 155p. (In Russian.)
Gilly, David. Handbuch der Land-Bau-Kunst. Berlin; Bei Friedrich Vieweg, 1798–1811. 3 vols. (Volume 3 edited by D. G. Friderici.)
Gimbutas, Jurgis. Das Dach des Litauischen Bauernhauses aus dem 19. Stuttgart; Gimbutas, 1948. 104p.
Goriachkin, Vasilli P. Agricultural Machinery. Moscow; Sel'Skok Mekhanikie, 1913. 96p. (In Russian.)
Goriatchkin, Vasilli P. Mowing, Reaping and Harvesting Machines. Moscow; Academia, 1919. 182p. (In Russian.)
Goriatchkin, Vasilii P. Theory of the Plow. Moscow; Akinoiroe "Proviat" Obuhistvo, 1927. 198p. (In Russian.)
Goriatchkin, Vasilii P., ed. Theory, Construction and Industry of Agricultural Machinery. Moscow; Mashgiz, 1935–1940. 5 vols. (In Russian)
Grandvoinnet, J. A. Traité Elementaire des Constructions Rurales. Paris; Librairie Agricole de la Maison Rustique, 1887. 2 vols.
Grisebach, H. Das Polnische Bauernhaus. Warsaw; Kaiserlich Deutsches Gen. Gov., 1917. 106p.

H

Hallie, J. Catalogue Raisonne et Illustre des Machines et Instruments D'une Application Usuelle en Agriculture. Bordeaux, France; Lafargue, 1860. 58p.
Heckl, Rudolf. Leitfaden fur das Landwirtschaftliche Bauwesen auf Grund der Wirtschaftlichen, Klimatischen und Siedlungs-Kundlichen. Wien; Scholle-Verlag, 1950. 204p.
Heine, J. A. Traite des Batiments Propres a Loger les Animaux: Qui Sont Necessaires a L'Economic Rural. Leipzig; Voss et Compagnie, 1802. 72p.
Heiss, Rudolf. Lebensmitteltechnologie. Munich; Verlag von J. F. Bergmann, 1950. 344p.
Heuser, Otto. Grundzuge der Praktischen Bodenbearbeitung auf Bodenkundlicher Grundlage. Berlin; P. Parey, 1928. 228p.
Hjelm, Lennart. Jordbrukets Byggnadskostnadsindex. Lund; H. Ohlssons Boktr., 1946. 126p.
Hogetsu, Keigo. Chusei Kangaishi no Kenkyu Hogetsu Keigocho. Tokyo; Meguro Shoten, 1950. 366p.
Holldack, Hans. Maschinenlehre fur Landwirte. Berlin; P. Parey, 1949. 509p.

J

Joao de Andrade Cervo. De Agua Para as Regas. Lisbon; Empreza Commercial e Industrial Agricole, 1881. 116p.
Jobst, Gerhard. Landliches Bauwesen. Berlin; W. Ernst, 1949. 156p.
Josse, Francois de Rennes. Chaleur de la Animales et de Ses Rapports. Paris; Gabon, 1801. 352p.

K

Kaerger, Karl. Kunstliche Bewasserung in den Warmeren Erdstrichen. Berlin; Gergonne et Cie, 1893. 183p.
Karek'Skikh, D. K. Theory and Construction and Tractor Use. Moscow; Gos. Nay.-Tekh. Izd-Vo Mach. Lit., 1950. 3 vols. (In Russian.)

Karel'Skikh, D. K., ed. Theory, Construction and Design of Tractors. Moscow; Gos. Nay.-Tekh. Izd-Vo Mach. Lit., 1950. 2 vols. (In Russian.)

Knapo, Alexander. Pol Nohospodarske Stavby. Bratislava; Slovenske Vydavatel'Stvo Technickej Literatury, 1914. 457p.

Kokovin, Evgenii V. Mechanization of Small Tasks. Moscow; Gos. Izd-Vo Sel. Lit-Pub Kiev, 1950. 350p. (In Russian.)

Konev, V. N. Fundamentals of Reclamation of Land Drainage and Irrigation. Moscow; Gos. Izd-Vo Sel. Lit-Pub Kiev, 1950. 380p. (In Russian.)

Koopstra, H., and D. B. Nieuwenhuis. Tractoren en Verdere Mechanisch drijfkracht voor de Landbouw. Haarlem; Uitg. V/H A. Kemperman, 1949. 362p.

Koren, Hanns. Pflug und Arl. Salzburg; O. Muller, 1950. 275p.

Korobov, V. A. Tractor, Automobile, and Agricultural Machine Engines. Moscow; Gox. Izd-Vo Sel. Lit-Pub Kiev, 1950. 447p. (In Russian.)

Kozenoer, G. Ya, F. V. Polyakova, and P. A. Skarin. Universal Tractor-1. Moscow; Gos. Nay.-Tekh. Izd-Vo Mach. Lit., 1950. 147p. (In Russian.)

Kuhne, Georg. Die Technik in der Landwirtschaft in den Vereinigten Staaten von Nordamerika. Berlin; P. Parey, 1926. 100p.

Kuhne, Georg. Handbuch der Landmaschinentechnik. Berlin; Springer, 1930. 187p. (1)

L

Lefeldt, W. Der Gegenwartige Stand der Abfuhr und Rationalisations-Frage in Grossbritannien. Berlin; Wiegandt, Hempel, & Baren, 1872. 102p.

Legris, L. La Nouvelle Mecanique Agricole. Paris; F. M. Maurice, 1825. 493p.

Lesagne, Maurice. Equipment Electrique des Exploitations Agricoles. Paris; Societe Pour le Developpement des Applications de L'Electricite, 1950. 138p.

Londet, Louis-Andre. Instruments Agricoles, Machines, Appareils et Outils Employes en Agriculture. Paris; Bouchard-Huzard, 1858. 303p.

Loy, A. Van. Landbouwmechanica. Antwerpen; N. V. Standaard-Boekhandel, 1945. 269p.

M

Mais, Hector. Le Travail du Sol et le Tracteur. Soissons; Diffusion Nouvelle du Livre, 1950. 189p.

Mangon, Charles-Francois-Herve. Etudes sur le Drainage au Point de Vue Pratique et Administratif. Paris; Carillian-Goeury et Vve Dalmont, 1853. 378p.

Mangon, Charles-Francois-Herve. Etudes sur les Irrigations de la Campine et les Travaux Analogues de la Sologne et D'Autres parties de la France. Paris; L. Mathias, 1850. 119p.

Mangon, Charles-Francois-Herve. Experiences sur L'Emploi des Eaux dans les Irrigations sous Differents Climats. Paris; Dunod, 1863. 132p. (1869 ed. as Experiences sur L'Emploi des Eaux dans les Irrigations sours Differents Climats et sur la Proportion des Limons Charies par les Cours D'Eau. 198p.)

Mangon, Charles-François-Herve. Travaux, Instruments et Machines Agricoles. Paris; Dunod, 1875. 840p.

Marty, Henri. Electricité dans l'Art Manager. Toulouse; Impr. Fournie, 1947. 320p.

Marty, Henri. L'Electricité dans l'Agriculture. Toulouse; Societe Pour la Diffusion et la Vulgarisation des Emplois de L'Electricita, 1948. 496p.

Midy, Felix. Le Drainage Radie, Nouveau Procede Simple et Economique de Drainage sans Tuyaux. Paris; Hatchette, 1858. 48p.

Milletti, Roberto. Nuovi Tipi di Costruzioni Rurali. Bologna; Edizioni Agricole, 1948. 169p.

Morelli, Luigi. Manuale del Casaro. Milan; Univ. Hoepli, 1942. 319p.
Mosolov, Vasilii. Agricultural Technology. Moscow; Gos. Izd-Vo Sel. Lit-Pub Kiev, 1948. 351p. (In Russian.)
Mossaeri, Victor M., and Ch. Audebeau. Les Constructions Rurales en Egypte. Carie; Impr. de l'Institute François de'Archaeologie Orientale, 1921. 172p.
Moysen, C. H. Nouveaux Instruments Aratoires Inventes et Decrits Avec des Gravures dans le Texte. Paris; Yve Bouchard-Huzard, 1854. 96p.
Muret, C. Topografia Aplicaciones Especiales a la Agricultura. Barcelona; Salvat, 1929. 482p.
Musil, Alfred. Die Motoren fur das Kleingewerbe. Braunschweig, Germany; F. Vieweg & Sohn, 1878. 120p.
Musil, Alfred. Die Motoren fur Gewerbe und Industrie. Braunschweig, Germany; F. Vieweg & Sohn, 1897. 311p.

N

Nadault de Buffon, Benjamin. Considerations sur la Regime Legal des Eaux de Sources Naturelles et Artifielles. Paris, Marescq. 414p.
Nadault de Buffon, Benjamin. Cours D'Agriculture et Hydraulic Agricole Comprenant les Principes Generaux de L'Economie Rural. Paris; Carilian-Goeury et V. Dalmont, 1852–1853. 3 vols.
Nadault de Buffon, Benjamin. Hydraulique Agricole Applications. 2d ed. Paris; Dunod, 1861–1863. 2 vols.
Nadault de Buffon, Benjamin. Hydraulique Agricole, Des Submersions Fertilisantes. Paris; Dunod, 1867. 492p.
Nadault de Buffon, Benjamin. Hydraulique Agricole, Des Alluvions Modernes. Paris; Dunod, 1873. 251p.
Natrus, Leendert van, Jacob Polly, and Cornelius van Vuuren. Groot Volkomen Moolenboek. Amsterdam; Johannes Covens en Cornelis Mortier, 1734–1736. 2 vols.
Nazarov, Georgii I. Electricity for Agricultural Operations. Moscow; Gos. Izd-Vo Sel. Lit-Pub Kiev, 1949. 359p. (In Russian.)
Niccoli, Vittorio. Prontuario dell'Agricoltore e dell'Ingegnere Agrario. 15th ed. Milano; Hoepli, 1945. 704p.
Norheim, Alfred. Landbruksbygninger. Olso; J. S. Cappelen, 1947. 184p.

O

Olalguiga, Ramon. Electrification Agricola. Madrid; Minister of Agriculture, 1948. 97p.
Ourches, Charles d'. Observations et Ameliorations sur Quelques Parties de L'Agriculture dans les Sols Sablonneux. Paris; Impr. de Mme Huzard, 1818. 175p.
Ourches, Charles d'. Traite General des Prairies, et de Leurs Irrigations. Nancy; Bontoux, Metz; J. B. Collignon, and Paris; Levrault, 1804. 184p. (2d ed., Paris; Chez A. J. Marchant, 1806. 224p.)

P

Pareto, Raphael. Irrigation et Assainissement des Terres. Paris; Roret, 1851. 4 vols.
Perels, Emil. Die Mahemaschinen. Ebend; Holzschn, 1870. 150p.
Perels, Emil. Handbuch des Landwirtschaftlichen Wasserbaus. Berlin; Wiegandt, Hempel & Parey, 1877. 692p.
Perels, Emil. Handbuch zur Anlage und Konstruktion Landwirtschaftlicher Maschinen un Gerathe. Jena; H. Costenoble, 1866. 2 vols. (2d ed., Emil Perels and Wilhem

Strecker. Perels' Ratgeber bei Wahl und Gebrauch Landwirtschaftlicher Gerate und Maschinen. Berlin; P. Parcey, 1897. 276p.)
Perthuis de Laillevaut, Leon. Traite d'Architecture Rurale. Paris; Deterville, 1810. 268p.
Petit, Antoine. Electricidad Agricola. Barcelona; Salvat Editores, 1928. 524p.
Petit, Antoine. Electricite Agricole. Paris; J. B. Balliere et Fils, 1909. 424p.
Philadelphy, Emit. Auto, Motocykel, Traktor. Turciansky Sv. Martin; Martin Marica Slovenska, 1948. 167p.
Philbert, J. Genie Rural. Paris; Vve C. Dunod, 1902. 422p.
Pogorelyi, I. P. Repair of Tractors. Moscow; Gos. Izd-Vo Sel. Lit-Pub Kiev, 1950. 631p. (In Russian.)
Provost, A., and P. Rolley. Practicas de Ingeneria Rural. Barcelona; Salvat, 1926. 455p.

R

Raggio, Juan L. Hidraulica Agricole. El Ateneo, Buenos Aires; Libreria y Editorial, 1947. 563p.
Ramelli, Agostino. Le Diverse et Artificiose Machines (Grinders, Windmills, et.). Paris; The Author, 1588. 338p.
Rau, L. Beschreibung und Abbildung der Nussbarsten Ackerwerckzeuge. Stuttgart; Ebner & Seubert, 1862. 40p.
Rau, Ludwig. Verzeichniss der Modell-Sammlung von Handgerathen and Pflugen Nach Ihrer Geschichtlichen Entwicklung. Frankfurt; Main, 1882. 16p.
Raug, Kurt. Entwicklungslinien im Landmaschinebau. Essen; W. Girardet, 1949. 141p.
Ringelmann, Maximilien. Genie Rural Applique aux Colonies. Paris; A. Challamel, 1908. 698p. (2d ed., Paris; Societe d'Editions Geographiques, Maritimes et Coloniales, 1930. 727p.)
Ringelmann, Maxmilien. Etude des Machines Destinees a la Preparation des Racines et des Tubercules. Nancy; Impr. de Berger-Levrault, 1883. 62p.
Ringelmann, Maxmilien. Habitations Rurales et Batiments de la Ferme de Regions Liberees. Paris; Librarie Agricole de la Maison Rustique, 1920. 86p.
Ringelmann, Maxmilien. Machines Employees en Agriculture pour l'Elevation des Eaux. Paris; Bureau de l'Industrie Laitier, 1889. 140p.
Ringelmann, Maxmilien. Le Materiel Agricole a l'Exposition de 1900. Paris; Dunod, 1901. 224p.
Ringelmann, Maxmilien. Traite des Machines Agricoles. Paris; Librarie Agricole de la Maison Rustique, 1985. 124p.
Risler, Eugene, and G. Wery. Irrigations et Drainages. Paris; J. B. Bailliere et Fils, 1904. 516p. (2d ed., 1909. 532p.) (1)
Risler, Eugene, and G. Wery. Riegos y Drenages. Barcelona; Salvat, 1919. 557p.
Ronna, A. Les Irrigations. Paris; Firmin-didot et Cie, 1889–1890. 3 vols.

S

Saint Felix, Armand-Joseph-Marie de. Architecture Rurale. Toulouse; J. M. Douladoure, 1820. 360p. (2d ed., 1826. 395p.)
Sarazin, J. N. Elements de la Mecanique Rationnelle de la Charrue. Nancy; Mlle. Gonet, 1853. 252p.
Schackt, Helge R. Landbrugets Mekanisering. Odense; Skandinavisk Bogforlag, 1947. 424p.
Schneider, Ernst, and H. H. Noth, eds. Landarbeit Leicht Gemacht. Berlin; Reichsnahrstandsverlag, 1943. 351p.

Schubert, Hermann. Theorie des Schlick'schen Massen-Ausgleichs Bei Mehrkurbeligen Dampfmaschinen. Leipzig; s.n., 1901. 132p.
Sedlacek, Jiri. Zemedelske Stroje a Jejich Obsluha. Prha; Brazda, 1949. 162p.
Semenov, Vadim M. Tractors and Automobiles. Moscow; Gos. Izd-Vo Sel. Lit-Pub Kiev, 1950. 295p. (In Russian.)
Sieck, K. A. Tractoren. Kobenhavn; Landbrugsteknisk Forlad, 1947. 376p.
Spitz, Georges. Sansanding: Les Irrigations du Niger. Paris; Societ d'Editions Geographiques, 1949. 237p.
Supino, Giulio. Irrigazioni. Bologna; R. Patron, 1949. 118p.

T
Thebis, Reinhold. Technisches Praktikum fur Kleinbauern und Siedler. Possneck; R. A. Lang, 1948. 128p.
Tramonte, Raffaele. Irrigazione in Puglia. Bari; Arti Grafiche A. Cressati, 1949. 103p.
Trefois, Clemens V. Ontwikkelingsgeschiedenis Van Onze Landelijke Architectuur. Antwerpen; De Sikkel, 1950. 297p.
Tresca, Alfred. Le Materiel Agricole Moderne. Paris; Firmin-Didot, 1893–1895. 2 vols.

V
Villeroy, Felix, and Adam Muller. Manuel de l'Irrigation. Paris; Dusacq, 1850. 380p.
Vuorela, Toive. Etela-Pohjanmaan Kansanrakennukset. Helsinki; Etela-Pohjalainene Osakunta, 1949. 271p.

Z
Zipfel, Matthias. Die Wirtschaftliche Stromversorgung der Landwirtschaft. Karlsruhe; H. L. Meyer, 1949. 146p.

Appendix: Journals and Annuals Recorded from Source Documents

This list demonstrates the range of subjects, types of publications, and diverse languages of the journals, annuals, and recurring proceedings volumes. Of the 590 titles, 467 were cited only once, a very common occurrence in citation analysis of specialized literature. The list has not been edited and is in brief title bibliographic style, having been used as a computer listing for tabulation. Those familiar with bibliographic descriptions in animal science should be able to decipher the abbreviations.

Adv fd res
Advances in Agronomy
Advances in Chem series
Advances in Irrigation
Ag Engineer
Ag Mech in Asia
Ag Water Mgmnt
Agr Tech
Agr Wastes
Agr & Environ
Agrar Ubersicht
Agric Algerienne
Agric aviation
Agric engng
Agric engng, Australia
Agric gaz N.S. W.
Agric Meteorology
Agric Pakistan
Agric Res Council, Res Rev
Agric Systems
Agric & Forest Meteorology
Agron trop
Agronomy
Agronomy Abs
Agronomy J
Agrotechnitar

Alimentozione Anim
Am J Ag Econ
Am J dimier nutrition
Am J Public Health
ASAE Grain & Forage Harvest., Proc
American Potato J
AMJ
Anal Chin Acta
Andhra J
Anim Prod
Animal Feed Sci & Tech
An Rev of Energy
Ann Rev Ferm Proc
Ann Rev microbio
Ann rev plan physio
Ann Sper Agr
Annali dell'Insituto Sperimentale per . . .
Annals bot
Annals chem
Annals de Recherches Veterinairies
Annals Inst Pasteur
Annals N.Y. Acad Sci
Annals Occup Hygiene
Annals of appl Biol
Annals of Vet Res
Annals Zootechnie

Antoine van Taewenhoek
Applied Animal Ethology
Applied Electrical Phenomena
Applied Env Microbio
Applied Microbiology
Applied Polymer Symp
Approp Technology
Aquaculture
Aquatic Botany
Arable fmr
Archiv fur Landtechnik
Archiv fur Tierzucht
Archiv Tieremach
Archives for Env Health
Arroz
Arroz Brazil
Atti Centro naz Mecc agric
ATZ Auto Zrectschrift
Australian J of agric res
Australian J of T Soil Rsch
Auto Engr
Automobiltechn Z
Avian Pathology
Avians Diseases
Azerb Agrartudom, egyet, Kozlem, Godollo, 1975

B Fiji Dept of agric
B Grain Tech (Burma)
B IMAG
B Indonesian Econ Stud
B Inform Rizicult, France
B Informacyjny Inst. Aootechniki, Zaklad Informacji
B Liaison Comm Mach Agric Outre-Mer
B Southern Coop Series
B U Cal Davis Div of Agr
B U of Maine
B Univ Osaka Prefecture, Osaka
B Vsesoyuznogo Ordena Lenina Instituta . . .
Bamidgeh
Bedrijfsonturkkeling
Berichte uber Landwirtschaft
Ber. Landtech. Kur Tech. Landw
Betterwe
Big Farm Mgmt
Biochem & Biophys Acta
Bioengineering
Bioengineering symp
Biometric
Bioscience
Biotech & Bioeng
Biotech & Bioeng symp
Boundary Layer Meteorology
British J of Nutrition
British poultry sc

British Sugar Beet Rev
Bull inf. CNEEMA
Bull Nat Inst Animal Ind
Bull Rech Agron Gembloux
Bull Swedish Inst Agric Engng
Bull Techn Inform (France)
Burma Med J

Calif agric
Can Agric Engeng
Can Chem Proc Ind
Can J Chem Eng
Can J of Animal Sci
Can J Plant Sci
Can J Res
Can J soil sci
Can Vet J
Cereal Chem
Cereal Sc Today
Ceres
Chem Eng
Chem Eng Progress
Chem Engng Program
Chem Engng Sci
Chem Engng Technol
Chem Metall Engng
Chemistry & Industry
Commerce Annual No. (India)
Commun in Soil Testing & Plant Anal
Computer J
Cornell Vet
Critical Reviews in Environ Control
Crop Science
Cult Dairy Prod J

Dairy Farmer
Das Papier
De Ingenieur
Deustche Tierarztlich Wochenshift
Deutsch Gartenbau
Deutsche Landwirtschaft
Deutsche Papierwirtschaft
Dev in Animal & Vet sci
Dev in Indust'l microbio
Development Digest
Development & Change
Die Grune

East African J Rur Dev
Eastern Anthropologist
Ecology
Econ Dev & Cultural Change
Econ & Pol. Weekly
Effluent & Water Treatment J
Effluent & water trmnt J
Electrochem Ind Process Biology

Electronics Australia
Empire J exp Agric
Energy
Energy International
Energy Policy
Energysprøtrum
Energy in Ag
Environment
Equine Vet
European J applied microbio
European Potato J
Expl Hort
Expl husb

Far eastern econ rev
Farm Building Progress
Farm implements and mach rev
Farm mech
Farmers Weekly
Farming in Zambia
Farming Japan
Fd Inds
Federal Register
Feedstuffs
Feldwirtschaft
Fertilizer News, New Delhi
Field Crops Research
Filtration and Separation
Food engineering
Food Policy
Food Res Inst Stud
Food Technology
Foreign Agric
Forest eco & mgmt
FORPRIDE Digest
Forsch Gebiete Agrikulturphys

Gartenbau & Gartenweld
Gartenbauwirtschaft
Gartenmeister
Geoderma
Geotechnique
German Chem Eng
Getreide Mehl Brot
Gidrotechnika i Melioracija
Grass Forage Sci
Groent en Fruit
Grower
Grundforbattring
Grundl Landtech

Haryana Agr U J of Rsch
Hidrotechnica
Highlights Agr res (Ala)
Hilgardia
Hort. Indy
Hort. Res

Hort. Sci.

IEEC process design & dev
Ill State Water Survey
Ind Engng Chem
Ind Fmg
Ind J vet Sci Anim Husbandry
Ind & eng chem
Indian agron
Indian Farming
Indian J Agric
Indian J Agric Econ
Indian J agron
Indian J of Agric Sci
Informatore Zootecnico
Inftore agrario
Ingenieur Archiv
Inorganic Chem
Inst of Pub Health Engrs
Int J Heat Mass Trans
Int Labour Rev
Invest Urology
Iowa St. J. of Research
IRC Newsletter
Israel Program for Sci Transl
Izchislitdna Tekh Automatizatsiya na Proizrod-stoenite Protsesi

J Agr Assoc of China
J Agr Res of China
J agric engng res
J agric food chem
J Agric Labour Sc
J agric sci, Cambridge
J Agric Sci, Netherlands
J Am Chem Soc
J Am Oil Chem Soc
J Am Optical C Assoc
J Am Soc Sugar Beet Techn
J Am Vet Med Assoc
J Amer Inst Chem Eng
J Analytic Chem
J Anim Scie
J appl chem
J Applied Bacteriology
J applied meteorology
J Applied Physics
J ASCE, Env Eng Div
J ASCE, Irr & Drain Div
J Asian Studies
J Assoc Official Anal Chem
J Atmospheric Sciences
J Aust Inst agric Sci
J Automative Engng
J Bacteriology
J biol chemistry
J biotech & Bioeng

Journals and Annuals

J Br Grassld Soc
J Chem Engng China
J Chem Tech & Biotech
J Chromotog Sci
J crystal growth
J dairy sci
J Development Stud
J econ Lit
J Electroanal Chem
J Engng Indy
J env qual
J Exp Botany
J Fd Sci
J Fermentation Tech
J Food Process & Preserv
J food sci and tech
J Franklin Inst
J Gen Microbiology
J Gen Physiology
J Gen & Applied Microbio
J Geophysical Res
J Heat & Vent Engrs
J hort sci
J hydrology
J Hygiene (Cambridge)
J Inst of Sewage Purif
J Inst Sewage Pur
J Instn agric engrs
J jap soc. grassld sci
J Kaisss Ent Soc
J Korean Soc of Agric Mach
J moder african stud
J Natn Inst Agric Bot
J nutrition
J Optical Sci
J Phys E Sci Instrumentation
J Proc Instn agric engrs
J R Soc Arts
J Rheology
J rural dev
J sci agric
J Sci Fd agric
J Sci & Tech (Burma)
J Scient Agric Soc, Finland
J Soc Agric Mach
J Soc Agric Mach Kansai Br
J Soc Agric Machinery (Japan)
J Soil Sci
J Soil Sci Soc Amer
J Soil Sci Soc Philippines
J Soil & Water
J stored products res
J Terramechanic
J Testing & Eval
J Textile Inst
J Vacuum sci & tech
J Water Poll Control Fed

J Weed Sci Soc Am
J Wind Engng & Ind Aerodynam
J world mariculture
J World Poultry Soc
J Yamagata Agric For Soc
Jap J Breeding
Japanese poultry sci
JCRR Taipai

La agric
Lab An
Laboratory An Sci
Lancet
Landb Onderz
LandbForsch-Volkenrode
Landboumechanisatie
Landbouwkundig Tijdschrift
Landtech Forsch
Landtechnik
Landwn Verstnen
Lav Arroz
Lav Arroz Brasil
Lenin Akad Selkhoz Nauk
Livestock Product Sci
Louisiana rur Econ

Macch Moi Agric
Macchine Mot agric
Machine Design
Madras agric J
Magyar Allatorvosok Lapja
Makhaniz. elektrif sots. sel'khoz
Malay Agric J
Marches coloniaux du monde
Mecc agric
Mechanizacae Zewmedelstoi
Medycyna Weterynayjna
Melioration und Landwirtschaftsbau
Memoir Fac Agric Kagshima U
Memoirs, Coll Agric, Nat Taiwan Univ
Mezogazd. gepesit. Tanul.
Microbiology
Mittdt LandwGes
Mitt. Schweiz. Landw
Monthly rev
Monthly Weather Rev
Motoris Technol Agric
Motschufte fur Veterinaumedizin

NachrBl dt Pflschut Berlin
Nat res forum
Nat Sci Council Monthly (Taiwan)
Naturaliste Can
Nature
Nauch Trudy ukransk Ordena Trudov Kransnogo Anameni Selsk Akad
Neth j agric sci

New times
New Zealand J of Agr Rsch
New Zealand J exper'l agric
Nogyo Oyobi Eng
Nordic Hydrology
NSDB Tech J (Philip gov't)
Nutr Rpts Int

Ohio Rpt
Outlook Agric

Personnel Mgmt
Phil Agric
Philippine agr engng J
Philippines Agric
Philippines Forests
Philippines J Agric
Philippines Lumberman
Physiol Planatarem
Phytopathology
Pig Farming
Pig Yearbook
Plant, Cell Envir
Plant Phys
Planta
Policy Sciences
Poljoprivredni fakultet
Polymer Engin & Sci
Pop Bull
Potato Res
Poultry Int
Poultry Notes
Poultry Sci
Poultry World
Power Farming
Process Biochem
Procs 14th World Poultry Sc Cong
Procs 14th World Poultry Sci Congress
Procs 1973 Livestock Waste Mgmt Conf
Procs 1st Intl Conf Water Poll Rsch
Procs 1st Intl Green Crop Drying Conf
Procs 1st Intl Symp on Livestock W
Procs 1st Symp on Food Phys
Procs 24th Purdue Indust Waste Conf
Procs 27th Industl Waste Conf
Procs 28th Industl Waste Conf
Procs 2d European Conf on Mixing
Procs 2d Intl Green Crop Drying Conf
Procs 2d Intl Livestock Env Symp,ASAE
Procs 2d Natn Dairy Housing Conf
Procs 2d Symp on Bioconversion & Biochem Eng
Procs 34th Purdue Indust Waste Conf
Procs 3d Intl Symp on Livestock Waste
Procs 3rd Intl Conf Energy Use, Mgmt

Procs 4th Intl Conf on the Phys Props of Agr Materials
Procs 4th Intl Conf Water Poll Rsch
Procs 4th Intl Symp Livestock Wastes
Procs 6th Conf Fluid Machinery (Budapest)
Procs 6th Int Cong Soil Sci, Paris, 1956
Procs 6th Intl Conf ISVS, Vienna, 1978
Procs 7th Tirenniel conf Eur assoc potato res, Poland, 1978
Procs 8th Int Conf agric mech, Zaragoza, Spain, 1976
Procs Agr res (Ala)
Procs Am Poll Control Ass
Procs Am soc hor sci
Procs amer assoc swine practitioners
Procs Amer Hort Soc
Procs ASAE Farmstead Engng Conf
Procs ASCE
Procs Auto Div Instn Mech E
Procs Brit Crop Prot Conf
Procs Brit Soc Animal Prod Symp on Processing of Roughages
Procs Conf on Ag Eng Queensland 1984
Procs Cornell Ag Wast Mgmt Conf
Procs Crop Prot., N. Britain
Procs Eur grassland fedn conf brighton, 1979, Bris soc Occ symp No. 11, 1980
Procs European symp on Anaerobic Waste Wat Trtmnt
Procs IMAG Symposium
Procs Inst Agric Engrs
Procs Inst Mech Engrs
Procs Int meeting animal produciton from temperate grassland, Dublin
Procs Int Ric Comm. 2d sess, Kuala Lumpur, Malaya
Procs Intl Conf on Plant and Veg Oils as Fuels, Am Soc Agr Eng
Procs Intl Seed Test
Procs IRC
Procs Nat Acad Sci
Procs Nutrition Soc
Procs of the Nutrition Soc
Procs pig Vet Sci
Procs RACI Cereal Chem Div An Conf
Procs Royal Society
Procs Soc Hort Sci
Procs soil sci soc am
Procs Soil Sci Soc Amer
Procs Symp on Farm Wastes
Procs Weed Sci Soc Am
Procs XIII Intnl grassland congr., Leipzig, 1977
Prog water tech
Progress in Indust Microbio

Przemydk Spzywezy
Public Admin rev
Public works
Publities Proefstn Rundveehoud
Pulp Paper Int.

Q bull mich st univ agric exp stn
Q J Int agric
Q J of Royal Meteorology Sci
Quaderni della Stazione Sperimentale di Risicolluroo

Rab. molodykh. unchkh Mekh elekt
Rec agric res
Revue des Fermentations et des Industries Alimentaires
Revue Suisse l'Agric
Rice J
Rice news
Rice News Teller
Rijks Univ Gent Med Fac Landbow
Riso
Rivista Zootecnologia
Riz et reziculture
Rsch and Training network

Schweinezucht Schweinmast
Science
Scientific Am
Seed Sci Tech
Seeds & Fertility
Sel'khozmashina
Sel'skogo Khozyaistva
Sewage & Indl Wastes
Soil Biol & Biochem
Soil Sci
Soil Sci Soc Am J
Soil Survey Horizons
Soil Tillage Rsch
Solar Energy
Span
Sta North Cent Agric Expt Sta, Liberia
Studie de Epurarea Apelor
Suinicoltura
Sveriges Lantbrouksuniv Inst for lantbroukets. . . .
Svinovodstro, Moscow
Symp Intl Livestock Envir
Symp on heat and mass transfer by comb forced & nat convection

Tag Ber Itakad Landw Wiss Berlin
Taiwan Irr
Techn sel Choz
Term Farm & Home Sc

Tetkahto Pyin Nya Padatha Sasong
Thai J of Agr Sci
Thin Solid Films
Tierenahrung & Futterung
Tierzucht
Tijdschr Diergeneesk
Timisoara
Trakt J
Trakt. Sel'khozmash
Trans Am geophys union
Trans Am Inst Chem Engr
Trans Int Cong Agric Mach, Paris
Trans Int Congr Agric Mach, Paris
Transactions, amer geogphys union
Transactions, amer soc agric engrs
Transactions, Br Mycolo Soc
transactions, inst chem engrs
Trop Agric
Trop Agric, Ceylon
Trop Agric, Trin
Trop Sci
Trudy Khar'kov Sel'khoz Inst
Trudy vses. nauchno-iiled Inst. sel-khoz. masch

Uganda J
Utah scie

VDI Berichte
Versl landbouwk. onderz. rijkslandb-Proefstn
Vestnik Selskokhozyaistvennoi Nauki
Vet Bull
Vet Record
Veterinaria Saraj
Vleerdistributie en Vleestechnologie

Wallerstein Lab Comm
Washburn Law J
Washington St U Ag Ext Newsletter
Water Engng & Mgmet
Water Industry
Water Poll Control
Water Resources Res
Water Rsch
Water Science & Technology
Water Services
Water & Sewage Works
Weed Research
Weed Sci
Wiadomosci Melioracyjse i lakarskie
Wirtschaftsergene Futter
World An Review
World Animal Rev

World Crops
World Development
World Farming
World Poultry Sci J
World Water

Yearbook Agric Engr
Yearbook Commercial Pig Prod

Zeitschrift fur Acker-und Pflanzenbau
Zemedelec, priloha AN
Zemedelski aktuality ze svets
Zemed. Tech
Zemeledelbchaskaya Mekhanika
Zert fur Kulturtechnik u Flurber
Zirocisna Vyroba
Zokrodnictvo

Index

Authors and titles in the core list of monographs (pp. 240–313), the core list of journals (pp. 325–328), the historically significant monographs (pp. 352–405), and Appendix A are *not* included in this index.

Academic Press 171
Accreditation Board for Engineering & Tech. 27, 32
AGRICOLA 99, 180, 219–223, 321
    Languages 223–224
Agricultural engineering:
    Citation analysis 226–239, 318–320, 345–351
    Courses 28, 41, 163–171
    Degrees 30
    Early history 3, 11, 23, 339–343
    Education 25–32
    First use in a book title vii, 340
    Journals and serials 318–331
        Citation analysis 230–239, 318–325
        Core lists 325–328
        Percentage of literature 224–225, 319
    Literature:
        Characteristics 213–225, 346–350
        Languages 223–224
        Percentage of agriculture 220
    Monographs:
        Citation analysis of 218–239
        Core lists 240–313
            Characteristics 312–317
            Compilation 227–234
            Median year 313
            Publishers 313
            Rankings 238–317
            Review and reviewers 234–239
            Top twenty 313–317
        Historically significant 352–405
        Percentage of literature 224–225
        Publishers 223
        Reports 330–331
        Publishing formats 224–225
        Reference works 332–337
    Students 30
    Subject concentrations 220–223, 235, 332, 340, 344
    Technologies 6
    Textbooks 163–174, 177
Agricultural Engineering 112, 178–179
Agricultural Engineering Abstracts 219–221, 227
Agricultural Engineering Index 22, 97–98, 107, 180, 342
Agricultural and Horticultural Engineering Abstracts 98, 342
Agricultural Mechanization in Asia, Africa, and Latin America 49–51, 180
Agricultural University, Wageningen 119
Agriculture and energy 59–95
AGRIS 219
Allahabad Agricultural Institute 34
Allyn and Bacon 172
American Fisheries Society 145
American National Standards Institute 205, 209
American Plywood Association 137
American Society for Engineering Education 105
American Society of Agricultural Engineers 22, 29, 31, 37, 62, 99, 105, 109, 135, 163, 213
    Centennial 181–183
    Conference proceedings 179–180
    Database 182

413

414 Index

American Society of Agricultural Engineers (*cont.*)
  Journals 176–178
  Publications 169, 171, 174–195, 303
    Database 182–196
    Index 183–196
    Standard deviation years 188
    Subjects 184, 186–187, 189–196
  Purpose 176
  Standards 180, 197–208
  Subject divisions 14, 17, 21, 97–98, 177
  Textbooks 167, 174, 181
American Society of Agronomy 317
American Society of Civil Engineers 105, 210
American Society of Mechanical Engineers 105, 210
American Society for Testing Materials 105, 209
American Standards Association 105
Appropriate Technology 49
Aquacultural engineering 144–160
Aquacultural Engineering 146, 157
Aquaria 149–150
Artificial intelligence 111
ASAE Standards 180
AVI Publishers 22, 125, 164, 169, 171–172, 313

Beasley, R. P., et al. 172, 247
Biltmore Forest School 127
Biological and Agricultural Index 180
Biological engineering and systems 19, 24, 31, 36, 39
Biomass 23, 59, 84–87, 223
Biotechnology 19
Books. See Monographs

CAB Abstracts 219–225
CAB and CABI 99, 180, 321
Canadian Institute of Forestry 134
Canadian Society of Agricultural Engineering 106, 177
Chemical Abstracts 180
Citation analysis 226–234
  Historical 343–350
Civilian Conservation Corps 133
Colorado State University 120
Commission Internationale du Génie Rural (CIGR) 105, 180, 343
COMPENDEX 99
Computers and Electronics in Agriculture 107
Conferences 115, 135, 156–157, 179–180
  Third World 51–54
Conservation 23
Core Agricultural Literature Project 228

Core journals and serials 325–331
Core literature:
  Citation analysis 226–239
  Reports 330–331
Core monographs 240–312
  Review process 234–237
Cornell University 14, 23, 120
Council on Forest Engineering 136
Crops 19
Cuenca, Richard H. 172
Czechoslovakia 231, 331

Dairy technology 121
Deere and Co. 169, 181, 314–315
Denmark 331
Dissertations 225
Drainage 223, 316
Drying crops 77, 81, 122, 125

Effluent in aquaculture 154–155
Electrical systems 21, 23
Electronics 23, 96–116
Elsevier 171, 313
Energy 59–93
  Alternatives 93
  Developing countries 90
  Environmental aspects 91
  Food system 62, 123
  International aspects 89–91
  Management 73
  Renewable 80
Engineering, colleges of 26
Engineering Council for Professional Development 31
Engineering Index 180
Europe 35

Farmall 15
Farm automation 96
Farm buildings 17, 21, 168–169, 223
Farm equipment 17, 168–169, 223
Farm mechanization 17, 23, 75, 168–169, 223
Farm power 17, 21, 168–169, 223
Ferguson Foundation 21, 163
Fertilizers 73
Food and Agriculture Organization 47–48, 232–233, 312–313, 317, 330
Food engineering 19, 23, 40, 117–125, 168–169
  First use in a textbook title 118
  Journals 124
Food processing 19, 23, 78, 223
Food safety 121
Food storage 223

Forest Engineering Research Institute (Canada) 140
Forest Products Research Society 134
Forestry engineering 39, 126–143

Geankoplis, Christie J. 172, 266
Germany 331
Gordon and Breach 172
Grain drying 77, 81, 122
Guide to Sources for Agricultural and Biological Research 332

Harper and Row 171
Historical literature 214–218, 339–405
   Citation analysis 343–351
   Identification 339–352
   Monographs 352–382
     Non-English 397–405
   Preservation 343–351
   Scholarly journals 382–384
   Trade and popular periodicals 384–397
Hummel, W. M. 214–218
Hunt, Donn 172, 273, 314–315

IMAG 331
Industry standards 202
Institute for Scientific Information 226
Institute of Food Technologists 117
Instrumentation 99–105
International Congress of Agricultural Engineering 105
International Crops Research Institute for Semi-Arid Tropics 47
International Federation of Automatic Control 109
International Rice Research Institute 47
International Standards Organization 199, 201, 210
Iowa State University 14–15, 27, 31, 120
   Press 169, 171–172
Irrigation 73, 223, 316
Irrigation and Drainage Abstracts 220

Journal of Agricultural Engineering Research 230–233, 320
Journal of the Japanese Society of Agricultural Machinery 51
Journal of the World Aquaculture Society 146
Journals 318–331, 406–412
   Analysis of 319–325
   Core list 325–328
   Historical 382–384
   Percent of literature 225

   Recent 337–338
   Subjects of 329

Kansas State University 14, 31, 120
Kepner, Robert A., et al. 172, 280, 314–315
Knowledge transfer 9
Krieger Publishers 172

Labor productivity 23
Languages in literature 223–224
Liljedahl, J. Bruce 172, 284, 314–315
Logging engineering 128–131, 137–141
Luthin, James N. 172, 285

McGraw-Hill 21, 163, 169, 172, 340–342
Massachusetts Institute of Technology 117
Mechanization 42, 75
   Third World 44, 47
Michigan State University 59, 106, 119
Midwest Plan Service 169, 171–172, 287
Mississippi State University 120
Mohensin, Nuri N. 172, 288, 314–315
Monographs:
   Citation analysis of 218–239
   Core lists 240–317
     Characteristics 313–315
     Evaluations and rankings 229–317
     Top twenty 314–317
   Historically significant 352–382, 397–405
   Percent of literature 225
   Reports 330–331
Morgan, B. A. 22, 99, 225, 329, 332

National College of Food Technology (Reading) 120
National Institute of Agricultural Engineering (U.K.) 98, 330, 342
Nelson, Gordon L. 172, 292
Netherlands 331
Norway 331

Ohio State University 31
Oils 86
Oregon State University 120, 131, 133

Pacific Logging Congress 129, 134
Pacific Northwest Forest Experiment Station 132
Patents 225
Pennsylvania State University 120
Ponds for aquaculture 150
Popular periodicals, historic 384–397
Prentice-Hall 169, 171–172

416   Index

Pulp and Paper Research Institute (Canada) 138, 140
Purdue University 119

Reclamation 17
Remote sensing 111
Reviewers 236–237
Reports 330–331
Robotics 109–110
Royal Indian Engineering College 126
Rural electrification 16–17, 23
Rural engineering 15, 23
Rutgers University 120

Science Citation Index 226, 320–322
Scottish Institute of Agricultural Engineering 331
Seawater aquaculture 151–153
Silsoe College 22
Society of American Foresters 134, 138
Soils 21, 168–169
Solar heating 80
Source documents 229–231, 343–345
Standards 197–212, 225
    Organizations 209–212
Stout, B. A. 314–317
Studijni Informace 231
Sweden 331
Symposia. *See* Conferences

Technologies 6
    Transfer 9
Texas A & M University 119
Textbooks 163–174
    First 119
Theses 225
Third World:
    Journals and serials:
        Core list 324–327
        Literature review 43–55, 230–231
        Rankings 322–323
    Monographs:
        Core lists 240–312
        Common titles 239
        Publishers 317
        Top twenty 315–316
        Reviews and reviewers 237

Trade periodicals, historic 384–397
Transactions of the ASAE 136, 146, 176–178, 320–321

UNIDO 47
University of:
    British Columbia 131, 133
    California, Berkeley 131
    California, Davis 71, 119
    Florida 119
    Georgia 120
    Guelph 119
    Idaho 131
    Illinois 120
    Kentucky 119
    Maine 120
    Maryland 120
    Massachusetts 119
    Minnesota 120
    Missouri 14, 119
    Nebraska 14, 27, 31, 120
    Oxford 126
    Reading 120
    Tennessee 119
    Washington 131
    Wisconsin 120, 176
USSR 331
U.S. Bureau of Reclamation 172
U.S. Bureau of Roads 14
U.S.D.A. Agricultural Energy Data Base 69
U.S. Department of Agriculture (USDA)
    Forest Service 128, 132–133
    Publications 330
    Southern Biomass Energy Research Center 59
Utah State University 14

Wageningen, Netherlands 119
Washington State University 120
Waste management 168–169
Water engineering 24, 168–169
Wiley Press 21, 163, 169, 171, 313, 341–342
Wind energy 83
World Bank 89
World War II 16, 43

Library of Congress Cataloging-in-Publication Data

The literature of agricultural engineering / edited by Carl W. Hall
   and Wallace C. Olsen.
     p. cm. — (Literature of the agricultural sciences)
   Includes bibliographical references and index.
   ISBN 0-8014-2812-2 (alk. paper)
   1. Agricultural engineering literature.   I. Hall, Carl W.
II. Olsen, Wallace C.   III. Series.
S674.3.L58   1992
630—dc20   92-24516

Powell's Literature of Agricultural (NoDJ)
$99.95 / 6.98   NDJ(H)
Science & Natural History   124168